CONSUMER MATH

Biblical Stewardship & Personal Finance

MASTERBOOKS®
CURRICULUM

Editorial Team:
Laura Welch
Craig Froman
Sony Elise
Willow Meek
Judy Lewis
Ally Teague

Design Team:
Diana Bogardus
Jennifer Bauer

First printing: June 2025

Master Books, P.O. Box 726, Green Forest, AR 72638
Master Books® is a division of the New Leaf Publishing Group, LLC.

ISBN: 978-1-68344-371-1
ISBN: 978-1-61458-889-4 (digital)
LOC: 2025933623

Printed in the United States of America

Please visit our website for other great titles: www.masterbooks.com

About the Author

Katherine Hannon, a homeschool graduate and homeschool mom, has been researching and writing about math for more than 20 years. She is the author of Master Book's bestselling *Principles of Mathematics* curriculum and other resources. Her heart is to point people to "seek the Lord, and His strength" (Psalm 105:4) in everything. Katherine resides with her husband and child near Atlanta, Georgia.

Disclaimer: Nothing in this course and these case studies should be construed as specific financial advice. Neither the author nor the publisher is liable for any financial decisions made. This course is not affiliated with, endorsed by, or sponsored by any other organization, institution, or entity. Likewise, this course does not endorse or represent any other parties or entities.

Table of Contents

Quarter 1: The Big Picture & Key Skills

Lesson 1: Introduction & Giving

Lesson 2: Introduction to Living Expenses

Lesson 3: Savings, Compound Interest, & Debt

Lesson 4: Taxes

Lesson 5: More Taxes & Review

Quarter 2: Budgeting

Lesson 6: Getting Started with a Budget

Lesson 7: More on Budgeting Income; Limiting Lifestyle (Food)

Lesson 8: Budgeting & Medical Expenses

Lesson 9: Budgeting for Big Expenses

Lesson 10: Review

Quarter 3: Financial Decisions

Lesson 11: Cars

Lesson 12: Career Preparation

Lesson 13: Housing, Part 1

Lesson 14: Housing, Part 2

Lesson 15: Decisions & Review

Quarter 4: More Applications, Investments, & Final Review

Lesson 16: Applying Math Skills to Daily Life

Lesson 17: Savings: Understanding Saving Options

Lesson 18: Savings: Mechanics & Retirement

Lesson 19: Estimating Taxes, Business Math, & Review

Lesson 20: Review

Welcome to Consumer Math!

This course is designed to equip you with essential skills to handle personal finances wisely. From budgeting for groceries to preparing for larger purchases like a car or home, this course connects math principles to real-life financial situations.

This course not only focuses on the practical aspects of financial management but also integrates a biblical perspective on stewardship. You will see how principles from Scripture can inform wise financial decisions, promoting generosity and responsible living. Engaging activities, case studies, and discussions will help solidify your understanding and make the learning experience both meaningful and enjoyable.

By course end, you will be ready to navigate financial decisions confidently, blending practical skills with a strong foundation in stewardship and responsibility.

Now we will embark on this journey towards becoming wise stewards of what God gives us.

Acknowledgments: A heartfelt thank you to all those who contributed to this course, including my mom, Cris Loop, who faithfully read drafts of the material; my family, who sacrificed to give me time to work on this; my friends who prayed with me through the process; the many men and women who have grappled with what the Bible says about money whose work helped me forward; and the incredible team at Master Books, who saw vision for this project and polished it into what it is today.

Quarter 1: The Big Picture & Key Skills

Welcome to *Consumer Math*. In this first quarter, we are going to look at the big picture of managing money.

In these first several lessons, we are going to be reviewing a lot of skills that you hopefully already know. They are skills that help with a lot of consumer math decisions, however, so are very important. As the course continues, we will move into much more application, applying those skills to everything from taxes to budgeting to choosing a career.

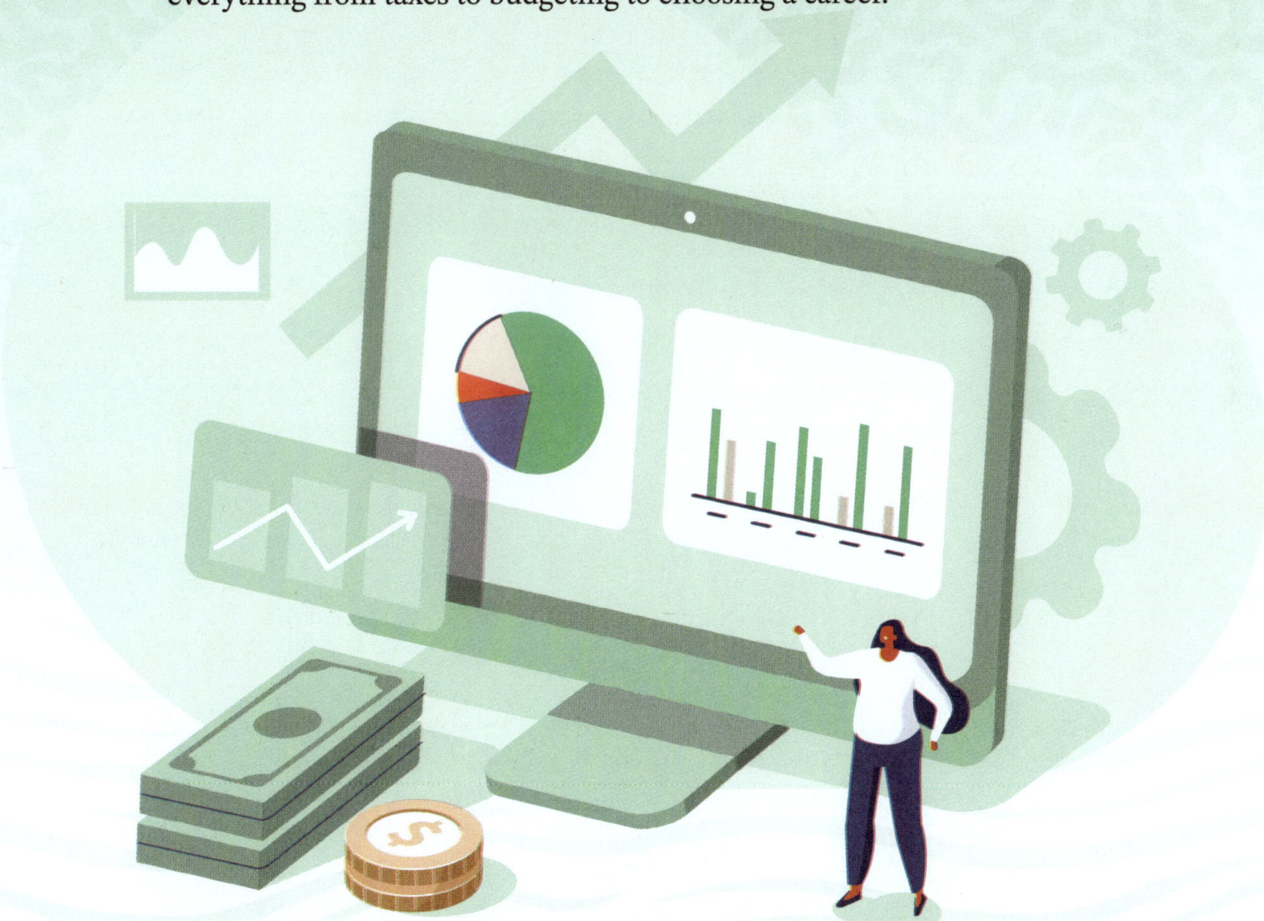

Page intentionally blank

Introduction & Giving

In this lesson, we will take a look at what consumer math is and at the biblical foundation for it. We will review some key skills we will need as we continue. We will see how money we get can either be given, saved, or spent. We will also explore God's heart for giving and review working with percent problems, which is a very useful skill we will use over and over again.

The Big Picture

We will begin by looking at what we are studying, what God has to say about it, and why consumer math is so important.

What Is Consumer Math?

Consumer = "a person who purchases goods and services for personal use."[1]

Consumer math is technically the math needed to make personal purchases, whether those be large purchases like a house or small ones like which type of spaghetti sauce to buy. We will broaden the term to include a big picture of all aspects of personal finances, including:

- income
- spending
- saving
- budgeting
- giving
- taxes

We are also going to focus throughout on God's principles for money.

What Does God Have to Say About Finance?

God has given us "all things that pertain to life and godliness" (2 Peter 1:3). His Word, the Bible, can equip us for every good work.

> *All Scripture is breathed out by God and profitable for teaching, for reproof, for correction, and for training in righteousness, that the man of God may be complete, equipped for every good work* (2 Timothy 3:16–17).

The Bible can guide us in every area of life, including finances. God has a lot to tell us about money — and I mean a lot. One man compiled 2,350 verses that deal with money in one way or another.[2] We will examine some key principles from God's Word to govern how we view and use money, starting today with two of them.

Principle #1: Everything Is God's.

Animals, people, money, and abilities — they all come from God and belong to Him.

> *The earth is the LORD's and the fullness thereof, the world and those who dwell therein* (Psalm 24:1).

While the Bible shows people setting aside money specifically for God's work, that does not mean that the other amount is ours to do whatever we want with. God rules over *all*.

In this course, we will use the word **steward** to convey the idea that we are really just managers of what God has given to us — His servant, overseeing His resources.

Principle #2: Loving Money Produces Slavery.

> *For the love of money is a root of all kinds of evils* (1 Timothy 6:10a).

Notice this verse does not say money itself is evil. But loving money is. God commands us to love *Him* with all our "heart and with all [our] soul and with all [our] mind" (Matthew 22:37).

The Bible is very clear: we can not serve both God and money.

> *You cannot serve God and money* (Matthew 6:24b).

If money holds our heart, God does not. The moment we start living for things here and now, we have stopped living as God's servant. Living for money (and what it can buy) does not truly satisfy. When we die, we can not take any of it with us. We were made to know our Creator.

- Money can easily become our master and put us in slavery to temporary things.
- Money can be one of God's many blessings when viewed appropriately and utilized like God intended.

Why Study Consumer Math?

Throughout your life, you are going to face a myriad of decisions, from how to spend money you are given for your birthday to which career, car, or house to choose. This course will help prepare you for financial decisions as you grow.

KEEPING PERSPECTIVE

Whether or not you like math, I hope you are excited about this course. It is my prayer that this will be a fascinating journey for you — and one that causes you to dig deeper into God's Word and seek Him for how to live as He desires.

Skill Review: Addition & Subtraction, Plus Abbreviations for Time

One great part about consumer math is that you do not need to be a math whiz to grasp it. The key is knowing when to use what operation. Here is a quick review of when to use addition and subtraction, as we will add and subtract throughout the course.

Addition

Addition shows us the result when we *combine* different values; it gives a *total* or a *sum*.

Example: If you buy items that cost $4.50, $6.79, and $3.99, what is the total?

We can calculate this with addition.

$4.50 + $6.79 + $3.99 = $15.28

Subtraction

Subtraction *takes away*; it shows what we have left after we *remove* some amount.

Example: If a service costs $49.99 and they offer a $10 discount, what is the new total?

Do not let the word total fool you. We have an amount, and we are *removing*, or *taking away*, the discount. This is subtraction.

$49.99 – $10 = $39.99

While a total can refer to the result of addition, in finance, the word *total* often just means the final owed/due, regardless of how it was obtained. **Do not assume just because you see that word that you need to use addition; think through what you are trying to do.**

We can also think of subtraction as showing us the *difference* between two values and helping us *compare* them.

Example: One item costs $4.99 and another $5.99. How much more is the second item?

We are looking here for the *difference* between the items. We calculate this using subtraction. After all, if we took away one amount from the other, it shows the *difference* between the two.

$5.99 – $4.99 = $1

Time Abbreviations & Relationships

Make sure you know these relationships between units of time and their abbreviations, as we will be using them throughout this course.

8 hours (hrs.) = 1 typical workday

5 days (ds.) = 1 typical workweek

12 months (mos.) = 4 quarters (qtrs.) = 52 weeks (wks.)
= 1 year (yr.) $\approx$ 365 days (ds.)

$\approx$ means approximately equal

1 quarter (qtr.) = 3 months (mos.)

$\approx$ 4.33 weeks (wks.) = 1 month (mo.)

(calculated by taking $52 \div 12 \approx 4.33$, as there are 52 wks. in a yr.)

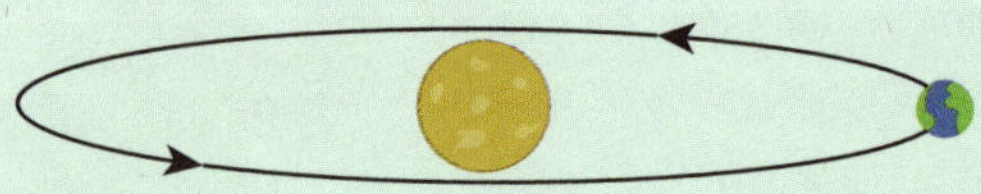

An actual year is 365.25 days because it takes a little longer for the earth to circle the sun. For the sake of ease, though, it is common to round to 360 or 365 in financial uses. We will round to 365 in this course.

KEEPING PERSPECTIVE

Your exercises will help you practice recognizing which operation to use. You will get more practice as we continue through this course because we will be using these operations over and over.

Have you ever wondered *why* addition and subtraction work? God's Word gives us an answer: It teaches us that God established and sustained "fixed laws" in creation. He is holding all things together in a consistent and predictable fashion — predictable enough for us to rely on 2 + 2 equaling 4. Math's ability to work is a testimony to us of God's never-changing nature and faithfulness.

> *Thus says the Lord: "If I have not established my covenant with day and night and the fixed order of heaven and earth, then I will reject the offspring of Jacob and David my servant ..."* (Jeremiah 33:25–26).

I have written this course with the desire to help you serve God by using money wisely without becoming a slave to it. But all the best money management and stewardship is pointless if you do not first have a relationship with God.

- Jesus said to love God "with all your heart ... soul ... and mind" (Matthew 22:37). We are all guilty of loving things in this world more than God at times, whether that is wealth, getting our own way, or being jealous of what other people have. We are all guilty of sin — of falling short of God's commands. This sin earns us eternal separation from God in hell.
- Jesus paid the penalty with death on the Cross "... for the wages of sin is death" (Romans 6:23). When we accept what Jesus did for us, God will make us new and restore our relationship with Him.

If you have not yet humbled yourself before God and received the gift of eternal life in Christ Jesus, I hope you will consider responding today to what Jesus has done.

Skill Review: Multiplication & Division

Now we will keep reviewing how to identify which basic math operation to use.

We are putting a lot of skill review at the beginning of this course to make sure you have all the math you need as we delve into other topics.

Multiplication

Multiplication is a quick way of *adding* the same amount repeatedly.

Example: If you spend \$5.99 a month on a service, what is the cost per year?

Since there are 12 months in a year, rather than adding \$5.99 12 times, we can simply multiply 12 × \$5.99.

$12 \times \$5.99 = \$71.88$

Example: If you earn \$200 a week at a job, how much will you make in a year?

Since there are 52 weeks in a year, we need to multiply 52 by \$200 to discover the yearly amount.

$52 \times \$200 = \$10,400$

Division

Division can be thought of as breaking apart, or *separating*, an amount into piles. In most consumer math applications, those piles are hypothetical piles — breaking a cost down into how much something is *per* month, ounce, or some other unit of measure.

Example: If I spend \$498 every 6 months on auto insurance, how much is it costing me *per* month?

We can discover the answer by division. We want to take that \$498 and hypothetically *separate* it into 6 piles — one for each month. That will tell us how much it is costing each month.

$\$498 \div 6 = \$83$

Note that we could also think of division as repeated subtraction, in that if we were to subtract \$83 from \$498 over and over, we would subtract it 6 times altogether.

Fractions as Division

It can be helpful to represent division using a fraction line. The previous example could be written as $\frac{\$498}{6}$. It can be helpful to read a fraction line as *divided by* or *per*. $\frac{\$498}{6}$ can be read as "\$498 divided by 6" or "\$498 per 6 [months]."

God gave mankind the ability to develop different ways to represent the same thing. There is more than one way to show division. Not only are fraction lines easier to write, but when using algebra or advanced math, they become indispensable. Because of this, **we will be mainly using fractions to show division.**

Multiplication and Division by 10 & Powers of 10

To multiply by 10 or a power of 10, move the decimal point to the *right* the same number of 0s as in the 10, 100, and so forth.

$10 \times 5 = 50$ (decimal point moved to the right 1 place, as 1 zero in 10)

$100 \times 5 = 500$ (decimal point moved to the right 2 places, as 2 zeros in 100)

100 is really 10 × 10, so moving the decimal point to the right 2 times, 1 time for each multiplication by 10, makes sense.

To divide or multiply by $\frac{1}{10}$, $\frac{1}{100}$, and so forth, move the decimal point to the *left* the same number of 0s as in the 10, 100, and so forth.

$50 \div 10 = \frac{50}{10} = 5.0 = 5$ *or* $50 \times \frac{1}{10} = 5$ (decimal point moved to the *left* 1 place, as 1 zero in 10)

$50 \div 100 = \frac{50}{100} = 0.50 = 0.5$ *or* $50 \times \frac{1}{100} = 0.5$ (decimal point moved to the *left* 2 places, as 2 zeros in 100)

Note that we can add 0s as needed when moving the decimal point over.

Dividing by 10 is the same as multiplying by 0.10, or 0.1. After all, 0.1 means $\frac{1}{10}$ ... which is 1 *divided by* 10.

The decimal point is to the right of the one's place, whether or not it is written out. You can add a .0 to the right of a whole number without changing the meaning.

5 = 5.0 and 4 = 4.0

Example: Multiply \$4.59 by 10.

10 × \$4.59 = \$45.90

Example: 100 × \$4

100 × \$4.00 = \$400 *or* \$400.00

Example: \$4 ÷ 10 *or* $\frac{\$4}{10}$ *or* \$4 × 0.10

\$4 ÷ 10 = \$0.40

KEEPING PERSPECTIVE

A big part of consumer math is learning to use the right operation. You might want to make yourself a reminder about what addition, subtraction, multiplication, and division do to consult as needed. There is also an overview you can reference at the beginning of Quarter 3 (page 175).

Income Distribution Equation

When you get money (i.e., income), you can either give, save, or spend it.

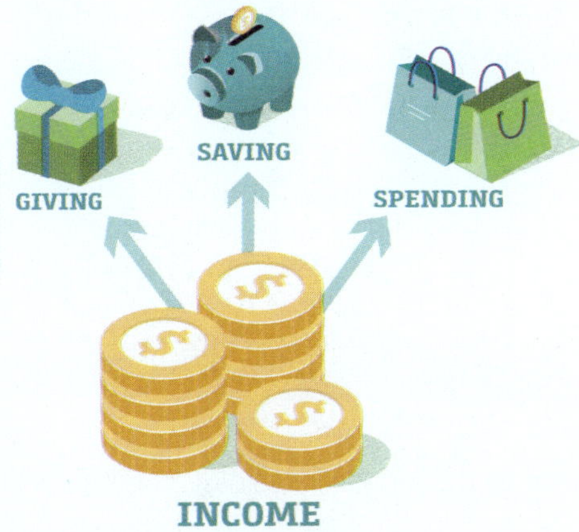

Income Distribution Equation

income = giving + saving + spending

Notice that we listed *spending* last. When you get money, what do you think of first? For many, the answer would be how to spend it. Pausing to think about whether you should *give* or *save* money *before* you start spending can help you make better long-term financial decisions.

Important! This Income Distribution Equation does not tell us how much is in savings already. Instead, it looks at how much of the income received in a period of time is going toward giving, spending, or *increasing* savings.

Look briefly at each value in the equation.

- ***Income*** is money you receive. Think of it as all the money that is coming into your hands — as the resources God is sending your way to steward.
- ***Giving*** is money you give away for others. God wants us to be "cheerful," generous givers (2 Corinthians 9:7).
- ***Saving*** is money set aside; you might save for the unexpected, for big expenses, for retirement, and for long-term goals.
- ***Spending*** is money you do not keep or give. You will need to spend some money on taxes. If you have a home mortgage, you might have to spend on repaying your debt.

  Much of your spending will likely be on living expenses.

spending = taxes + debt + living expenses

Applying the Equation

The Income Distribution Equation can help you figure out how much you have left for one use of money after taking care of the others.

ALGEBRA Connection

We could write the equation algebraically like this:

income =
giving + saving + spending

$I = G + S_a + S_p$

You can think of letters in math as emojis; just like an emoji stands for an emotion, letters stand for quantities, such as *I* for *income.*

ALGEBRA Connection

The Income Distribution Equation is a formula. A **formula** is simply a mathematical relationship, often expressed in symbols instead of words. To use it, plug in the values and simplify.

ALGEBRA Connection

Because God governs creation consistently, you can add in any order, perform the *same operation* using the *same amount* to both sides of an equation without changing the equality, and know that adding $0 will not change the value (thus allowing us to simplify $0 + *saving* to *saving*). We subtracted $2,800, as that was the amount needed to get *saving* on a side by itself.

Example: If you make $3,000 a month, give $500, and spend $2,300, how much are you saving?

The *income* (what you make) is $3,000, *giving* is $500, and *spending* is $2,300.

income = giving + saving + spending

Insert values: $3,000 = $500 + *saving* + $2,300

Simplify addition: $3,000 = **$2,800** + *saving*

Subtract $2,800 from both sides of the equation:

$3,000 = $2,800 + *saving*
−$2,800 −$2,800
$200 = $0 + *saving*
$200 = *saving*

As Christians, we want to be known as givers (we serve a giving God), so we have listed *giving* first and separately, as we want it to be a priority, not an afterthought. While many other Christian resources list giving separately, most apps and budgeting tools will clump giving as a part of spending.

If you are not given a separate giving amount in a problem, assume it is incorporated inside of *spending* and view *income* as equaling your *spending* plus *saving*.

Example: If you make $3,000 a month and want to save $600 of it, how much of it can you spend?

Since nothing is mentioned about *giving*, we will put a $0 for it in the equation, but think of it as being part of the *spending* amount we calculate.

income = giving + saving + spending

Insert values: $3,000 = $0 + **$600** + *spending*

Subtract $600 from both sides of the equation:

$3,000 = $0 + $600 + *spending*
−$600 −$600
$2,400 = $0 + $0 + *spending*
$2,400 = *spending*

KEEPING PERSPECTIVE

If you look just at your income for a period of time, you are either giving, saving, or spending it. Understanding this relationship can help you as you steward the resources God gives you. As He sends you income, you will have choices about how much to give, save, and spend — and where to do so.

Skill Review: Rounding & Percent

Before we continue, there are a couple of math concepts we will use throughout the course that we are going to review.

Rounding

Rounding helps throughout consumer math.

- As you solve problems on the calculator, you will get answers that have many digits after the decimal place. All those digits can get tedious to keep track of. In this course, you will typically **round your calculator's answer to the nearest cent.**
- You will also find rounding helpful when shopping. Have you ever noticed that retailers rarely use whole number prices? Mentally rounding $1.99 to $2 can help you not get fooled into thinking you are only paying $1 when it is really almost $2.
- Rounding can also help you approximate answers and spending.

In Matthew 10:30 we learn God keeps track of every hair everyone has ... "And even the very hairs of your head are all numbered."

»» To round, look at the digit to the right of where you want to round to. If it is a 5 or greater, round up. If it is less than a 5, round down. This is because if the digit to the right is 5 or more, the value is closer to the next whole dollar, cent, or other value we are rounding to; if it is less than 5, it is closer to the previous one.

closer to 0	closer to 1
0	0.5 … 1

Example: Round $4,914.561767 to the nearest dollar.

$4,914.561767
dollar's place ↑ ↑ digit to the right = 5 or greater (so round up)

$4,915 (rounded the 4 up to a 5 and removed all the digits to the right)

Example: Round $4,914.561767 to the nearest cent.

$4,914.561767
cent's place ↑ ↑ digit to the right = less than 5 (so round down)

$4,914.56 (kept the 6 as a 6 and removed all the digits to the right)

When *rounding down*, we keep the digit we are rounding to as it is but remove all digits to the right of it.

Example: Round $5.295 to the nearest cent.

$5.295
cent's place ↑ ↑ digit to the right = 5 or greater (so round up)

$5.30 (rounded $0.29 up to $0.30)

Notice that this particular problem resulted in two digits changing.

Percent

A percent shows what portion of 100 something represents, making it an easy way to see what portion of a whole something represents. The list of consumer math percent applications goes on and on.

- **Stores** run percent-off sales.
- **Interest rates** are given as a percent.
- **Taxes** are a percent of income.
- **Tips** are calculated as a percent of the total.

For example, it is easier to compare what portion of the people surveyed supported a new pool when written using a percent.

If in a survey about a pool, we had 332 people out of 401 vote yes, that would be the same portion of the whole as 82.79 out of 100. **A percent shows us the portion of the whole.**

332 out of 401, or $\frac{332}{401}$ people voted yes

Percent notation: 82.79% voted yes

82.79% could also be written as $\frac{82.79}{100}$; it means 82.79 *per* 100.

Writing as a Percent

The percent sign, %, just means "*per* hundred." Notice that it even has two tiny 0s, just like 100 has two 0s.

>>> To rewrite a fraction or decimal as a percent:

Step 1: If a fraction, remember that a fraction means division, and complete that division.

$$\frac{332}{401} = 332 \div 401 = 0.827930174564 \approx 0.8279$$

Step 2: Once in decimal form, move the decimal point two places to the *right* (i.e., multiply by 100) and add a % sign (to show that the amount is really per 100).

$$0.8279 = 82.79\%$$

>>> In this course, when rounding values, round to the 4th decimal place before writing as a %, as we did with 0.8279. This will result in 2 decimal places in the %, as in 82.**79**%.

Rewriting a Percent as a Decimal

The word *per* is a way of saying "divided by." To rewrite a percent as a decimal, we simply complete the division. Since we are dividing by 100, we move the decimal point 2 places to the left.

$$20.12\% = 20.12 \div 100 = \frac{20.12}{100} = 0.2012$$

KEEPING PERSPECTIVE

Rounding and percents are both concepts you have hopefully learned in the past, but they are worth reviewing, as we will use them over and over again.

Giving & Calculating a Percent of a Value

income = ***giving*** + *saving* + *spending*

While giving is an afterthought for many people, we are starting with it because, biblically speaking, it is important. Think about it: God owns everything, yet He gave up heaven to become a man in order to give us eternal life. We serve an incredibly giving God. As we seek to reflect God's image here on Earth, giving is one of our greatest privileges. We should be asking God to work in our hearts a spirit of generosity and a love for others that flows out of our love for God. Generosity is part of exhibiting the Fruit of the Spirit love. As God fills our hearts with love for Him and others, we will want to use the resources at our disposal to bless those around us.

Giving is specifically called out in Romans 12:8 as one of the gifts of the Holy Spirit. While some will excel at this gift more than others, we are *all* called to give.

For you know the grace of our Lord Jesus Christ, that though he was rich, yet for your sake he became poor, so that you by his poverty might become rich (2 Corinthians 8:9).

Some find 10%, the percent the Israelites brought as a tithe, as a helpful starting amount to give. Just keep in mind that God really owns everything, and He may call you to give more.

How Much Should We Give?

The Israelites were to bring a tithe of 10% of all they earned to God, along with other sacrifices and offerings. Altogether, they were really giving more than 20%. Back at this time in Israel, this giving would have included some of what we pay in taxes too, as the Levites also handled some of the responsibilities our government handles today.

The word *tithing* does not appear in the New Testament. Instead, there is a theme of giving regularly and generously. In 1 Corinthians 16:2, Paul instructed that money be set aside each week. There are examples of believers selling what they had and giving the entire "proceeds to all [who] had need" (Acts 2:45). Early believers did not limit themselves to giving 10%. In 2 Corinthians 9:7 it states "... for God loves a cheerful giver." He desires each of us to give abundantly, just as He does to us.

Calculating a Percent of a Value

You may find it helpful to set a percent of your income to give before you figure out how much is left to save or spend. That way, you guard against letting other priorities keep you from the joy of giving.

So how can you find a percent of your income? For example, what is 20% of $1,599?

>>> To determine a percent of another number, simply multiply the percent by that number. The word *of* proves helpful when calculating a percent. *Of* indicates multiplication. You multiply the percent by the value *of* which you want to take that percent.

20% is the same as 0.20.

Example: Calculate 20% of $1,599.

20% *of* $1,599

0.20 × $1,599 = $319.80

»» **Calculating 10% of a value is easy.** Multiplying by 10%, or 0.10, is the same as multiplying by $\frac{1}{10}$, which we reviewed earlier is really *dividing* by 10. So, simply move the decimal place to the *left* one digit to discover 10%.

Example: Calculate 10% of $670.

10% *of* $670

0.10 × $670 = $67.0 = $67

KEEPING PERSPECTIVE

The point is this: whoever sows sparingly will also reap sparingly, and whoever sows bountifully will also reap bountifully. Each one must give as he has decided in his heart, not reluctantly or under compulsion, for God loves a cheerful giver (2 Corinthians 9:6–7).

Notice these verses compare giving to sowing seeds. If you only sow a few seeds, you only get a small harvest, while more seeds yield more plants and thus more harvest. Some have interpreted this and a few other verses to mean that God promises a financial reward for giving, but the Bible never promises that godliness will lead to material wealth. Just look at Paul — he wrote most of the New Testament, yet he spent a lot of his life in prison. Jesus' Kingdom is "not of this world" (John 18:36).

The idea that giving leads to financial wealth here can stem from (and lead to) the love of money the Bible warns about. There is indeed a blessing in giving — but that blessing may be an eternal one. The harvest that we yield may be an eternal harvest, as what we give impacts souls for eternity. God promises, though, that He will give us what we need to complete the work He has given us to do.

And God is able to make all grace abound to you, so that having all sufficiency in all things at all times, you may abound in every good work (2 Corinthians 9:8).

Percentage Given

When you calculate a percent *of* a value, you end up with an amount that represents that portion *of* the value.

30% *of* $400 is $120

$$0.30 \times \$400 = \$120$$

Now, we will generalize and write a formula for working with percent problems. To do so, we will use names to refer to the different parts of the equation. In the calculation above, $120 *is* what we call the **percentage**; it is the number that is 30% *of* $400. That is, if we look at $400 as a whole made up of 100 parts, $120 would be 30 of those parts.

Percent Formula

percent **of** *base* **is** *percentage*

% × *base* = *percentage*

or

P × *base* = *percentage*

Do not let the terms *percent* and *percentage* confuse you. A *percent* is a portion of 100 (it will be written as a % or decimal), while a *percent**age*** [same word but with "age" at the end] is the *amount* that is that same portion of some value — it is the *result* of calculating a percent of a value.

Knowing this formula, we can then plug in values to calculate the percent (%) when we know the other values.

> Books vary in the terms they use for the different values in a percent problem (or even rearrange the equation to solve for different values). A few common terms are shown. Whatever terms or method you have learned is fine to use.
>
> **% or rate *of* base or whole *is* percentage or amount or part**

Example: What percent of $13,000 is $400?

The words *of* and *is* help us figure out where to plug in the values. When showing examples of solving a problem, we will use a *P* instead of %, since that is easier to handwrite.

$$P \times \$13{,}000 = \$400$$

We need to divide *both sides* by the same amount ($13,000). We can show this with a fraction.

$$\frac{P \times \cancel{\$13{,}000}}{\cancel{\$13{,}000}} = \frac{\$400}{\$13{,}000}$$

Notice the division cancels out the multiplication on the left, leaving us with the equation below.

$$P = \frac{\$400}{\$13{,}000}$$

We can then complete the division, round, and write the answer as a percent.

$$P = \$400 \div \$13{,}000 \approx 0.0308 = 3.08\%$$

ALGEBRA Connection

In algebra, we look at consistencies of God's created order that hold true for any value and use that to find unknowns. Any value divided by itself equals 1. Thus, $\frac{\$13{,}000}{\$13{,}000}$ equals 1. And any value multiplied by 1 equals itself, which allows us to know that $P \times 1$ will just be *P*. (Note that we did not show the step of $\frac{\$13{,}000}{\$13{,}000}$ equaling 1; we just crossed it out and wrote *P*, as that is the end result.)

>>> When you calculate a percent, round your answer to the fourth decimal place, then rewrite as a percent.
Example: 0.2012415 ≈ 0.2012 = 20.12%

Notice that we ended up dividing the value after the "*is*" by the value after the "*of*" to calculate the *percent*. We are working with the same basic relationship we have been, only discovering a different value in that relationship.

Example: If you have an annual income of $50,000 and give $8,000, what percent of your income did you give?

We need to figure out what percent *of* $50,000 *is* $8,000.

P *of* $50,000 *is* $8,000

Replace the *of* with a multiplication sign and the *is* with an equal sign.

$$P \times \$50{,}000 = \$8{,}000$$

Divide both sides by $50,000.

$$\frac{P \times \cancel{\$50{,}000}}{\cancel{\$50{,}000}} = \frac{\$8{,}000}{\$50{,}000}$$

Complete the division and write the answer as a percent.

$$P = \frac{\$8{,}000}{\$50{,}000} = 0.16 = 16\%$$

As we go forward, we often will not show every step of a percent problem. For example, we might show the above example like this:

$$P \times \$50{,}000 = \$8{,}000$$

$$P = \frac{\$8{,}000}{\$50{,}000} = 0.16 = 16\%$$

Remember that we got from the first line to the second by dividing both sides by the same amount ($50,000).

KEEPING PERSPECTIVE

Giving is truly a gift. It is a way we can experience the joy of helping others. And believe it or not, that is way more wonderful than getting all we want.

In all things I have shown you that by working hard in this way we must help the weak and remember the words of the Lord Jesus, how he himself said, "It is more blessed to give than to receive" (Acts 20:35).

The Methods & Heart of Giving

Since giving is important, we will take another look at some mechanics and further explore the heart of giving.

Methods

There are a lot of different ways to give — and I mean a lot. Below are a few of the methods, along with some considerations of each:

Method	Explanation	Considerations
→ **Cash or Check**	You can put a check or cash in your church's offering plate (or box) or mail it to the church or charity.	Be careful mailing cash, as it is easy to steal.
→ **Electronically**	You can often give via a credit card or electronic withdrawal from your bank account. You may even be able to automate giving this way, agreeing to pay a certain amount per month, which allows the church or charity to better plan.	If you use a credit card, the church or charity will have to pay a small credit processing fee, or you may opt to pay the fee. With electronic withdrawals, make sure you trust the security of the website or app before giving out your bank account information.
→ **Directly from Paycheck**	Some employers will allow you to give money to a charity directly from your paycheck.	You may be limited in where you can give.
→ **Via a Payment App or Other Money Service**	Payment apps or money services are particularly helpful when giving to individuals. If your account does not display your name, it can even be anonymous.	Gifts that are not to a qualifying nonprofit are not tax-deductible, but this may still be part of where God leads you to give. Be discerning and do your due diligence.

Donor-Advised Funds

- Donor-advised funds can help you give to multiple charities all in one place.
- Donor-advised funds may help you save on taxes by letting you count donations in different tax years.
- Donor-advised funds may help you give non-monetary gifts to a charity in a way that can help you save on taxes.

A few notes about donor-advised funds:

- There is usually a monthly fee for donor-advised funds, making it better suited for those giving a lot.
- Check the fund out thoroughly before you sign up. If the fund goes under, any money you have in there will too, so pick one that is respected.
- Funds may suggest deadlines before December 31 for end-of-the-year donations, so plan ahead.

Get a parent's permission before making any payment online or within an app.

Other Ways

There are many other ways to give, such as at the end of your life through your will or life insurance. You can also give your time, talents, or non-monetary resources. (So even before you have an income, there are plenty of ways to give.)

Calculating Amount Left to Reach a Giving Goal

>>> To calculate the amount left to reach a giving goal, first discover the percent you have given. Then do one of the following:

1. Determine the amount (i.e., percentage) you want to give and subtract the amount (i.e., percentage) you have given from it.
2. Subtract the percent you have given from the percent you want to give and calculate that amount (i.e., percentage).

Example: You check and see that you have given $1,200 and have made $20,000 so far this year. What percent of your income have you given, and how much more would you have to give to give 10%?

% *of* **base** *is* **percentage**

Calculate the percent you have given:

$P \times \$20{,}000 = \$1{,}200$

$P = \dfrac{\$1{,}200}{\$20{,}000} = 0.06 = 6\%$

METHOD 1

Amount (i.e., percentage) you want to give: $0.10 \times \$20{,}000 = \$2{,}000$

Subtracting what you have given: $\$2{,}000 - \$1{,}200 = \$800$

METHOD 2

Percent left to give: $10\% - 6\% = 4\%$

Amount (i.e., percentage): $0.04 \times \$20{,}000 = \$800$

You can add and subtract numbers written as a percent just like you would other decimal numbers — just keep the percent sign in your answer.

KEEPING PERSPECTIVE

You might be tempted to think that what little you can give does not make a difference. The truth, though, is that the *act* of giving will make a difference *in you*. God can do amazing things with small offerings, as the feeding of the 5,000 reminds us. In the Book of John and the Book of Matthew, Jesus took just five loaves of bread and two fish and used it to feed 5,000 men, plus women and children. Jesus did not need that boy's lunch — He could have made food appear from nothing. He does not need our money either. Yet He invites us to have the privilege of giving.

Start being a generous giver now. Do not wait until you have more money.

Review & Case Study

Your exercises will have you begin reviewing Lesson 1. Remember, when we get income, we either give, spend, or save it. Each paycheck gets broken down like we have learned:

income = giving + saving + spending

As we continue to explore this equation, we will apply the math we are reviewing to many other applications.

The Heart of Giving

Since we talked about giving in this lesson, we will look at some of what Jesus had to say about it. **Jesus talked about not giving for recognition from others.** The Pharisees gave for others to see, but that is not what God is after. Jesus said not to let our left hand know what our right hand is doing — in other words, to not give to impress others.

> *But when you give to the needy, do not let your left hand know what your right hand is doing, so that your giving may be in secret. And your Father who sees in secret will reward you* (Matthew 6:3–4).

Jesus also talked about how **God cares about our hearts, not the value of what we give**. One day, Jesus pointed to a widow who put a very small amount into the treasury, yet Jesus said it was more than all the others because she put in all she had to live on.

> *Jesus looked up and saw the rich putting their gifts into the offering box, and he saw a poor widow put in two small copper coins. And he said, "Truly, I tell you, this poor widow has put in more than all of them. For they all contributed out of their abundance, but she out of her poverty put in all she had to live on"* (Luke 21:1–4).

God does not need our money, but He wants our hearts. He wants us to have the joy of stepping out in faith and trusting Him with our finances by giving abundantly, even sacrificially like that widow. Giving is an opportunity for us to grow in faith — to partner with God and His work. And it can start *now*, even if you do not have very much. The following case study brings this point home.

Case Study: A Handful of Rice

This case study comes from Generous Giving. Generous Giving is a nonprofit that helps people on their journey in biblical generosity. They have been kind enough to grant us permission to share a synopsis of some of the modern-day true stories of people living out godly stewardship that they offer in video form on their website.

How could a state in India that has been called the "most backward state in India" and the "poorest of the poor"[3] possibly give to advance God's Kingdom? Mizoram Christians (in northeastern India) have every excuse not to give, yet their giving is astonishing.

Back in 1910, the church realized they needed "Bible women" to spread the gospel. They started a Handful of Rice ("Buhfai Tham") ministry to support these women. The Handful of Rice ministry was simple: every time those participating cooked a meal, they set aside a handful of rice to take to the church. The church then sold the rice to get income for ministry.

More than 100 years later, the Handful of Rice ministry has grown, and giving has expanded even beyond a handful of rice. In 2009–2010, the equivalent of 1.5 million dollars was given through the Handful of Rice ministry, and 13 million dollars altogether. This poor state is able to support their own ministry and other gospel ministries.

The Mizorams' giving attitude reminds us that we can *all* experience the joy of giving, no matter what our financial situation. One woman only makes the equivalent of less than a dollar a day,[4] yet she is part of giving through a simple handful of rice. Giving has become a habit that does not depend on wealth. It is one you can start now and build on as you grow.

"As long as we have something to eat every day, we have something to give God everyday."[5]
— Rev. Zosangliana Colney, Executive Secretary, Mizoram Presbyterian Church

"There have been times when some churches have thought that we need to get blessings from God ... and the attitude has been what can we get when we become part of the church. But here, the Handful of Rice offering inspires us that God has called us to share what we have with God, for God's ministry."[6] — Rev. Dr. Roger Gaikwad, Principal, Aizawl Theological College

From Mizoram Women:

"I give because I believe it is a way of showing my gratitude to the Lord."[7]

"Whenever I do this, I believe that I cook a meal for the Lord as well."[8]

Introduction to Living Expenses

In this lesson, we are going to begin exploring living expenses. Since a lot of purchases fall under living expenses, we will discuss living expenses many times in this course. For now, we are going to take a high-level look at how decisions you make will affect how much you will spend. We will also focus on reviewing some math skills that help when making these decisions.

income = *giving* + *saving* + ***spending***
spending = *taxes* + *debt* + ***living expenses***

Living Expenses = Lifestyle Choice (Needs or Wants)

As we start looking at living expenses, there are two key principles to keep in mind.

Key #1: Wants Are Not Needs.

I need that. Without it, I can not be ____________ [happy, successful, satisfied, popular — you fill in the blank].

Advertisers and marketers carefully craft everything from ads to product placements to try to convince you to buy their offerings. It is tempting to think you *need* everything you see and to spend a lot.

For those of us living in the United States, we live in luxury compared to many people in the rest of the world. We think we need a steak dinner; others are hoping they at least get a bite of anything to eat that day.

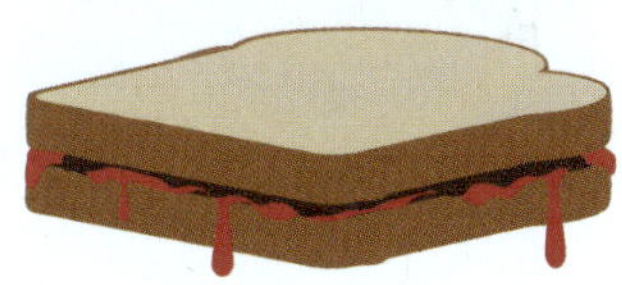

The Apostle Paul says all we really *need* is food and clothing.

> *But if we have food and clothing, with these we will be content* (1 Timothy 6:8).

Paul had learned the secret of contentment is not found in getting everything we want. **Contentment is found in living for and in Christ.**

> *Not that I am speaking of being in need, for I have learned in whatever situation I am to be content. I know how to be brought low, and I know how to abound. In any and every circumstance, I have learned the secret of facing plenty and hunger, abundance and need. I can do all things through him who strengthens me* (Philippians 4:11–13).

The Bible tells us that true joy comes from God's presence, not temporal things that pass away (Psalm 16:11).

As you face decisions regarding living expenses, **watch your heart.** Do not buy into the lies that you *need* things to make you happy.

You may *want* a steak dinner ... but all you really *need* is food to fill your stomach.

Key #2: What You Spend Makes a Difference.

Many living expense decisions seem minor, but they add up. For example, you spend $2 extra at the store each week. A little multiplication can show us what that looks like over longer periods of time.

Multiplying weeks in a year (52) by $2:

$$52 \times \$2 = \$104$$

Spending $2 extra each week means you are spending $104 extra a year. After 10 years, you will have spent $1,040 extra.

$$10 \times \$104 = \$1{,}040$$

This number could be even more if you increase spending by more than $2 each week. What if you spend $25 extra each week at the store?

Extra spent after 1 year = 52 × $25 = **$1,300**

Extra spent after 10 years = 10 × **$1,300** = $13,000

Just $25 a week extra means you will spend an extra $13,000 over 10 years.

KEEPING PERSPECTIVE

How much you spend on living expenses can vary greatly. Keep in mind that the more you spend, the less you have available to give and save. Knowing how to stretch your money can be helpful.

- Spending adds up ... as do small savings.
- While you may not have much to spend now, thinking through what you are spending is something you can start doing today.

Mental Math: Introduction and Addition & Subtraction

As you shop, the decisions you make add up.

- Being able to mentally perform some math as you go can help you quickly evaluate decisions.
- Quick mental math can also help you check to make sure you are being charged the accurate price — mistakes happen. Plus, there could be fees you did not know about that get added that would cause you to change your purchase.
- Mental math can also help you with bigger decisions that you might pull a calculator out for. It is easy to accidentally plug the wrong amount into a calculator, but if you can approximate an answer mentally, you will know if your answer is way off.

Hopefully, you have already learned a lot of mental math. Just in case, though, we are going to look at a couple of strategies and then practice.

Mental Problem-Solving Strategies

- Round one of the values, then adjust at the end.
- Break the problem down by digit or by dollars and cents if needed.
- Turn subtraction into addition by adding from the amount you are subtracting.

Round One Value and Then Adjust

Example: For items that cost \$1.99 and \$1.20, what is the total?

Round \$1.99 up \$0.01 to \$2:
\$2 + \$1.20 = **\$3.20**

Since rounded up by \$0.01, we have to subtract \$0.01:
\$3.20 – \$0.01 = \$3.19

Breaking Down by Dollars and Cents

Example: If you buy items that cost \$1.50 and \$2.75, how much will you spend?

Break down by dollars and cents: **\$1.50** + **\$2.75**

Add the cents, thinking through that we need \$0.50 of the \$0.75 to make \$1 (as \$0.50 + \$0.50 = \$1), and then we would still have \$0.25: **\$0.50** + **\$0.75** = **\$1.25**

Add the dollars: **\$1** + **\$2** = **\$3**

Add together: **\$1.25** + **\$3** = \$4.25

Turning Subtraction into Addition

In this example, we have also thought in terms of adding actual coins (a nickel and pennies) to go from the total to what was given.

Example: If you are buying something that costs $0.15 and have $0.23, how much will you have left?

You are trying to calculate $0.23 – $0.15.

Subtract:	$0.15	
Add $0.05 (1 nickel) to get to $0.20 (the nearest multiple of $0.10):	$0.05 (1 nickel)	→ $0.20
Add $0.03 to get to $0.23:	+ $0.03 (3 pennies)	→ $0.23
We have to add a total of $0.08 to get $0.23.	$0.08	

When asked to approximate a value, so long as your answer is fairly close to the actual, it will be counted as correct.

Finding an Approximate Answer

Sometimes all we need is an approximate answer. In that case, you can round one or *both* values.

Example: *Approximately* how much would items that cost $1.99 and $1.20 total?

We could approximate this as $2 + $1.20 = $3.20 (rounding one value) or as $2 + $1 = $3 (rounding both values).

KEEPING PERSPECTIVE

- **Do not stress over what strategy you use;** so long as you can get the right answer (or close if approximating), that is fine.
- **Do not panic if some of the mental math feels a bit challenging.** It is okay if you get some problems wrong or have to use paper. If you keep practicing, you will improve your skill.
- **Remember that the purpose is to help** you quickly assess costs and make sure you are being charged accurately.
- **Be patient** as you practice.

Lifestyle Choice & Unit Price

- Sarah heads out to eat at a steak restaurant, where she orders prime rib.
- Melody picks out supplies from a grocery store for a gourmet steak dinner.
- Jeremy selects some clearance chicken that he plans to use for making a simple stew.

All these people are buying food for dinner. The amount they are spending, however, will vary greatly.

How much you spend on living expenses is a lifestyle choice that makes a difference. Think about the Income Distribution Equation: *income = giving + saving + spending*. If *spending* increases, either *giving* or *saving* has to decrease; if *spending* decreases, one or the other (or both) will increase. If your *spending* gets out of control, then you are hampering your ability to give generously or save for priorities God has given you (a car, home for your future family, travel, or some other goal). If you spend $500 a month less, that is $500 more you could either give or save.

If you want to avoid the never-ending pull of wanting more and more, have a plan ahead of time for how much you are willing to spend on living expenses (and categories of living expenses, such as food and clothing) each month. Decide beforehand (when you are looking at the big picture and not just awestruck by something you want) how much you will spend on different expenses. We will look more at how to do this using a *budget* in the next quarter.

Daily Choices & Unit Price

How do you live within whatever amount you set for living expenses? You will make a lot of smaller decisions. **Basic math skills can help you evaluate these decisions.** Which pair of socks is cheaper, and by how much? How much more would one bag of chips cost than another?

You can find the answer using basic division by looking at the unit price. After all, you want to hypothetically break apart, or divide, the cost by the units you are buying.

>>> The **unit price** is found by taking the total $ and *dividing* it by the total units.

$$\textit{unit price} = \frac{\textit{total }\$}{\textit{total units}}$$

Notice how the unit prices of the bottles show us that the largest one, while it costs the most, costs the least *per ounce (oz).*

Common Unit Abbreviations

gallon (gal.)
quart (qt.)
pint (pt.)
fluid ounce (fl. oz.)

Dry Measurements

ounce (oz.)
pound (lb.)

Notice the use of the word ***per***. As we learned when looking at ***per***cent, ***per*** means "divided by." In the case of the ratio examples of the bottles, we have divided the dollars by the ounces, so our answer is in dollars ***per*** ounce.

$3.29 14 oz.	$4.29 20 oz.	$4.99 32 oz.	$5.29 38 oz.
$\frac{\$3.29}{14 \text{ oz.}} \approx \$0.24$ per ounce	$\frac{\$4.29}{20 \text{ oz.}} \approx \$0.21$ per ounce	$\frac{\$4.99}{32 \text{ oz.}} \approx \$0.16$ per ounce	$\frac{\$5.29}{38 \text{ oz.}} \approx \$0.14$ per ounce

Keep the Heart in Mind

While math can help us understand and evaluate a decision, it can not tell us which is the best decision to make. There may be times when spending that extra $2 is good. God has given us all we have and wants us to enjoy it — but not to set our hope on it.

> *As for the rich in this present age, charge them not to be haughty, nor to set their hopes on the uncertainty of riches, but on God, who richly provides us with everything to enjoy* (1 Timothy 6:17).

We need wisdom to make wise choices. We get that by depending on the Lord and renewing our minds in God's Word.

> *Do not be conformed to this world, but be transformed by the renewal of your mind, that by testing you may discern what is the will of God, what is good and acceptable and perfect* (Romans 12:2).

The Bible also puts a huge priority on our heart. We need to keep it in the right place if we want to make wise decisions. Remember, the love of money will lead us astray. If we love things here on earth (which money helps us get), it is going to affect our decisions. If we love God above all else, we will be able to see clearer to make wise decisions.

> *Keep your heart with all vigilance, for from it flow the springs of life* (Proverbs 4:23).

KEEPING PERSPECTIVE

How much you spend on living expenses varies greatly based on the choices you make. Someone shopping at a name-brand store is going to spend a different amount for clothing than someone wearing thrift-store or off-brand outfits (which can sometimes be just as nice). Consider your priorities, such as how long you plan to keep clothing, the quality, and the cost. Having a plan (and knowing your priorities) *before* you shop can help keep your spending in check so you do not wind up buried in credit card debt. Math skills can help you figure out how much something costs so that you live within that plan.

Mental Math: Multiplication & Division

We will continue exploring mental math by discussing some strategies for multiplication and division.

For calculating change with money, we worked on finding an exact value with addition and subtraction. Now we are going to switch focus to approximating. **In many mental math scenarios, you are simply looking for an approximation.**

For example, if you are trying to figure out which of these boxes of cereal is less expensive per ounce, rather than pulling out a device to find an exact unit price, a mental approximation would tell you what you need to know.

$\frac{\$4.70}{20\text{ oz.}}$

$\frac{\$3.25}{12\text{ oz.}}$

Mental Problem-Solving Strategies

- **Round or use a nearby value that is easy to work with** (especially if approximating).
- **Break the problem down into problems you can find easily** (such as multiplying or dividing by 10 or dividing by 2) or into dollars and cents or digit by digit.
- **Ignore units of measure until the end.**

Example: Approximate the unit price of the cereal box that is $4.70 for 20 ounces.

Round: $\frac{\$4.70}{20\text{ oz.}} \approx \frac{\$5}{20}$ Think of 20 as 10 × 2: $\frac{\$5}{10 \times 2}$

Now divide by 10 and then by 2:

By 10: $\$5 \div 10 = \$0.50$ By 2: $\$0.50 \div 2 = \$0.25$

Answer: $0.25 per oz.

Example: Approximate the unit price of the cereal box that is $3.25 for 12 oz.

Think of 12 oz. as close to 10 and divide:

$\frac{\$3.25}{12\text{ oz.}} \approx \frac{\$3}{10} = \$0.30$ per oz

Notice we ignored the ounces until the end and just thought about finding $5 divided by 10 instead of $5 divided by 10 oz. But we specified that it was $0.25 per oz. at the end.

The actual and approximate unit prices for the cereal boxes are shown. Notice that either way, the 20-oz. box is less per ounce.

$\frac{\$4.70}{20\text{ oz.}}$
approximate unit price: $0.25
actual unit price: $0.24

$\frac{\$3.25}{12\text{ oz.}}$
approximate unit price: $0.30
actual unit price: $0.27

»» Sometimes, rather than rounding, you might need to approximate as a value that perfectly divides by the other.

Example: Approximate the unit price of something that is \$8.99 for 5.

Approximate \$8.99 as \$10 so it will evenly divide by 5: $\frac{\$8.99}{5} \approx \frac{\$10}{5} = \$2$

Now, the actual answer will be slightly less than \$2, as we rounded up (\$1.80 to be exact).

These next two examples deal with finding an exact answer rather than an approximation; they break the problem down by digit.

Example: Approximately how much would 6 of a \$1.99 item cost?

If we round to \$2, it is easy to see the answer is about \$12, as 6 × \$2 = \$12.

Example: You are considering signing up for a software that costs \$43 per month. If you keep it for 5 months, what will that cost you?

Calculate:	5 × \$43
Break down by digit:	\$43 = **\$40** + **\$3**
Multiply 10's place by 5:	5 × **\$40** = **\$200**
Multiply 1's place by 5:	5 × **\$3** = **\$15**
	\$200 + **\$15** = \$215

KEEPING PERSPECTIVE

While an approximation is often all you need when mentally multiplying or dividing, we will continue to have more addition and subtraction problems asking for an exact answer, as you need an exact amount when making change.

As you solve problems, **pay attention to whether the problems ask for an approximation or estimation or not**, and keep the mental strategies we covered here in mind. If approximating, remember that you can round to help you. Approximating will help you estimate costs.

Skill Review: Percent Off & Mental Math: Finding Half

Today, we are going to continue reviewing some math skills and mental math that come in handy in daily life-spending scenarios.

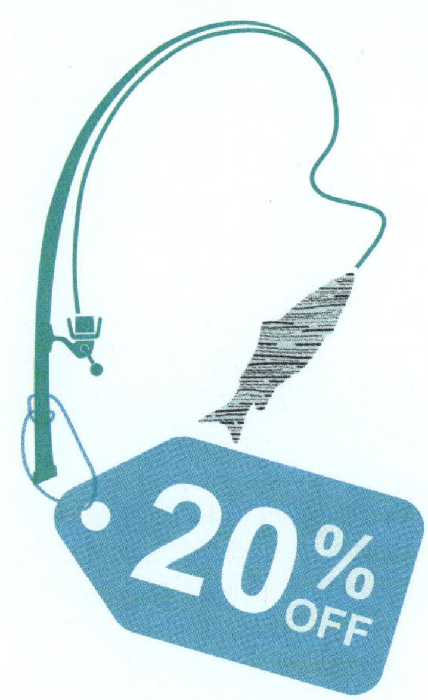

Skill Review: Percent Off

"Wow, everything is 20% off!" you exclaim as you notice a 20% off sign while walking into the store. You see an item you have been wanting. The price tag reads $15.99. You count the money in your pocket. You have $11. Is it enough to buy the item after the discount?

To answer that question, you might think you should start by calculating 20% *of* $15.99.

% *of* base *is* percentage

$$0.20 \times \$15.99 \approx \$3.20$$

Now, this did not really tell you what you wanted to know, did it? You just found 20% *of* $15.99. That is the amount of the discount. But you really want to know your cost *after* this discount.

> ⋙ There are two different ways to calculate the total after a percent off:

1. Subtract the amount off (i.e., the percentage) from the original cost.

   $\$15.99 - \$3.20 = \$12.79$

2. Calculate the percent you will end up paying by subtracting the percent off from 100%, then calculate that percentage to determine that amount (i.e., percentage).

   A percent is a portion *of* 100, so 100% is the original value. If something is 20% off, you will end up paying 100% – 20% = 80%.

   $0.80 \times \$15.99 \approx \$12.79$

> ⋙ **You can add or subtract numbers written as a percent just like you can if they were in decimal notation.**
> 100 – 20 = 80, and 100% – 20% = 80%.

Calculating the cost after the discount would let you know that the $11 you have is not enough. Keep in mind that sales tax will likely be added to the cost after the discount too.

Mental Math: Finding Half

You and your friend go out to eat. You both decide to split the bill — that is, to each pay half of it. How can you calculate how much you should each pay?

Save Time – Memorize what 1, 3, 5, 7, and 9 (the odd numbers through 9) divided by 2 equal. It will help you with mental arithmetic.

>>> To split or halve something, just divide it by 2 using the mental math strategies we discussed in the previous section.

Example: If you and your friend plan to split a $33 bill, how much will you each pay?

Splitting a bill is the same as dividing it by 2. We do this by dividing the 10s and 1s places separately, thereby breaking the problem into ones we can more easily do mentally.

Calculate:	$33 ÷ 2
Break down by digit:	$33 = **$30** + **$3**
Divide 10's place by 2:	**$30** ÷ 2 = **$15**
Divide 1's place by 2:	**$3** ÷ 2 = **$1.50**
Add together:	**$15** + **$1.50** = $16.50

We could also have thought through that $16 + $16 = $32, while $17 + $17 = $34, so $33 would be halfway in between the two at $16.50. **There is often more than one way to solve a problem.**

>>> If you only need an approximate answer, remember that you can round.

Example: If you and your friend plan to split a $33 bill, approximately how much will you each pay?

Since we only need an approximate answer, we can round to $30.

$30 ÷ 2 = $15

KEEPING PERSPECTIVE

He who loves money will not be satisfied with money, nor he who loves wealth with his income; this also is vanity (Ecclesiastes 5:10).

As you learn math to help make living expense decisions, remember that if having more and more to spend becomes your love, it will never be enough. The skills you are learning will instead help you to live within a boundary you prayerfully set.

Mental Percent Problems

Since we find a percent of a value when calculating sales, sales tax, giving, leaving a tip, and more, knowing how to work with percent problems mentally will save you a lot of time ... and it is not as hard as you might think.

Rounding and Thinking in Terms of 10%

If the percent you are trying to find is easy to calculate from 10%, then you can use the knowledge that 10% can be found by moving the decimal place to the left (refer to Lesson 1, Day 6) to easily solve.

>>> **Round and think of how the percent you are calculating relates to 10%.**

Example: If you want to leave a 20% tip on a $159.98 bill, about how much should you leave?

1. Round the original price.	*Round $159.98 to $160.*
2. Discover 10% of the original price by moving the decimal place to the left.	*10% of $160 is $16.*
3. Since 20% is twice 10%, we need to take the 10% twice.	*$16 + $16 = $32*

Rounding and Thinking in Terms of 50%

Thinking in terms of 10% works well for calculating a percent close to 10%. What about a percent like 50% or 60%? While they can still be found by taking 10% and adding it several times, it is often easier (and more accurate) to think in terms of 50%.

>>> **Calculating 50% of a number is the same as calculating half of a number or dividing it by 2.**

After all, $50\% = \frac{50}{100} = \frac{1}{2}$. Like 10%, it is not hard to calculate 50% mentally.

Example: Calculate 50% of $45.99.

1. Round to the nearest whole dollar.	**$46**
2. Divide by 2 to calculate 50%.	*40 divided by 2 is 20, and 6 divided by 2 is 3, so 46 divided by 2 is 20 + 3 = 23.*

50% of $45.99 is about **$23**.

Different Ways to Approximate: There are different ways to approximate an answer. On the examples, we could have multiplied the value for 10% by 5 for 50%, 4 for 40%, and 6 for 60%. However, that answer would not have been as accurate since we would be multiplying a rounded value. Feel free to use whatever methods you like; your answer just needs to be reasonably close to the actual.

Example: Calculate 40% of $45.99.

Follow steps 1–2 in the example on page 39 to calculate 50%.

3. Think through how 50% compares to the percent you want to find. 50% – 10% = 40%

   10% *of* **$46** = 0.10 × $46 = $4.60 ≈ $5

   $23 – $5 = $18

Example: Calculate 60% of $45.99.

Follow steps 1–2 in the example on page 39 to find 50%.

3. Think through how 50% compares to the percent you want to calculate and adjust as needed.

   50% + 10% = 60%

   10% *of* **$46** = 0.10 × $46 = $4.60 ≈ $5

   $23 + $5 = $28

Calculating the Amount Left to Pay

Sometimes, we want to know the amount left to pay, not just a percent of a value. You can approximate this mentally by adding another step.

Example: Calculate how much something would cost if it is on sale for 15% off $87.50.

1. Round the original price. *Round $87.50 to $88.*
2. Calculate 10% of the original price by moving the decimal place to the left. *10% of $88 is $8.80.*
3. (optional) Round the 10% to the nearest whole dollar to make it easier to work with. *$8.80 rounds to $9.*
4. Think about how that compares with the percent you want to find. *15% is 10% + 5% (which is half of 10%). So, we would have $9 + $4.50 (half of $9). We can round $4.50 to $5, giving us $9 + $5, or $14.*
5. **Added Step:** Subtract the discount amount from the rounded original price. *$88 – $14 = $74*

We could also have calculated 85% of $87.50 to solve this problem. Since the sale is 15% off, the item will end up costing 100% – 15% = 85%.

KEEPING PERSPECTIVE

- Calculating a percent off mentally may seem involved, but it gets quite easy with practice.
- Notice that we are applying other mental math skills we have looked at.

As you learn to work with coupons, keep in mind that just because something is on sale does not mean you need to buy it. Stores often run sales to try to get you to buy things you do not really need. While buying items you have planned for on sale can save you money, beware of letting sales trick you into things you do not need and really can not afford. Remember, "... godliness with contentment is great gain" (1 Timothy 6:6), not all the things money can buy.

Subscriptions

Cell phone plans, internet plans, streaming services — these are all examples of subscriptions. In a subscription, you agree to pay a fixed amount every month, year, or other time period for some kind of service.

Beware of the Dangers

While subscriptions can be hard to avoid, there are some key dangers to be aware of.

Danger #1: They Sound More Inexpensive Than They Are.

Subscriptions can sound less expensive than they are — $10.99 a month does not sound like much ... until you do the math and see how it adds up over time.

Cost after 1 year: 12 × $10.99 = **$131.88**

Cost after 5 years: 5 × **$131.88** = $659.40

Cost after 10 years: 10 × **$131.88** = $1,318.80

Imagine if you had four or five monthly subscriptions of $10.99. They would add up.

Danger #2: It Is Easy to Forget Them.

It is easy to sign up for something and then forget about it. In that case, you are spending money for something you are not even using — not just once, but over and over again.

Danger #3: Prices Might Increase or Competition May Lower.

Sometimes, places will offer a free trial or a reduced rate for a time if you subscribe. After that free trial or reduced-rate time period, though, the subscription price increases. Companies might also increase their rates — or competitors might lower them, making your existing subscription no longer the value it was at first.

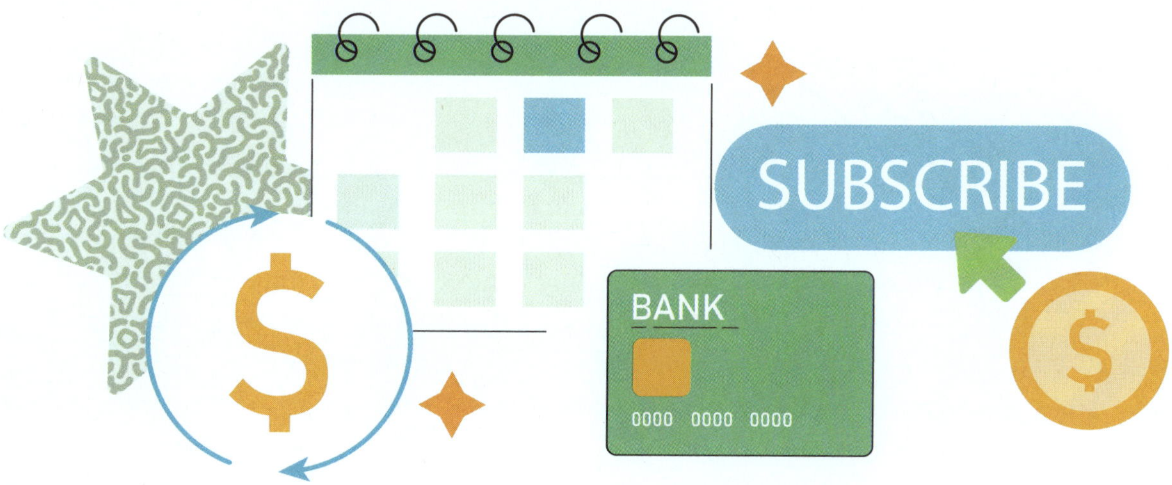

Comparing Costs

Sometimes, you may need to compare the cost of a subscription with the cost of a different subscription or the cost of a one-time purchase. You can do so using basic math.

> *Example:* A certain subscription service is $35 a month or $80 a year. After how many months would it cost more to pay monthly than yearly?
>
> To answer this, we need to divide $80 by $35. That will tell us how many months of the monthly subscription cost the same as the yearly price.
>
> $\frac{\$80}{\$35} \approx 2.29 = 3$ months

With monthly subscriptions, parts of a month often mean you have to sign up for an entire additional month, even if the value would round down.

Notice we wrote the answer as 3 months instead of rounding down to 2 months. We can not really pay for part of a month, can we? After 2 months, we would only have paid $70, so that is still cheaper than $80. But after the third month, we will have paid more than $80. Thus, the answer to after how many months it would cost more to pay monthly is 3, not 2, months.

Cautions

Be cautious when signing up for a subscription. Companies use them because they know that they are a great way to get recurring money, which adds up over time.

- **If you plan to cancel after the free or reduced-rate trial,** make sure you set reminders so you do not forget to cancel.
- **If you plan to keep your subscription,** check to make sure you understand the true cost over a year of what you are saying yes to. Compare that with other options.
- **Re-evaluate your recurring subscriptions from time to time.** Make sure that you understand the true cost, that you are still using them, and that they are still the best choice for the task.

KEEPING PERSPECTIVE

The Bible does not give formulas for exactly how to spend money. God wants our hearts to be His. He wants us seeking Him and viewing all we have as His. He wants us to care about eternity and what He cares about. As we do that, it is going to affect the financial decisions we face.

Utility Bills

So far, we have looked at living expenses over which you have a lot of control. Utilities are ones over which you have less control. A **utility** is "a service (such as light, power, or water) provided by a public utility."[9] Utilities do not come labeled with price tags. Because they are provided publicly, there is often only one company to use and thus no way to shop around for the best price. They are billed based on usage. Use more, and you pay more. Use less, and you pay less.

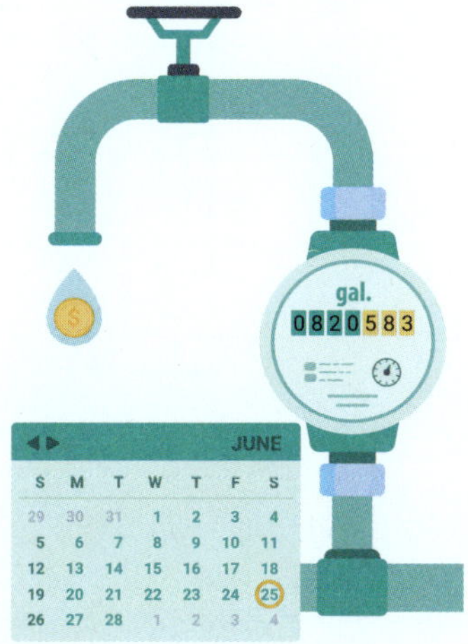

Water

Homes have water meters that keep track of the water that is used. The reading on the meter is checked each month and compared with the reading the previous month. The water usage can be reported using different units.

- **Gallons** – Sometimes, the meters measure in *1,000 gallons* used. To determine the *gallons* used, multiply the amount used on the bill by 1,000. The first table shown on the right is for a usage of 4,700 gallons.
  4.7 × 1,000 = 4,700
- **CF** – Other water companies may measure in units called CF, which stands for cubic feet. Each cubic foot of water holds about 7.48 gallons. Since each CF is about 7.48 gallons, we can determine the gallons by multiplying the CFs by 7.48:
  600 × 7.48 = 4,488.
  The first table shown on the right is for a usage of 4,488 gal.

Water companies typically charge a set amount per gallon/thousand gallons/CF/HCF used. You may also find charges on your water bill for sewage. Sewage is a charge for the water that goes down the pipes. Sewage cost is also tied to your usage.

Gas

Gas, like water, is billed by what you use. And, like water, it is measured in different units, which we will not go over.

Just keep in mind that **even if you do not know anything about a unit of measure, you can still use basic math to explore your bill.**

Example: How much are you paying per ***therm*** (a unit gas can be measured in) if you are charged $33 for 60 therms?

This is a unit price question. We want to calculate the price *per* therm — and we can do so just like we would if calculating the unit price of an item.

$\frac{\$33}{60 \text{ therms}} = \$0.55$ per therm

Used = Ending – Beginning

Examples:
326.2 – 321.5 = **4.7**
10,600 – 10,000 = **600**

Beginning Reading	Ending Reading	Used
321.5	326.2	**4.7**

Measurements given in 1,000 gal.

Beginning Reading	Ending Reading	Used
10,000 CF	10,600 CF	**600** CF

If you are ever trying to compare water bills from different cities, make sure you are comparing the same units. Most of the time, though, you are not going to need to compare. The key is to know that **the more you use, the more you pay,** regardless of what unit it is measured in.

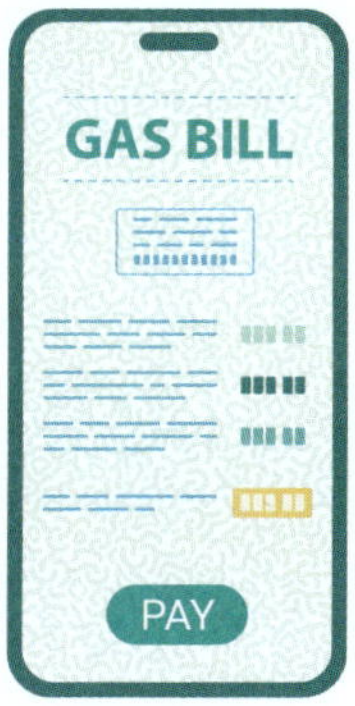

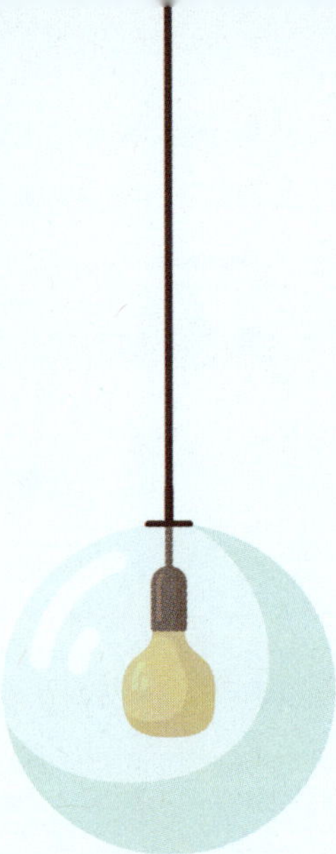

Example: By how much would your bill increase if you pay $0.55 per therm and use 30 extra therms a month?

We want to add up $0.55 for all 30 extra therms — repeated addition means we need to multiply.

$30 \times \$0.55 = \$16.50$

Electricity

Electricity is billed by kWh (kilowatt-hour). As with water, the bill is based on the difference between the beginning reading and the ending reading. And as with other utilities, the more you use, the more you pay.

Extra Charges

Get to know your utility bills so you can make wise choices.

In addition to the fees for the usage, utilities often add other charges. There might be a base fee for the service that gets charged no matter what. There could be taxes added. Extra may be charged for power used during peak times — or a discount given for non-peak usage.

Substitutes for Utilities

There are an increasing number of options available as substitutes for public utilities, such as solar panels for electricity. Math can help you explore these options too.

>>> Figure out how long it will take to recuperate a cost by dividing the cost by the savings for some period of time.

There may be tax savings for using certain renewable energy sources, thereby lowering the cost (and thus the years to recuperate that cost).

Example: If it costs $16,800 to install solar panels but it saves you $2,400 a year, how many years will it take to recuperate the cost?

We need to *divide* the cost by the yearly savings: $\frac{\$16,800}{\$2,400} = 7$

Some homes may use well water instead of city water. In that case, your cost would be the cost needed to maintain your well — likely *much* less than what you would pay the city. However, you might still want to watch what you use so you do not waste your well water.

There can be more reasons than financial to watch what you use of resources.

KEEPING PERSPECTIVE

When my husband and I moved into our first home, we were warned to be careful how much we ran our sprinklers, as we had a large yard and a generous sprinkler system. We listened at first but then forgot and ran the sprinklers extensively in an effort to get greener grass. A significantly higher water bill greeted us the next month. Be mindful of how much water, gas, and electricity you are utilizing.

Review

As you navigate life's little choices, remember that **godliness with contentment is what is great gain, not simply getting all you want.**

> *But godliness with contentment is great gain, for we brought nothing into the world, and we cannot take anything out of the world* (1 Timothy 6:6–7).

The lifestyle God calls us each to may be different, but we need to make sure the choices we make are in obedience to Him and not simply because we have fallen into materialism and the love of money. Buying a pricey shirt is not a sin, but if we are buying a pricey shirt because we have bought into the lie that we need it to be happy or attractive (or with money we either do not have or should be spending elsewhere), we need to take a good look at our heart.

> *Trust in the Lord with all your heart, and do not lean on your own understanding. In all your ways acknowledge him, and he will make straight your paths* (Proverbs 3:5–6).

> *You say, "... If I had a little more, I should be very satisfied." You make a mistake: if you are not content with what you have you would not be satisfied if it were doubled.*[10]
> — Charles Spurgeon, British preacher

Page intentionally blank

Savings, Compound Interest, & Debt

So far, we have seen that when we get income, we either give, save, or spend it. We also mentioned that spending can be broken into categories, including debt.

income = *giving* + ***saving*** + *spending*

spending = *taxes* + ***debt*** + *living expenses*

In this lesson, we are going to explore both saving and debt. We are looking at them together because both are connected to interest — a fee earned or paid for the use of money.

- **When you save money, it can *earn* you interest** (if you put it in a bank).
- **When you have debt, you have to *pay* interest.**

The Importance of Saving

When you get your first paycheck, it may be tempting to spend all of it. It is not hard to think of things you would like to buy. But is spending all you make wise?

Prepare for the Lean Years.

God warned Joseph that there would be 7 years of plenty followed by 7 years of famine. Joseph advised Pharaoh to prepare for the years of famine by storing $\frac{1}{5}$th of the harvest during the years of plenty. That savings would carry them — and many surrounding nations — through the years of famine (Genesis 41).

> *Moreover, all the earth came to Egypt to Joseph to buy grain, because the famine was severe over all the earth.* (Genesis 41:57).

There is wisdom in saving up during the times of plenty to help with life's expected and unexpected harder times. We live in a fallen world.

- People get sick.
- Cars break down.
- We age.

If you spend all your paycheck each week:

- You may not have money to pay for that unexpected missions trip.
- You may not have money to give to someone in need.
- You may not have extra cash to fix your car when it breaks down.

>>> **A reasonable amount of savings helps us make wise preparations for the future and frees us up to respond to opportunities that arise.**

The prudent sees danger and hides himself, but the simple go on and suffer for it (Proverbs 22:3).

In general, it is a good idea to get 3–6 months of your monthly expenses in savings that you could easily access if needed. There are a variety of purposes for which you might save additional money or for which you might end up using what you have saved. Here are a few:

As a Bill Buffer – If you have some savings, when a bill is due before your paycheck, you are not in a bind.

Medical Expenses – Savings set aside to cover medical expenses.

Emergencies – We live in a fallen world where emergencies (car breakdowns, layoffs, and other unforeseen hardships) happen.

Big Expenses – Some big-ticket items or events (cars, home down payments, vacations, wedding, college, and other expensive purchases) may require saving up for.

Retirement – We will look at saving for when you can not earn money anymore in a later lesson.

When you save money, you might not have a specific purpose in mind. **General savings** set aside for whatever need arises can later get used for one of the purposes above.

There Is a Balance.

Savings should never replace trusting God. We can not ever save enough to cover every emergency, nor should we use a desire to save as an excuse to not be generous and help others right now.

Our goal in savings should not be:

- to get all we want
- to relax, eat, drink, and be merry
- to protect ourselves from all hardship

The wise store up choice food and olive oil, but fools gulp theirs down (Proverbs 21:20; NIV).

The goal should be Christ and His glory and Kingdom.

Part of living for Christ's Kingdom, though, is avoiding the constant pull of materialism that urges us to spend everything we have on what we do not truly need. Having a more long-term perspective can free us up to live generously and not be caught in constant crises.

KEEPING PERSPECTIVE

None of us knows what the future holds. From the Bible, though, we know life includes trials. Wise stewardship saves when there is excess to help through those times.

Understanding Interest

Rather than keeping your savings in a piggy bank, ideally, you want to be earning money on your savings so that it grows. Through growth and time, a little bit of savings can grow into a lot.

There are many ways you could invest your savings. For now, we will look at the general idea of interest.

When a bank has your money, they invest it. In return for the use of your money, you can earn interest on it. **Interest** is a fee earned or paid for the use of money. It is calculated as a percent, called the **interest rate**.

Types of Interest

- **Simple interest** is earned on your initial investment.
- **Compound interest** is earned on your initial investment *and* previous interest earned.

For example, you have $5,000 and invest it where it can earn 10% interest.

With simple interest, it earns $500.

$$10\% \textit{ of } \$5{,}000 = 0.10 \times \$5{,}000 = \$500$$

If you keep it in the investment another year, it is going to earn another $500. It will keep earning interest on *just the initial $5,000.*

With compound interest, the first year it will earn $500. But the next year, it will earn 10% of the initial $5,000 *and* the $500 interest from the previous year.

$$10\% \textit{ of } (\$5{,}000 + \$500) = 0.10 \times \$5{,}500 = \$550$$

The year after that, it will again earn on both the initial investment *and* all the earnings.

$$10\% \textit{ of } (\$5{,}000 + \$500 + \$550) = 0.10 \times \$6{,}050 = \$605$$

In other words, every time the interest is **compounded** (or calculated), you earn interest on the initial amount *and* all the interest. Over time, this can add up, especially as interest is typically calculated more often than the one time a year we showed here.

Investing Terms Defined

- **Deposit**: Putting money in
- **Withdrawal**: Taking money out
- **Balance**: The amount of money held or owed. For example, your bank balance is the money the bank is holding for you. Your **beginning balance** is the starting amount; the **ending balance** is the ending amount.

The parentheses show to perform the addition first and then multiply. We will review parentheses further in the next section.

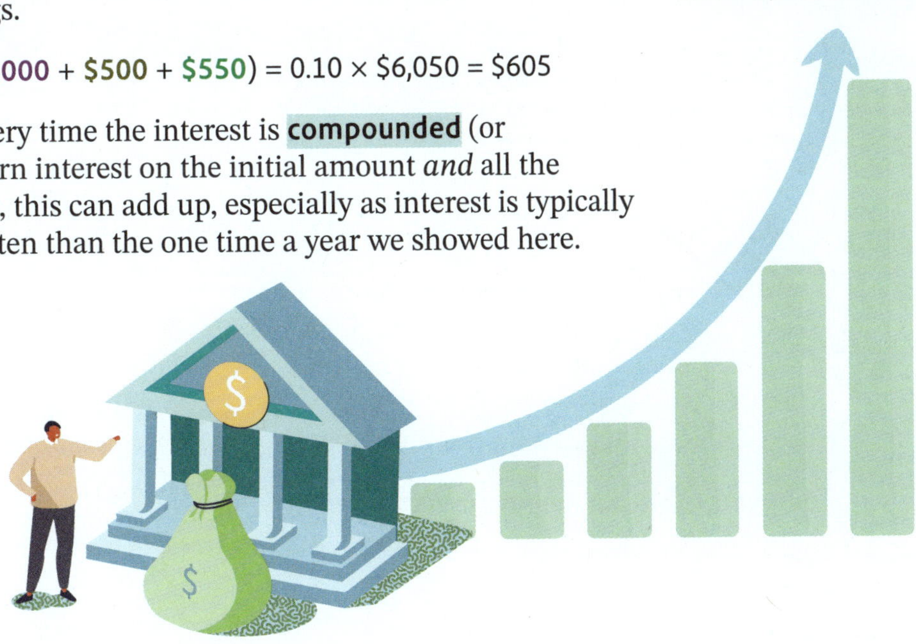

The graph shows how a beginning balance of $5,000 earning 10% interest calculated yearly will grow. The horizontal axis (i.e., line) marks the years and the vertical one tells the balance.

Notice that 30 years is not labeled. We see 32 along the horizontal axis, but not 30. However, we can see that there is a marker every 4 vertical lines that increases by 8 years (8, 16, 24, 32). To find the years per each vertical line, we need to divide 8 by 4.

$$8 \div 4 = 2$$

Each vertical line tracts 2 years, which means that 30 years would be the vertical line before the 32 marker.

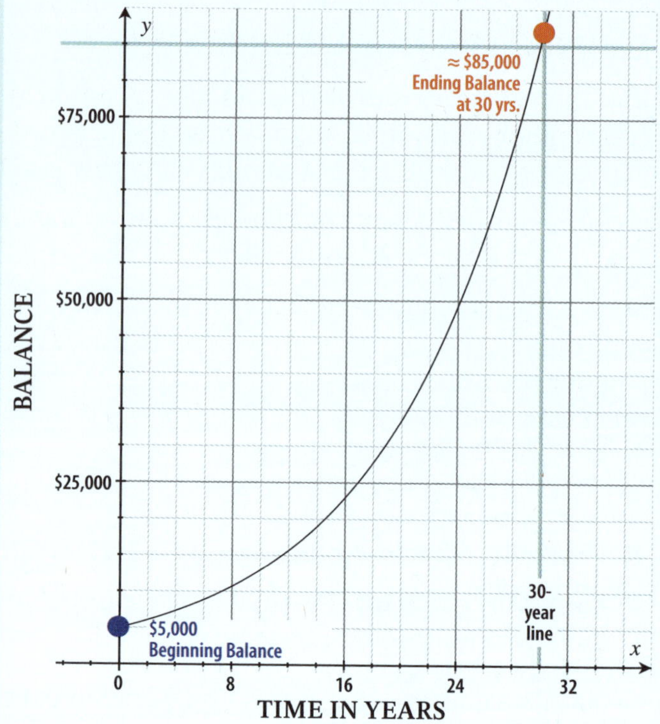

>>> To determine how much each line is showing an increase of, divide the increase between two markers by the number of lines between those markers.

Example: How much does each horizontal line show an increase of?

There is a marker every 5 horizontal lines that increases by $25,000.

$25,000 ÷ 5 = $5,000

Each horizontal line shows an increase of $5,000.

Depending on the current interest rates, you might not earn 10%. Smaller interest rates, while they will not see as much growth, will still add up to significant amounts over time.

We can use this knowledge to help us see the value after 30 years. Notice that the 30-year line intersects the curve just below the horizontal line that is 2 above the $75,000 label. If each line represents $5,000, that would make it $75,000 + $5,000 + $5,000 = $85,000. $5,000 will grow to more than $85,000 in this scenario. Start saving while you are young so that savings has time to really grow.

We rounded to $85,000, as that was the nearest horizontal line, but the actual would be closer to $87,000. In this course, if the curve intersects between horizontal lines, round to the nearest line to help keep grading easier.

KEEPING PERSPECTIVE

In the parable of the talents, the first two servants invested their money, making more. The third did not. The master in the parable of the talents pointed out that the third servant could have at least put his money with bankers to earn interest. While it is not the main point of the passage, there is wisdom in earning money on the savings you have been given.

Then you ought to have invested my money with the bankers, and at my coming I should have received what was my own with interest (Matthew 25:27).

Skill Review: Order of Operations/Parentheses, Plus Compound Interest Formula

As you hopefully already know, the **order of operations** is to solve problems from left to right,

1. **solving anything inside parentheses first**, following the rest of the order of operations within the parentheses,
2. then **exponents and roots**,
3. then **multiplication and division**,
4. then **addition and subtraction**.

Knowing the order of operations can help you solve problems more concisely — and, as we will see in a minute, will help with calculating compound interest.

Example:

$4 + 5(4 + 2)^2$

$4 + 5(6)^2$

$4 + 5(36)$

$4 + 180$

184

Note that we did not write out the × sign. **Putting a value next to a parenthesis automatically means to multiply.** 5(36) means 5 times 36.

Example: You are buying 2 items at the store. One costs \$5 and the other is 2 for \$8. You have a coupon to save 20% off your total purchase. How much will you owe?

Calculate the cost of the 2 for \$8 item: $\frac{\$8}{2} = \$4$

Add the costs together: $\$5 + \$4 = \$9$

Calculate the percent that you would actually pay: $100\% - 20\% = 80\%$

Multiply the percent times the cost: $0.80 \times \$9 = \$7.20$

Rather than writing out separate steps, we can show all of this in a single step using parentheses and the order of operations.

$(100\% - 20\%)\left(\$5 + \frac{\$8}{2}\right) = 80\% \times \$9 = 0.80 \times \$9 = \$7.20$

We followed the order of operations within the parentheses, dividing the \$8 by 2 *before* adding to the \$5.

After we simplified everything inside the parentheses, we removed the parentheses and simplified the rest of the problem.

Compound Interest Formula

To calculate how much money will grow if earning compound interest, use the Compound Interest Formula. Notice the formula has parentheses.

Compound Interest Formula

$$P = P_i(1 + r)^t$$

$$\textit{ending balance} = \textit{beginning balance}(1 + \textit{growth rate})^{\textit{number of times compounded}}$$

- The *growth rate* is the rate of growth per compound period — the growth that occurs every time the interest is compounded, or calculated. For now, it will be the interest rate you are given.
- Parentheses are used to show that we need to add 1 to the interest rate *first*.

ALGEBRA Connection

The Compound Interest Formula has an exponent (t) — a superscript value that tells you how many times the value to its left needs to be multiplied by itself. For example, 4^3 means $4 \times 4 \times 4$. In algebra, you will explore mathematical relationships with exponents further.

calculated = compounded

- This is finding the *ending balance*, not the *total interest*. To find the total interest, we would need to find the ending balance and then subtract the beginning from it.
- We could have written $P = P_i \times (1 + r)^t$, but the $\times$ sign is not necessary next to parentheses.
- Use the formula by plugging in values and simplifying.

Example: Calculate the ending balance after the interest is calculated 50 times if the beginning balance is $1,000 and the interest is calculated at an interest rate of 2.5% each time it is calculated.

Plug in values: $P = \$1{,}000\,(1 + 0.025)^{50}$

If your calculator has ( ) buttons, input the right side of this equation and press enter. If not, then continue.

Solve inside the parentheses: $P = \$1{,}000\,(1.025)^{50}$

Simplify exponents and roots: $P = \$1{,}000(3.437109)$

Complete multiplication and division: $P \approx \$3{,}437.11$

If you wanted to determine the total interest, you could now do so.

total interest = ***ending balance*** – ***beginning balance***
= **$3,437.11** – **$1,000** = $2,437.11

To calculate exponents on a calculator, locate a button that looks like y^x or ^ (if your calculator does not have either, you will need to ask a parent to help you search online for how to enter exponents on it or use one that does). In the example we would input 1.025, then press either y^x or ^ and then the exponent of 50.

If your calculator has ( ) buttons, you could input the entire expression at once. If you input $\$1{,}000\,(1 + 0.025)^{50}$, the calculator will automatically follow the order of operations.

KEEPING PERSPECTIVE

Parentheses and the order of operations are part of the language side of math. God gave man the ability to develop standard ways of writing to make it easier to communicate. Once you learn the language, it makes solving problems simpler.

More with Calculating Compound Interest

Compound Interest Formula: $P = P_i(1 + r)^t$

When using the Compound Interest Formula remember the following:

- The r must be the ***interest rate used to calculate the interest each time***.
- The t must be the ***number of times the interest is compounded.***

Finding the r

Different investments advertise different rates. Sometimes, you will need to adjust the rate before plugging into the formula. Many times you will be told the yearly or annual rate. In this course, unless told otherwise, **assume a rate is a yearly rate.**

≫ **To find the r for the formula, divide the yearly, or annual, rate by the number of times the interest is compounded in a year.** That will tell you the growth that occurs each time the interest is compounded.

See Calculator Review on next page for an explanation of the E^{-4}.

COMPOUNDED MONTHLY

Example: Find the r to use in the formula for 5.1% compounded monthly.

Divide by 12 since 12 months in a year:

$$r = \frac{0.051}{12} = 0.00425$$

COMPOUNDED DAILY

Example: Find the r to use in the formula for 5.1% compounded daily.

Divide by 365 since 365 days in a year:

$$r = \frac{0.051}{365} \approx 1.397260274E^{-4}$$
$$= 0.0001397260274$$

Finding the t

The t in the formula is the number of times the interest is compounded.

≫ **You can find t by multiplying the number of years by the number of times in a year the interest is compounded.**

COMPOUNDED MONTHLY

Example: Find the t to use in the formula for interest compounded monthly after 10 years.

Multiply years by 12 since 12 months in a year:

$$t = 12 \times 10 = 120$$

COMPOUNDED DAILY

Example: Find the t to use in the formula for interest compounded daily after 10 years.

Multiply years by 365 since 365 days in a year:

$$t = 365 \times 10 = 3{,}650$$

Calculator Review

When calculating the rate used each time, you might get odd-looking results on your calculator, such as

$$1.397260274E^{-4} \text{ or } 1.397260274E\text{-}4 \text{ or } 1.397260274^{-4}$$

The *E*, or exponent, is a calculator's way of abbreviating scientific notation, where numbers are written as a multiplication by 10 raised to a power. All of the expressions shown above, when on a calculator, mean this:

$$1.397260274 \times 10^{-4}$$

While we will not review scientific notation here, know that when the *E* or an answer is followed by a negative exponent on a calculator, you can rewrite it in the decimal notation we typically use by simply moving the decimal point the same number of places to the left as the negative exponent, adding zeros as needed. This is essentially multiplying by 10 raised to that negative power.

$$1.397260274 \times 10^{-4} = 0.0001397260274$$

Decimal point moved 4 places to the left.

Putting It All Together

By inputting everything into the calculator at one step, you did not have to worry about reading the rate in scientific notation. **Input parentheses around the exponent if needed for the calculator to treat the exponent as 12 × 20, not just 12.**

Example: Calculate the ending balance after 20 years at an interest rate of 3% on an initial balance of $5,000 if the interest is compounded monthly.

Formula: $P = P_i(1 + r)^t$

Insert values: $P = \$5{,}000\left(1 + \frac{0.03}{12}\right)^{(12 \times 20)}$

Notice that we are dividing the annual interest rate (0.03, or 3%) by 12 to calculate the interest rate used each time and multiplying the years (20) by 12 to find the number of times the interest is calculated and added to the balance.

Input into a calculator and solve: $P \approx \$9{,}103.77$

>> Whenever using a formula, always make sure you understand what the values are really representing. Otherwise, it is easy to end up inserting the wrong values, thus getting an incorrect answer. In the case of compound interest, watch what you plug in for *r* and *t*.

KEEPING PERSPECTIVE

Interest, especially compound interest, can help savings grow. Even if you are only able to save a small amount of money each month, saving a little now while you are young can result in sizeable savings by the time you need it. Earning interest is a way to grow the money God has entrusted to you.

The Dangers of Debt

We will now switch gears from from savings to **debt**, or something (for our purposes, money) *owed*. While savings is generally a wise thing, debt is something to avoid whenever possible.

- Borrowing makes you a slave now and in the future.

> *The rich rules over the poor, and the borrower is the slave of the lender* (Proverbs 22:7).

Lending indicates having money to spare; borrowing indicates being indebted or enslaved to others. Rather than being free to use money you earn as you need to, you are obligated to pay off your debt. And if you do not get the money you planned on, you may have to start working extra or selling things to pay it off. In a very real way, debt enslaves.

> *For the LORD your God will bless you, as he promised you, and you shall lend to many nations, but you shall not borrow, and you shall rule over many nations, but they shall not rule over you* (Deuteronomy 15:6).

- Borrowing can easily become a way to avoid living within what God has provided, even if that is not the lifestyle you want. Watch out for borrowing from a discontented heart.
- When you owe money, instead of *getting* interest, you end up *owing* interest.

We will now take a look at how interest on debt works. Checking out the numbers will better illustrate the dangers of debt.

>>> We can use the Compound Interest Formula for interest *owed* as well as interest *received*.

Interest in Reverse

Compound Interest Formula: $P = P_i(1 + r)^t$

$$\text{ending balance} = \text{beginning balance}(1 + \text{growth rate})^{\text{number of times compounded}}$$

With debt:

- The *beginning balance* is the initial *amount borrowed*.
- The *ending balance* is the *total debt* with interest that you will end up paying.

This rule is the same as we looked at on page 52 for interest earned, only we are using borrowing wording. Many finance formulas are the same idea, just reworded.

Interest rates on money borrowed are often given in **APR**, or **annual percentage rate**. Credit cards, for example, will list their APR. The APR is an annual rate, and you can use it in the Compound Interest Formula. We will adjust the wording of the formula to clarify how to use it for debt.

Reworked for debt:

$$\underline{\text{total debt}} = \underline{\text{amount borrowed}}(1 + \text{growth rate})^{\text{number of times compounded}}$$

When Romans 13:8 says to "Owe no one anything," it is in the context of paying taxes. When taken in context of the whole of Scripture and considering other verses that discuss borrowing, Romans 13:8 does not seem to be prohibiting all borrowing, but rather emphasizing showing love by fulfilling obligations and making payments as agreed upon. However, some Christians have personal convictions not to borrow at all.

This formula assumes *no payments are made* on the debt along the way.

Example: Calculate the total owed, including interest, after 10 years on a $100 credit card debt if the interest rate is 18% APR compounded daily and no payments are made.

Insert values: $P = \$100\left(1 + \frac{0.18}{365}\right)^{(365 \times 10)} \approx \$604.70$

>>> Calculate the total interest owed by subtracting the beginning amount from the ending amount. In the case of debt, that is the *amount borrowed* from the *total debt.*

Example: Calculate the total interest for the previous example.

total interest = ***total debt*** – ***amount borrowed***

total interest = **$604.70** – **$100** = $504.70

We can also use graphs instead of the formula to explore interest paid or gained.

ALGEBRA Connection

Graphs like this are generated on a calculator by inserting values for all but ***P*** and ***t*** in the Compound Interest Formula (the ***P*** gets input as a ***y*** and the ***t*** as an ***x***, as those are the letters calculators use).

Example: Use the graph to find the *interest paid* on a $10,000 loan at just 5% APR compounded monthly over 20 years, assuming no payments are made.

All we have to do is identify the total due and the amount borrowed on the graph and subtract to calculate the interest paid.

$27,000 – **$10,000** = $17,000

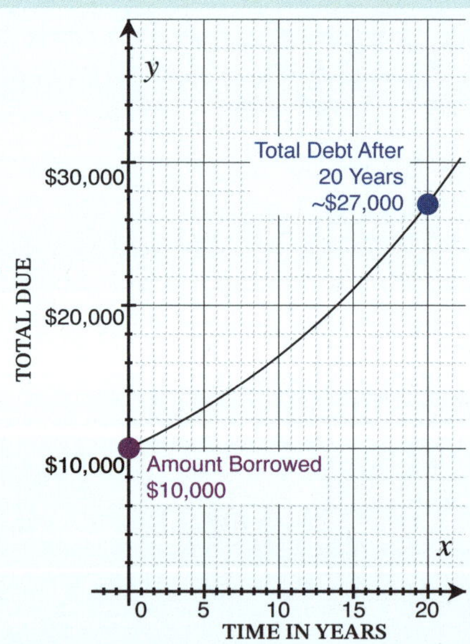

KEEPING PERSPECTIVE

Savings:	**Debt:**
you *gain* interest	you *pay* interest
prepares for the future	*enslaves* the future (to pay the debt)

It is wise to pray long and hard (and check your heart) before taking on debt.

Credit Cards & Credit Card Debt

A credit card is exactly what it sounds like — a card that gives you credit, or the ability to buy something and pay for it later.

↑ **Credit cards can be a convenient way to pay for items.** Rather than carrying around lots of cash or a checkbook, you can use a credit card, making one payment to the credit card company each month instead of tons of little payments.

↓ **Credit cards can also be an easy way to end up in debt** if you are not careful. If you do not pay your credit card bill on time, then whatever you have charged to that credit card automatically becomes a loan (i.e., money owed).

Credit Card Debt Adds Up

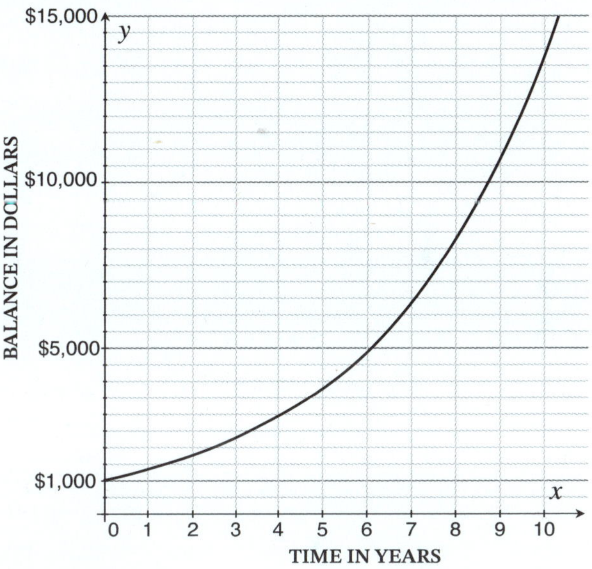

Credit cards, on average, can charge as high as 27.80% APR or more.[11] That is way more than a bank would charge. At this rate, debts pile up very quickly.

Now round to 30% APR. That would mean that *every year*, you are paying 30% of whatever your unpaid balance is. The graph shows a hypothetical credit card debt of $1,000 over time at 30% APR if no payments are made on it and the interest is only compounded once a year. If you do not pay the interest off, then the next year you will owe interest on that $300 too — and so the debt can keep growing. Debts can quickly become astronomical.

Credit card companies require minimal payments. However, if you keep charging other items to the card, it can offset that minimal payment, and your debt can keep growing.

Credit Card Fraud Cautions

We live in a fallen world with fallen people. Credit card fraud is sadly common. The good news is that credit card companies are good at removing fraudulent charges and not charging you.

However, it is wise to be careful and avoid the hassle of the fraud in the first place. Here are some suggestions:

- Keep your card in a safe place. If you lose it, report the loss right away by calling the credit card company.
- Guard your credit card number.
  - Make sure the websites you enter the card number into are legitimate and secure.
  - Do not give it out over the phone unless you are sure you know who you are giving it to.
  - Shred documents that have your account number on them.

Beware of credit card debt; it is a dangerous slope. In general, **it is wise to pay off your credit cards every month, thus avoiding the slavery of debt and the exorbitant interest rates credit cards charge.** If you find you are not able to use credit cards without stumbling into debt, consider finding other payment options.

- Beware of spam calls, texts, and emails asking for money over the phone or email.
- Do not give your credit card information when using public or unsecured Wi-Fi.
- Keep your online passwords secure.
- Check your credit card statements each month, looking for suspicious charges. (Otherwise, you might pay for something you did not buy.)

Credit Card Considerations

When selecting a card, here are some things to keep in mind:

- **APR** – APR makes a big difference if you have a balance, but if you pay your card off each month, the APR does not matter.
- **Rewards** – Many cards offer rewards such as cash back or airline miles. Look for a card with rewards you will use.
- **Acceptability** – Not every card is accepted everywhere; it is good to have a backup card, especially if you use a less common card.
- **Annual Fee** – Some credit cards charge an annual fee. Only go for these if you are sure you will get more out of them than the cost of the fee. Be careful about signing up because the first year is free; if you do, you will need to make sure you cancel in time. Canceling the card temporarily negatively affects your credit score, so it is not something to do often.
- **Limit** – Credit cards put a limit on how much they are willing to let you charge. If you show yourself trustworthy, that limit will increase.
- **Quantity** – Whatever credit cards you have, you will want to pay them off every month. Otherwise, you will face late fees, interest, and a lowered credit score.

KEEPING PERSPECTIVE

We live in a world that is constantly trying to get us to live for the here and now, convincing us we need that new outfit, latest video game, or item that is being advertised. Beware of the love of money (and what money can buy). Buying on credit promises the ability to buy everything you want, until you find yourself drowning in debt you can not pay back.

At the same time, you may find credit cards a convenient mode of paying in our digital age. So long as you pay your credit card off each month when the bill comes in, you avoid debt and demonstrate your ability to pay what you have promised, which can help you qualify for bigger loans, such as home loans.

Home Debt

While housing probably is not an issue for you right now, when you get older, you will have to decide where to live. We will explore housing in much more detail later. For now, here is a bird's-eye view of home debt.

Since most people can not afford the prices of homes right away, buying a home typically involves a type of loan called a mortgage. In a **mortgage**, the lender can claim property if the borrower does not make their payments.

Home mortgages are one kind of debt that many Christian financial advisors will recommend in many situations, with the caveat to be careful about how much of a mortgage you take on and pay it off as quickly as possible. Why?

- **It may be better financially than other options.** Unless you can afford to buy a place to stay with cash, if you do not get a mortgage, the alternative is renting. In both renting and a mortgage, you make monthly payments. But with renting, your money all goes to the landlord, while with a mortgage, some of each payment goes toward paying off the loan, leading to eventually owning the home debt-free.
- **It is a debt you can typically repay.** You typically can not sell what you buy with credit cards to pay off your credit card debt if needed, while you hopefully could sell your home if you had to for enough to pay off your mortgage.

Understanding Amortization Schedules

Calculating the interest in home mortgages (as well as car loans or any kind of loan that has a monthly payment and a fixed amount of time) is a little more complicated than the Compound Interest Formula. You can use an amortization schedule (there are many available online that you could use with a parent's permission) to see how a loan is paid off over time. Each row shows a payment and how that payment affects the loan.

Amortization schedule for a 30-yr., $100,000 loan at 4% yearly interest rate

	Fixed amount you pay each month	*Amount of payment going toward interest*	*Amount of payment going toward paying off loan*	*Amount still left on the loan after principal portion of the payment deducted*
Month	Payment	Interest	Principal	Loan Balance
1	$477.42	$333.33	$144.09	$99,855.91
2	$477.42	$332.85	$144.57	$99,711.34
3	$477.42	$332.37	$145.05	$99,566.29

Notice that each month, the interest goes down slightly. That is because the interest is calculated each month on the current balance ... which decreases a little with each payment.

>> Every month,

- the loan balance decreases,
- the interest decreases, and
- the **principal** (i.e., amount going toward paying off the loan) increases.

The interest amount in the schedule shown on page 59 comes from **taking 4% and dividing it by 12 to calculate the monthly interest** and then multiplying that by the **previous loan balance** (for the first month, this was the original loan amount of $100,000).

MONTH 1: $\frac{0.04}{12} \times \$100{,}000 \approx \$333.33$

The rest of the monthly payment goes toward paying off the loan (i.e., toward principal).

payment amount – interest = principal
$\$477.42 - \$333.33 = \$144.09$

The principal portion of the payment is then deducted from the previous loan balance.

previous loan balance – principal = ending loan balance
$\$100{,}000 - \$144.09 = \$99{,}855.91$

The next month, the interest is calculated off this lower amount ... and is thus less.

MONTH 2: $\frac{0.04}{12} \times \$99{,}855.91 \approx \$332.85$

Month	Payment	Interest	Principal	Loan Balance
358	$477.42	$4.74	$472.68	$950.08
359	$477.42	$3.17	$474.25	$475.83
360	$477.42	$1.59	$475.83	$0

Now look at the last several months of this 30-year loan. Notice how much lower the loan balance is and, as a result, the interest. The portion of each payment going to principal (i.e., to pay off the loan) has greatly increased as a result. (There are additional costs that come with a mortgage, like property taxes and insurance.)

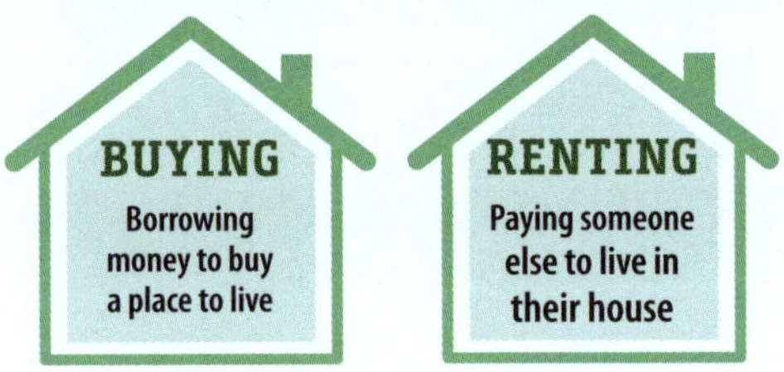

Interest vs. Renting

Now, the interest adds up over the life of a loan. Some amortization schedules add up all the interest amounts for you, showing the **cumulative interest**. Cumulative interest of more than $70,000 sounds like a lot, but after 360 payments (there are 360 payments in a 30-year loan, as $30 \times 12 = 360$), you will have a home you could stay in without payments. If instead you spend the same $477.42 a month on renting for 360 months, you will have spent $360 \times \$477.42 = \$171{,}871.20$ ... and at the end, you will not own anything.

Month	Payment	Interest	Principal	Loan Balance	Cumulative Interest
359	$477.42	$3.17	$474.25	$475.83	$71,867.92
360	$477.42	$1.59	$475.83	$0	$71,869.51

KEEPING PERSPECTIVE

- There are times when borrowing money may make sense.
- Borrowing typically has a cost in the form of interest attached to it.

Credit Scores

While, in general, debt should be avoided, there are times (such as with housing) where you may choose to take out a loan. Thus, it is important to have good credit.

Your **credit score** is a measure of how likely you are considered to make good on debt.

- It can affect if a lender will give you money.
- It can affect how high of an interest rate you pay.

>> Good credit (a higher credit score) is established by:

- borrowing money (also called getting a **line of credit**) and paying it back as promised
- doing this consistently over time

Credit can not be built in a day; lenders want to see that you have consistently kept your word over time. The time to start building your credit is not when you want to buy a home; it is years before that.

Think about it. Would you want to lend $300,000 to someone who has not proven himself responsible to pay anything back before? No — you are going to want to see a pattern of faithfulness in small things first.

> *One who is faithful in a very little is also faithful in much, and one who is dishonest in a very little is also dishonest in much* (Luke 16:10).

Consider looking for little ways you can build your credit without actually paying any interest.

- Have a credit card or two that you consistently pay off the whole balance of each month, thereby avoiding interest but showing that you can pay back debt.
- Buy something on an interest-free loan, paying it back before the interest starts. For example, as a young adult, a store offered me a 6-month no-interest loan on a piano I was buying, so I took the loan and paid it off before the interest started in order to help my credit. Just make sure you keep money to pay the loan off so you do not end up being unable to.

Check out the chart summarizing what increases and decreases your credit score.

Increase Credit Score	Decrease Credit Score
Have multiple lines of credit.	Do not have any lines of credit.
Do not use all the credit available to you.	Max out every line of credit (such as credit cards).
Pay off debt on time consistently.	Miss or be late on a payment.
Have the same credit card or other line of credit for a long period of time.	Constantly cancel credit cards.
Slowly build credit over time.	Apply to a lot of lines of credit all at the same time.

KEEPING PERSPECTIVE

God is trustworthy. As His servant, you should be trustworthy too, which means paying what you promise to pay. And as a bonus, doing so will help you build good credit. (So be careful about the debt you sign up for.)

If a man borrows anything of his neighbor, and it is injured or dies, the owner not being with it, he shall make full restitution (Exodus 22:14).

Review & Spiritual Debt

We have talked a lot in this lesson about interest. It can work for us in savings or against us with debt.

As we continue through the course, we will continue to see how helpful savings is.

Savings can:

- help you obtain goals that God may have for you (like a future home)
- free you up to not wind up in debt or panic when unexpected financial strains arise (which they do in a fallen world)
- help you make wise preparations for the future

>>> Prioritize saving now so your savings have time to grow.

Debt can:

- make you a "servant to the lender" (Proverbs 22:7)
- become a way to buy more than you can really afford rather than resting on God's provision, even if that provision does not include that new TV you really want

Before entering into debt, it is a good idea to do a heart check. Remember, God warns us about the love of money and calls us to live content in Him.

>>> Be careful with debt.

Spiritual Debt

Since we have explored debt so much, we will look at a parable Jesus told in Matthew 18:23–35 that revolved around debt.

- A man owed a huge debt — one he could never have paid back, even if he worked his whole life. But the king forgave him.
- The forgiven servant began choking another servant over an insignificant debt.

Jesus was using the parable to teach about forgiveness.

- Just like the servant was forgiven an impossible-to-pay debt, if we trust in Jesus Christ, we have been forgiven an impossible-to-pay spiritual sin debt, as Jesus' death paid the debt for us.
- Now we are to extend forgiveness to others who sin against us rather than acting like this servant did with his fellow servant. Whatever others have done to us, it is nothing compared to what we have been forgiven by God.

> *Jesus paid it all,*
> *All to him I owe.*
> *Sin had left a crimson stain,*
> *He washed it white as snow.*
>
> — *Christian Hymnal*[12]

Have you had your sin debt paid for you through Jesus' death and Resurrection?

- If not, He is ready to pardon you through Jesus if you ask.
- If you have, extend the forgiveness you have received to others — and tell others the incredible news that their sin debt before God can be forgiven too.

To share the good news of spiritual debt forgiveness, you may need to start by sharing that we all have a sin debt that we can not pay on our own. The news that your debt is canceled is only good news if you realize you have a debt first.

Taxes

income = giving + saving + ***spending***

We have looked at giving, saving, and some spending. It is time now to learn more about spending we all have to plan for: taxes.

spending *=* ***taxes*** *+ debt + living expenses*

Taxes are money that the government collects in exchange for the services and protection it renders. You can think of taxes as a fee for living in the city, state, county, or country. Since you will pay a lot of different types of taxes throughout your life, it is important to understand them.

In this lesson, we will focus on general principles and processes that tend to stay consistent from year to year. Specific amounts, including brackets and deductions, change from year to year. In this course, we are using example amounts based on a sample tax year.

Flat Tax (Sales, Social Security, & Medicare)

There are **two main types of taxes**: a flat tax and a progressive tax.

Flat Tax

A **flat tax** is taxed at the same rate for everyone, no matter their income. Sales tax is an example of a flat tax — when you buy something, you are taxed at the *same rate* (typically expressed as a percent), regardless of how much you make or how much you buy. While you will pay more tax on a more expensive item, you will pay it at the same % as you would on a less expensive one.

>>> With **flat taxes**, calculate the percent *of* the income or item that is being taxed.

Example: Calculate the sales tax on a $1,000 item if the sales tax rate is 5%.

0.05 × $1,000 = $50

% Stays the Same

5% *of* $1,000 *is* $50

5% *of* $10,000 *is* $500

>>> To calculate the total with tax, use one of these methods:

Method 1: Add the tax to the subtotal (i.e., the total without the tax).

Method 2: Multiply the subtotal by 1 + *the sales tax rate written as a decimal.*

Example: Calculate the total with sales tax on a $130.99 item if the tax rate is 5%.

Method 1: **0.05** × **$130.99** = **$6.55**
$130.99 + **$6.55** = $137.54

Method 2: **$130.99** × (1 + **0.05**) = $130.99 × 1.05 = $137.54

Medicare and Social Security taxes are not completely flat taxes. For incomes over certain high thresholds, Medicare adds an additional tax on just the portion over the threshold. Social Security does not tax above a certain income.

Medicare tax and **Social Security tax** are two other predominately flat taxes that all Americans have to pay. While confirmed when you file your federal income taxes (which we will look at in the next section), these taxes are taken directly from paychecks if you have an employer. The employer also has to pay part of the tax.

>>> Medicare sample tax year rates: 2.9% (1.45% for both the employer and employee)

>>> Social Security sample tax year rates: 12.4% (6.2% for both the employer and employee)

Example: Calculate an individual's portion of Medicare tax on a paycheck of $1,500 for the sample tax year rates.

Since both the employer and employee pay 1.45%, the individual portion will be 1.45%.

0.0145 × **$1,500** = **$21.75**

KEEPING PERSPECTIVE

When Jesus was asked about paying taxes, He pointed to Caesar's inscription on the money of the day and said, "Render to Caesar the things that are Caesar's, and to God the things that are God's" (Mark 12:17). The message is clear — we are to pay taxes due.

Now, the Roman government was quite corrupt. One could have argued that taxes were being used for corrupt purposes. Notice, though, how that did not make a difference. Jesus still said to give to Caesar what belonged to him. Paul echoes this message.

Let every person be subject to the governing authorities. For there is no authority except from God, and those that exist have been instituted by God.... Pay to all what is owed to them: taxes to whom taxes are owed, revenue to whom revenue is owed, respect to whom respect is owed, honor to whom honor is owed (Romans 13:1, 7).

God ordains the government, even if corrupt, and taxes should be paid.

Progressive Tax

When people talk about their taxes being due each April, they are referring to their **federal income tax**. This tax is collected by a government agency called the Internal Revenue Service, or **IRS**.

You will not have to deal with filing federal income tax until you make more than the minimum amount. However, since it is the tax that is the hardest to file and pay (and yet needs to be dealt with every year once you make more than the minimum), we are going to spend a lot of time learning about it.

Federal income tax and many state income taxes are not flat taxes. Instead, they are what we call progressive. In a **progressive tax**, the rate you pay changes for different ranges of income.

These ranges of income are called **tax brackets**. The table represents a sample of the first few tax brackets for single filers in the sample tax year.

Taxes change each year, so we will work with some samples.

Consider a hypothetical situation for a single person named James with a taxable income of $48,045. In the chart, $48,045 falls between the $44,725 and $94,375 of the 22% tax bracket. This means James is "in" the 22% bracket.

Sample Tax Brackets

Tax Rate	Single Filing Status
10%	$0 to $11,000
12%	$11,001 to $44,725
22%	$44,726 to $95,375

This table represents only part of tax brackets for the sample tax year.

Now, you might think this means that James will owe 22% *of* $48,045 in taxes, or $10,569.90 (since 0.22 × $48,045 = $10,569.90). This is *not* the case with progressive taxes.

>>> While we refer to a person as being "in" whatever bracket their income falls into, the portion of income that falls within each bracket is taxed at that bracket.

Saying that James is in the 22% tax bracket just means that his taxable income falls inside the income range that is taxed 22%. However, the first $11,000 of his income is going to be taxed at 10%, the amount from $11,000 to $44,725 at 12%, and the amount above $44,725 at 22%. The tax rate *progresses* at different levels of income.

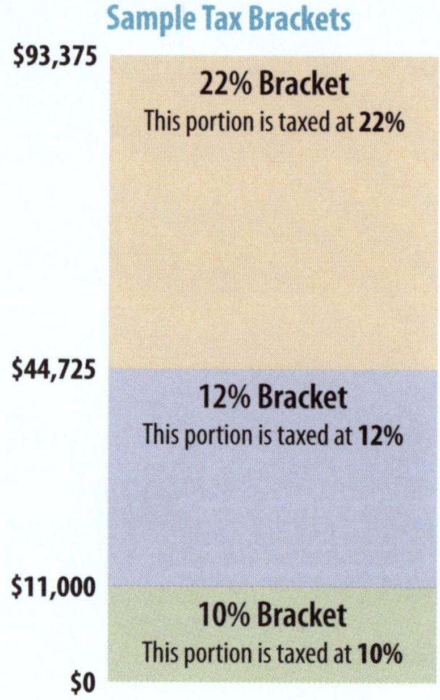

The graphic is based on the IRS graph.

Tax on the portion greater than $44,725

$48,045 – $44,725 = $3,320 (the amount taxed at 22%)

0.22 × $3,320 = **$730.40**

Tax on the portion between $11,000 and $44,725

$44,725 – $11,000 = $33,725 (the amount taxed at 12%)

0.12 × $33,725 = **$4,047**

Tax on the portion between $0 to $11,000

$11,000 – $0 = $11,000 (the amount taxed at 10%)

0.10 × $11,000 = **$1,100**

Total taxes = **$1,100** + **$4,047** + **$730.40** = $5,877.40, which rounds to $5,877.

James would owe $5,877 of federal income tax.

>>> We found the portion taxed at each level by subtracting the greatest value in the bracket from the greatest value in the previous bracket (or from $0 in the case of the first bracket)

For example, the 12% bracket ends at $44,725, while the previous one ends at $11,000. We thus subtracted $11,000 from $44,725 to find the income in that bracket. For the last bracket, we subtracted the greatest value in the previous bracket from the income ($48,045 – $44,725).

Now, do not worry: the government has tables, and there are other resources to make calculating taxes easier by doing a lot of this math for you. But it is important to understand that progressive taxes are calculated at different rates for different ranges of income.

>>> **Taxes and numbers input on federal income tax forms get rounded to the nearest dollar.**

We will typically have you round to the nearest dollar when working with income tax.

KEEPING PERSPECTIVE

The rate in progressive taxes change for different income levels. Understanding this will help you plan for taxes. For example, if you knew that you were in the 22% tax bracket and expected to get a $5,000 bonus, you should plan to pay 22% of that bonus in taxes. You should not, however, plan to be taxed 22% on everything you have made, as the first part of your income will be taxed at lower tax rates.

Different Incomes on Federal Taxes

Now we will dig a little deeper into federal *income* tax. Based on its name, you might assume you will be taxed on your income. And you are ... but what you might think of as your income is not actually the income used to calculate your taxes. The **1040 tax form**, the main tax form for calculating an individual's income tax, lists three different incomes.

Here is an overview of these different incomes on the 1040 form:

- **Total Income** (or Gross Income) (line 9)
- **Adjusted Gross Income** or **AGI** (line 11)
- **Taxable Income** (line 15)

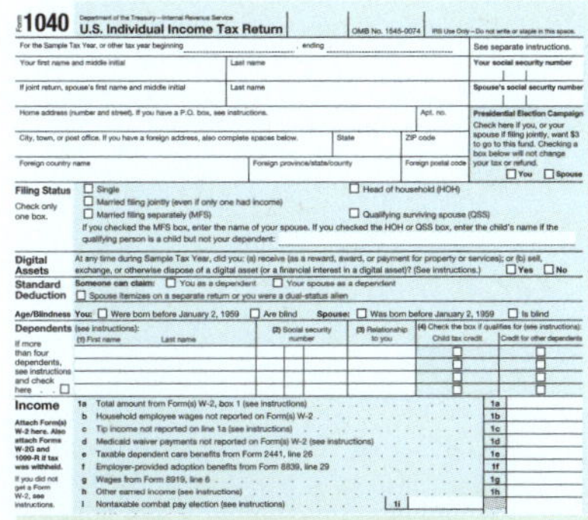
Form 1040 Department of the Treasury—Internal Revenue Service
U.S. Individual Income Tax Return
OMB No. 1545-0074 | IRS Use Only—Do not write or staple in this space.
For the Sample Tax Year, or other tax year beginning , ending | See separate instructions.
Your first name and middle initial | Last name | Your social security number
If joint return, spouse's first name and middle initial | Last name | Spouse's social security number
Home address (number and street). If you have a P.O. box, see instructions. | Apt. no. | Presidential Election Campaign Check here if you, or your spouse if filing jointly, want $3 to go to this fund. Checking a box below will not change your tax or refund. ☐ You ☐ Spouse
City, town, or post office. If you have a foreign address, also complete spaces below. | State | ZIP code
Foreign country name | Foreign province/state/county | Foreign postal code
Filing Status (Check only one box.) ☐ Single ☐ Married filing jointly (even if only one had income) ☐ Married filing separately (MFS) ☐ Head of household (HOH) ☐ Qualifying surviving spouse (QSS)
If you checked the MFS box, enter the name of your spouse. If you checked the HOH or QSS box, enter the child's name if the qualifying person is a child but not your dependent:
Digital Assets: At any time during Sample Tax Year, did you: (a) receive (as a reward, award, or payment for property or services); or (b) sell, exchange, or otherwise dispose of a digital asset (or a financial interest in a digital asset)? (See instructions.) ☐ Yes ☐ No
Standard Deduction: Someone can claim: ☐ You as a dependent ☐ Your spouse as a dependent ☐ Spouse itemizes on a separate return or you were a dual-status alien
Age/Blindness You: ☐ Were born before January 2, 1959 ☐ Are blind Spouse: ☐ Was born before January 2, 1959 ☐ Is blind
Dependents (see instructions): (1) First name Last name | (2) Social security number | (3) Relationship to you | (4) Check the box if qualifies for (see instructions): Child tax credit | Credit for other dependents
If more than four dependents, see instructions and check here ☐
Income (Attach Form(s) W-2 here. Also attach Forms W-2G and 1099-R if tax was withheld. If you did not get a Form W-2, see instructions.)
1a Total amount from Form(s) W-2, box 1 (see instructions) 1a
b Household employee wages not reported on Form(s) W-2 1b
c Tip income not reported on line 1a (see instructions) 1c
d Medicaid waiver payments not reported on Form(s) W-2 (see instructions) 1d
e Taxable dependent care benefits from Form 2441, line 26 1e
f Employer-provided adoption benefits from Form 8839, line 29 1f
g Wages from Form 8919, line 6 1g
h Other earned income (see instructions) 1h
i Nontaxable combat pay election (see instructions) 1i

Check out the 1040 on page 341. Notice that **the lines on the tax form are numbered**; tax instructions use that number to refer to them (and we will too).

Total Income, or Gross Income

The 1040 starts income calculations with your total income. Different incomes you might have are reported on lines 1z, 2b, 3b, 4b, 5b, 6b, 7, and 8 on the 1040 form and then summed up in row 9 to find what is called the total income. **Total income**, or **gross income**, is what you might think of as your income: it is essentially the total of all your income over the year. You can *estimate* this income by adding up all the income you know of for the year.

> ***Example:*** Estimate your total income for federal taxes if you plan to earn $50,000 from one job, $100 in interest, and $1,506 on the side by helping people move.
>
> $50,000 + $100 + $1,506 = $51,606

Adjusted Gross Income (AGI)

The IRS has some *adjustments* that you might be able to make to your income. These adjustments are detailed on a different form called Schedule 1 and are summed on line 26 of that form.

>>> The income after *subtracting* any adjustments from your *total, or gross, income* is called your **adjusted gross income** or **AGI**.

adjusted gross income = total income – adjustments

Because of the changing nature of adjustments, do not worry about learning the specifics of adjustments but rather the principle: there may be some adjustments made to your total income. The result after these adjustments is your adjusted gross income.

Some Adjustments to Total (Gross) Income

- Educator expenses
- Certain business expenses
- Health savings account (HSA) deduction
- Half of self-employment taxes
- Moving expenses for members of the Armed Forces

Example: Calculate your adjusted gross income if your total (or gross) income was $51,606 and your adjustments were $1,000.

$51,606 – $1,000 = $50,606

>>> Adjusted gross income is an important term to note, as your state taxes will probably have you copy over your adjusted gross income. You may also need it for other purposes, such as if applying for college aid.

Taxable Income

>>> **Taxable income** is the income actually used to calculate taxes; it is the result of subtracting something called deductions from your *adjusted gross income.*

taxable income = adjusted gross income – deductions

According to the IRS, "A deduction reduces the amount of a taxpayer's income that is subject to tax, generally reducing the amount of tax the individual may have to pay." Thus, a **deduction** is basically an allowable subtraction to your adjusted gross income.

Example: If your adjusted gross income is $50,606 and you qualify for $15,000 in deductions, what is your taxable income?

$50,606 – $15,000 = $35,606

Sample Tax Brackets

Tax Rate	Single Filing Status
10%	$0 to $11,000
12%	$11,001 to $44,725
22%	$44,726 to $95,375

$35,606 falls in this bracket

It is important to understand the meaning of taxable income. In the example, you will only pay taxes on the *taxable income* of $35,606, *not* on the total income of $51,606 or the adjusted gross income of $50,606. Assuming you are single and the sample tax year brackets shown apply, you would be in the 12% bracket, *not* the 22% bracket, even though you earned $51,606 that year.

KEEPING PERSPECTIVE

If you use a tax software or hire an accountant, your total (or gross), adjusted gross, and taxable income will all be calculated for you. You do not need to memorize how to calculate each. The important takeaway is that there are different incomes, and the one you use on your taxes is significantly less than what you might think of as your income.

Filing Status

Not only does your taxable income affect how much taxes you pay, but so does something called your **filing status**, which is a category of filing based on your life station. Filing statuses allow the government to give tax breaks to those in certain situations, such as those married or caring for others.

The Filing Status section from the sample tax year 1040 form is shown.

Filing Status Check only one box.	☐ Single ☐ Married filing jointly (even if only one income) ☐ Married filing separately (MFS)	☐ Head of household (HOH) ☐ Qualifying surviving spouse (QSS)

If you checked the MSF box, enter the name of your spouse. If you checked the HOH or QSS box, enter the child's name if the qualifying person is a child but not your dependent: ....

The table shows sample tax year filing statuses and their meanings.

Sample Filing Statuses

Single	Unmarried individuals who do not qualify for head of household
Married Filing Jointly	Married people filing one return which shows their joint income
Married Filing Separately (MFS)	Married people each filing their own return that lists their individual income
Head of Household (HOH)	People caring for others but not married, such as single parents
Qualifying Surviving Spouse	Someone who has lost their spouse within a certain timeframe and has not remarried

The filing status chosen makes a difference in the tax bracket in which someone falls, as you can see in the sample tax year table.

Sample Federal Tax Brackets

This table represents information from the sample tax year.

Tax Rate	Single	Married Filing Jointly or Qualifying Surviving Spouse	Married Filing Separately	Head of Household
10%	$0 to $11,000	$0 to $22,000	$0 to $11,000	$0 to $15,700
12%	$11,001 to $44,725	$22,001 to $89,450	$11,001 to $44,725	$15,701 to $59,850
22%	$44,726 to $95,375	$89,451 to $190,750	$44,726 to $95,375	$59,851 to $95,350
24%	$95,376 to $182,100	$190,751 to $364,200	$95,376 to $182,100	$95,351 to $182,100
32%	$182,101 to $231,250	$364,201 to $462,500	$182,101 to $231,250	$182,101 to $231,250
35%	$231,251 to $578,125	$462,501 to $693,750	$231,251 to $346,875	$231,251 to $578,100
37%	$578,126 and up	$693,751 and up	$346,876 and up	$578,101 and up

$100,000 is in the 24% bracket for the single filing status

$100,000 is in the 22% bracket for Married Filing Jointly Filing Status.

Notice that the same taxable income falls in different brackets depending on the filing status.

Example: In the sample tax year, which bracket would someone with a taxable income of $80,000 be if married filing jointly?

To find the answer, look at the ranges in the married filing jointly column. You will see that $80,000 falls inside the 12% row, meaning he or she would be in the 12% bracket.

While married people can file separately, it will typically result in the most savings for those married to file jointly.

KEEPING PERSPECTIVE

The status you file under makes a difference. Filing using the most advantageous status can help you save on taxes, freeing up more money to give, save, or spend elsewhere.

Regardless of your filing status, the more you make in income, generally speaking, the more your taxes are. **Owing taxes means God allowed you to make money.** Owing more means God entrusted you with more resources to steward.

Deductions

Deductions get subtracted from your *adjusted gross income* to calculate your *taxable income*. You get to choose between the following:

- a **standard deduction**, which is a specified amount the government sets each year based on your filing status
- **itemized deductions**, the sum of individual qualified expenses you had throughout the year

There is also a deduction for qualified business expenses you can take either way if it applies to you.

11	Subtract line 10 from line 9. This is your **adjusted gross income**	**11**	
12	**Standard deduction or itemized deductions** (from Schedule A)	**12**	
13	Qualified business income deduction from Form 8995 or Form 8995-A	**13**	
14	Add lines 12 and 13	**14**	
15	Subtract line 14 from line 11. If zero or less, enter -0-. This is your **taxable income**	**15**	

Since tax is calculated as a percent of your taxable income in each bracket, **the more you can deduct to lower your taxable income, the less taxes you will pay**. While we want to pay taxes we owe, we do not need to pay more taxes than what are owed. The tax code gives individuals choices as they file, and some choices may result in owing less than others.

This table represents standard deduction information from the sample tax year.

After looking up the current year's deduction in a chart, you can subtract it from your adjusted gross income to find your taxable income.

Sample Tax Year Standard Deduction

Filing Status	Sample Standard Deduction
Single	$13,850
Married filing jointly; surviving spouse	$27,700
Married filing separately	$13,850
Head of household	$20,800

Reduction that applies to the Finleys.

Example: If Mr. and Mrs. Finley's adjusted gross income was $50,000 for the sample tax year and they are married filing jointly and claim the standard deduction, what is their taxable income?

$50,000 – **$27,700** = $22,300

»» The standard deductions vary by the year and by filing status.

Itemized Deductions

Instead of claiming a standard deduction, the government allows for adding up qualifying expenses and using them as deductions. The chart shows some deductions from the sample tax year; remember, tax laws change each year, so these may no longer apply — and others might.

Hint: **Record-Keeping Makes Tax Time Easier.**

If you think you might be able to itemize deductions, **keep track of your medical (including dental) expenses throughout the year.**

Examples of Deductions (for the Sample Tax Year)

- **Certain medical and dental expenses** (Can only deduct the portion that is more than 7.5% of the adjusted gross income.)
- **Interest on a home mortgage** (*Not* the entirety of mortgage payments, but rather only the *interest* portion of them.)
- **Contributions to tax-deductible charities** (including churches)
- **Certain taxes** (such as property tax, state tax, and local taxes)
- **Self-employed health insurance premiums**

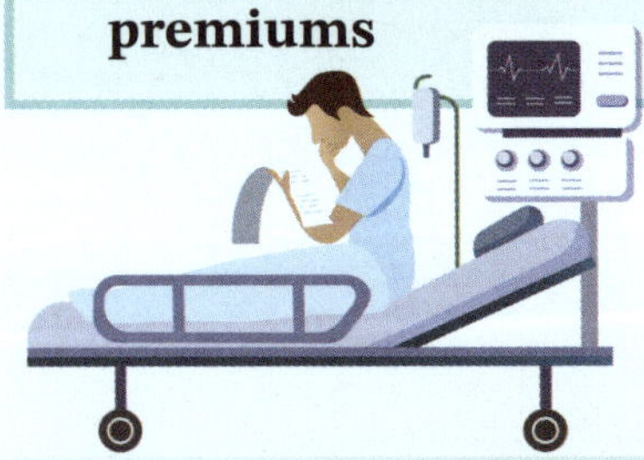

1. Figure out how much of medical expenses count (you can only deduct those over 7.5% of adjusted gross income [AGI]).
2. Add all the deductions together, using just the portion of medical expenses you calculated in step 1.
3. Compare the itemized deduction with the standard to see which is greater. The greater the deduction, the more savings.

Example: Consider Mr. and Mrs. Finley again, with an adjusted gross income of $50,000 in the sample tax year. Their medical and dental expenses for the year total $4,000 (but only the amount greater than 7.5% of their income count for a deduction); they donated $5,000 to charities; and they paid $2,000 in property tax, $5,000 on their home mortgage interest, and $3,000 in state taxes. Which will save them more: itemizing deductions or taking the standard deduction?

1. Calculate countable medical expenses:
0.075 × $50,000 = **$3,750** (calculating 7.5% of AGI)

total medical expenses – 7.5% *of AGI* =
$4,000 – **$3,750** = **$250**

2. Add deductions together:
$250 + $5,000 + $2,000 + $5,000 + $3,000 = $15,250

3. Compare with standard deduction:
In this case, the amount we calculated in Step 2 is less than the standard $27,700 deduction listed on the chart for married filing jointly. They will thus save more by taking the standard deduction.

Medical expenses often can not be deducted at all because they do not reach the 7.5% threshold — which is often a praise, as it means you and your family did not have major medical expenses.

KEEPING PERSPECTIVE

Do not let taxes control how you spend money. God may lead you to give somewhere that does not qualify as a charitable deduction you could use to itemize deductions. However, deduction knowledge can still prove helpful in stewarding well. For example, it may help you time your giving or certain expenses to have more deductions in one year so you can itemize deductions that year. More deductions lowers the amount of income on which you are taxed, lowering your taxes.

- Do what you can to lower your taxable income with deductions you qualify for.
- Cheerfully pay what you owe, praising God for blessing you with income on which to pay taxes.

Tax Tables

Up to now, we have been having you use the tax brackets to calculate federal income taxes. We did this to help you understand progressive taxes.

Sample Tax Rate Schedules and other tables are on pages 339–340.

However, the IRS puts together easy tables that do a lot (or all) of the math for you. Understanding how to read tax tables is important, so we are going to explore two differently formatted tax tables you might encounter.

The schedule shown, for single filers, comes from instructions on estimating federal income tax.

To better understand the table, we will revisit James' taxes that we looked at on page 67. As a reminder, he had a taxable income of $48,045 and was filing as single. We calculated his taxes using the brackets to be about $5,877.

Using the table, we could have found James' tax more easily. We would just need to find the row for his income. $48,045 falls in between $44,725 and $95,375, so we will use that row. That row says to take $5,147 + 22% of the amount over $44,725. The $5,147 is really the amount due for the previous brackets, leaving us to do the calculations for the final tax bracket.

Sample Tax Rate Schedules for Estimating Taxes

Schedule X — Use if your filing status is **Single**

If line 3 is: *Over—*	*But not over—*	**The tax is:**			*of the amount over—*
$0	$11,000	---------	+	10%	$0
11,000	44,725	$1,100.00	+	12%	11,000
44,725	95,375	**5,147.00**	+	22%	44,725
95,375	182,100	16,290.00	+	24%	95,375
182,100	231,250	37,104.00	+	32%	182,100
231,250	578,125	52,832.00	+	35%	231,250
578,125	---------	174,238.25	+	37%	578,125

Calculate amount over $44,725:

$$\textit{taxable income} - \$44,725$$
$$= \$48,045 - \$44,725 = \$3,320$$

Total taxes:

$$\$5,147 + (0.22 \times \$3,320)$$
$$= \$5,147 + \$730.40 = \$5,877$$

→ *22% of amount over $44,725* (refers to $730.40)

Tax on Previous Brackets

10% Bracket
($11,000 of income inside):
$0.10 \times \$11,000 = \$1,100$

12% Bracket
($33,725 of income inside):
$0.12 \times \$33,725 = \$4,047$

Total = $4,047 + $1,100 = $5,147

The table does the calculations on the amount in all the previous brackets so all the taxpayer needs to calculate is the tax on that final bracket.

Sample Tax Year 1040 Instructions Tax Table

1. Find the Section. This section has $48,000 up to $49,000.

2. Find the Row. $48,045 is between $48,000 and $48,050.

3. Find the Filing Status. This column lists the numbers for singles.

If line 15 (taxable income) is--		And you are--			
At least	But less than	Single	Married filing jointly*	Married filing sepa-rately	Head of a house-hold
		Your tax is--			
$48,000					
48,000	48,050	5,873	5,323	5,873	5,449
48,050	48,100	5,884	5,329	5,884	5,455
48,100	48,150	5,895	5,335	5,895	5,461
48,150	48,200	5,906	5,341	5,906	5,467
48,200	48,250	5,917	5,347	5,917	5,473
48,250	48,300	5,928	5,353	5,928	5,479
48,300	48,350	5,939	5,359	5,939	5,485
48,350	48,400	5,950	5,365	5,950	5,491
48,400	48,450	5,961	5,371	5,961	5,497
48,450	48,500	5,972	5,377	5,972	5,503
48,500	48,550	5,983	5,383	5,983	5,509
48,550	48,600	5,994	5,389	5,994	5,515
48,600	48,650	6,005	5,395	6,005	5,521
48,650	48,700	6,015	5,401	6,016	5,527
48,700	48,750	6,027	5,407	6,027	5,533

In the sample tax year 1040 instructions, the IRS does even more of the calculations for taxable incomes less than $100,000.

You just have to find the section and row based on your taxable income and then look at the correct column for your filing status. To find James' taxes, we would first find the section where all the $48,000s are listed. We would then locate the row where $48,045 falls and check out the single column there.

The tax table is slightly different due to the way the IRS calculates the tax for these tables, but it is about the same as what you would get if you used either the tax brackets or the table given with the estimation form.

Remember, the 1040 is the main federal tax form for individuals.

Due to the slight difference between this table from the 1040 Instructions and one you would get using other tables, know that taxes found using other tables are technically only estimates. **Use the 1040 Instruction table (or a software) when filing.**

KEEPING PERSPECTIVE

While tax tables can simplify the calculations, do not forget that federal income taxes are progressive.

- If you increase your income drastically, you may end up paying a higher percent of that additional income than you currently pay.
- Just because someone is in a high tax bracket, it does not mean they are paying that high percent of all of their income.

Deductions and Credits

Credits are similar to deductions in that they are subtractions that lower your taxes. However, while deductions come off the adjusted gross income, thus lowering the taxable income on which the taxes are calculated, credits come directly off your taxes.

For example, if you owe $4,000 in federal taxes but qualify for a tax credit of $300, your taxes will get lowered by $300.

$4,000 – $300 = $3,700

The government tends to change up the credits they offer. Your accountant or tax program will help you identify ones you qualify to take.

Sample Tax Year Federal Tax Credits Examples

- Child credit (for each child)
- Education credit (for certain educational expenses)
- Earned income credit (for low tax brackets)

Credits make more of an impact on your taxes than deductions. A $300 deduction lowers your taxable income by $300. If you are being taxed in the 10% tax bracket, that means it saves you 10% of $300, or $30. But if you have a $300 credit, that saves you $300.

>>> Credits lower your taxes directly dollar for dollar, rather than just lowering your income like deductions do.

Example: Calculate how much a credit of $2,000 and a deduction of $2,000 would save if you are in the 10% tax bracket and your taxes are $5,000.

CREDIT

$5,000 – $2,000 = $3,000	*Credits get deducted from taxes.*

DEDUCTION

0.10 × $2,000 = $200	*If you are taxed 10% on $2,000 less than you would be otherwise, then the deductions save you 10% of $2,000.*

While the idea of credits is simple enough, they are complicated a little by the fact that credits can be either refundable or nonrefundable.

- **Nonrefundable credits** only reduce your total tax — they do not result in a refund.
- **Refundable credits** can lower your taxes to where the government owes you.

Taxes	$800
Nonrefundable Credit	$1,000
Your taxes now equal	$0

Taxes	$800
Refundable Credit	$1,000
The government returns	$200

To understand the difference, suppose you qualify for a $1,000 nonrefundable credit but only owe $800. Because the credit is nonrefundable, you can only take $800 worth of the nonrefundable credit, lowering your taxes to $0. You can not use the credit to get a refund.

If instead the $1,000 credit were refundable, then the government would give you back $200.

Refundable and nonrefundable taxes get calculated at different places on the 1040 form. The non-refundable credit lines on the sample tax year 1040 are shown here, along with lines about other taxes that get calculated on different forms and added to your federal income tax in the same part of the 1040. We will look at the refundable credit lines in the next section.

Tax and Credits	16	**Tax** (see instructions). Check if any from Form(s): **1** ☐ 8814 **2** ☐ 4972 **3** ☐ ________	16	
	17	Amount from Schedule 2, line 3	17	
	18	Add lines 16 and 17	18	
non-refundable	19	Child tax credit or credit for other dependents from Schedule 8812	19	
	20	Amount from Schedule 3, line 8	20	
	21	Add lines 19 and 20	21	
	22	Subtract line 21 from line 18. If zero or less, enter -0-	22	
	23	Other taxes, including self-employment tax, from Schedule 2, line 21	23	
	24	Add lines 22 and 23. This is your **total tax**	24	

KEEPING PERSPECTIVE

There are a lot of steps to calculating taxes, which is why most Americans use tax software or an accountant. Do not get lost in the details. The key thing to remember is that credits lower your taxes due, dollar for dollar.

As we look at how taxes are calculated and at some key terms used, it is to help prepare you to know how much of your earnings belongs to the government so that you can pay honestly what is owed, while at the same time minimizing what you truly owe where possible.

After all, the Bible tells us to be honest and truthful. This includes on our taxes.

Lying lips are an abomination to the Lord, but those who act faithfully are his delight (Proverbs 12:22).

Total Due or Refund

Taxes are required to be paid throughout the year, not just on April 15.

- If you work for an employer, they will **withhold** (i.e., take out of your paycheck and send to the government automatically) taxes for you.
- If you are self-employed, you will need to make your own tax payments throughout the year.

If the sum of your prepayments and refundable credits is:

- less than what you were required to pay that year, you will need to pay the difference
- more than what you were required to pay that year, you will get a refund

When you file your taxes on April 15, the 1040 form not only calculates how much you owe for the year in taxes, but it also takes into account how much you have already prepaid throughout the year.

Reporting Prepayments

The 1040 has a place to report how much you have prepaid throughout the year in federal income tax. If you are using software or an accountant, you will be asked for this information. This information will either be:

- sent to you from your employer, or
- you should have a record of payments you have made on your own.

Do not worry about understanding each line of the form. Just know it adds up payments and refundable credits.

The 1040 form then adds your prepayments together with refundable credits.

Payments	**25**	Federal income tax withheld from:		
	a	Form(s) W-2	**25a**	
payments	**b**	Form(s) 1099	**25b**	
	c	Other forms (see instructions)	**25c**	
	d	Add lines 25a through 25c		**25d**
If you have a qualifying child, attach Sch. EIC.	**26**	Estimated tax payments and amount applied from previous return		**26**
	27	Earned income credit (EIC)	**27**	
	28	Additional child tax credit from Schedule 8812	**28**	
	29	American opportunity credit from Form 8863, line 8	**29**	
refundable	**30**	Reserve for future use	**30**	
	31	Amount from Schedule 3, line 15	**31**	
	32	Add lines 27, 28, 29, and 31. These are your **total other payments and refundable credits**		**32**
	33	Add lines 25d, 26, and 32. These are your **total payments**		**33**

Taxes Owed or Refunded

The next part of the Payments section tells you exactly what to subtract or add to calculate your refund or the amount you owe (or your software or accountant will do it all for you).

Refund	**34**	If line 33 is more than line 24, subtract line 24 from line 33. This is the amount you **overpaid**	**34**	
	35a	Amount of line 34 you want **refunded to you.** If Form 8888 is attached, check here . . ☐	**35a**	
Direct deposit? See instructions.	**b**	Routing number ______ **c.** Type: ☐ Checking ☐ Savings		
	d	Account number ______		
	36	Amount of line 34 you want **applied to your 2024 estimated tax** **36**		
Amount You Owe	**37**	Subtract line 33 from line 24. This is the **amount you owe.** For details on how to pay, go to *www.irs.gov/Payments* or see instructions	**37**	
	38	Estimated tax penalty (see instructions) **38**		

≫ To see how much you owe or will get refunded without filling out the 1040 form:

- Use subtraction to compare the sum of your prepaid taxes and refundable credits with your taxes.
- If the sum is higher, you will get a refund; if the taxes are higher, that is what you owe.

Sum of prepaid taxes and refundable credits: $9,000

Taxes: $10,000

Difference: $10,000 – $9,000 = $1,000

Taxes were higher, so amount owed.

Sum of prepaid taxes and refundable credits: $5,000

Taxes: $4,000

Difference: $5,000 – $4,000 = $1,000

Prepaid taxes and refundable credits sum higher, so amount refunded.

KEEPING PERSPECTIVE

While it may seem exciting to get a refund on your taxes, is it really a good thing?

- **Refunds actually mean you let the government borrow your money interest-free.** That refund could have been earning you interest throughout the year.
- On the flip side, you do not want to owe too much, or you could have steep fees for underpaying your taxes.

It is best to prepay your taxes as accurately as you can.

Review

We have sure covered a lot of ground. While whole courses could be written on just one aspect of taxes, hopefully this general overview gives you a big picture of how taxes work in the United States.

Remember, **there are tax programs and accountants who can walk you through the process**, so you do not have to do it all alone if you do not want to. Knowing the basics, though, can help you:

- understand terms you will hear
- make decisions that affect your taxes
- complete a general check of your taxes to make sure they seem reasonable.

Remember also that:

- owing taxes means you made money, which is a cause for giving thanks.
- as a child of God, you are called to live honestly and to submit to the government, which includes paying taxes owed.

The table on the next page gives an overview of some taxes. Federal and state income taxes are the main taxes individuals have to file themselves. We have focused on federal income tax, as it is the most complicated. We will discuss state taxes briefly in the next lesson.

Two Things Are Sure: Death and Taxes

There is an old saying that there are two things that are certain in life: death and taxes. We are taking a look at taxes so you are prepared to file them; it is even more important that you are prepared for death.

The only way to be prepared is to have your sins covered through Jesus Christ (Acts 4:12). Make sure you are ready by placing your faith in what Jesus did to pay for your sins, trusting that He rose from the dead and has the ability to save all those who call upon Him (Romans 10:13).

Type of Tax	About	Type of Tax *Sample Tax Year Information Given*	How Billed
Federal income tax	Taxed by the IRS on income earned	**Progressive**	Individuals have to file
Medicare tax	Taxed by the IRS on income earned	**Flat** percent of income for most; progressive on high incomes. *For the sample tax year it was 2.9% (1.45% for both the employer and employee) with an additional employee percent of 0.9% on income greater than a certain amount.*	Automatically deducted from paycheck; self-employed have to file
Social Security tax	Taxed by the IRS on income earned	**Flat** percent of income. *For the sample tax year it was 12.4% (6.2% for both the employer and employee) on up to $160,200 of income.*	Automatically deducted from paycheck; self-employed have to file
State income tax	Taxed by a state on income earned. Not all states have income tax.	**Varies**; can be either progressive or flat	Individuals have to file
Sales tax	A state tax on items purchased in a state. Some counties or cities add additional tax to the state sales tax.	**Flat**; varies in amount	Automatically charged on purchases
Property tax	A county tax on property owned (such as a house)	**Flat**; varies in amount	On home property tax, bill is sent to home. Some locations may require reporting other property over a certain value.

Fun Fact: The First Tax Recorded in the Bible

The first tax recorded in the Bible is the one Joseph established in Egypt. When the people ran out of money and livestock to buy grain, Joseph bought their land in exchange for the grain, instituting a $\frac{1}{5}$, or 20%, flat tax on future crops.

"And at the harvests you shall give a fifth to Pharaoh, and four fifths shall be your own, as seed for the field and as food for yourselves and your households, and as food for your little ones" (Genesis 47:24).

More Taxes & Review

This lesson is going to be different than the lessons we have had so far, as it is the end of the first quarter of material. Each quarter ends with a shorter lesson that includes lots of review leading up to a quarterly test. We will also be looking at some more tax concepts.

> *income* = *giving* + *saving* + ***spending***
> ***spending*** = ***taxes*** + ***debt*** + ***living expenses***

Effective & Marginal Tax Rates

As we learned in the last lesson, tax tables make calculating taxes super easy. For example, Jamie is a single filer with a taxable income of $30,000. You can simply look at the table shown below and see that his tax is $3,383.

You could also look at a tax bracket table and discover that his taxable income falls in the 12% bracket.

Without understanding progressive taxes, you might conclude that Jamie is paying 12% of his taxable income in taxes. In reality, though, he is paying 10% on some and 12% on some.

Today, we are going to look at a term used to describe the actual percent of taxable income that someone is paying, and one that describes the percent they would pay on the next dollar earned.

Sample 1040 Tax Table

If line 15 (taxable income) is—		And you are—			
At least	But less than	Single	Married filing jointly*	Married filing separately	Head of a household
		Your tax is—			
$30,000					
30,000	30,050	3,383	3,163	3,383	3,289
30,050	30,100	3,389	3,169	3,389	3,295
30,100	30,150	3,395	3,175	3,395	3,301
30,150	30,200	3,401	3,181	3,401	3,307
30,200	30,250	3,407	3,187	3,407	3,313
30,250	30,300	3,413	3,193	3,413	3,319

Sample Tax Brackets

Tax Rate	Single Filing Status
10%	$0 to $11,000
12%	$11,001 to $44,725
22%	$44,726 to $95,375

Effective Tax Rate

What percent of his taxable income is Jamie paying?

Tax programs will often list what is called the **effective tax rate**. To calculate the effective tax rate, divide the amount of taxes by the taxable income.

$$\text{effective tax rate} = \frac{\text{amount of taxes}}{\text{taxable income}}$$

$$\text{effective tax rate} = \frac{\$3{,}383}{\$30{,}000} \approx 0.1128 = 11.28\%$$

Jamie is effectively paying approximately 11.28% of his income in taxes.

Rather than memorizing this formula, think of the principle we have already looked at of figuring out the percent of a value. We want to figure out what percent of the total income is the taxes paid.

% *of* base *is* percentage

$$P \times \$30{,}000 = \$3{,}383$$

The effective tax rate gives you a guide of about how much you need to plan on for taxes the next year, assuming you expect your taxable income to be the same. It essentially treats Federal Income Tax as a flat tax to see what percent of your taxable income you are paying overall.

To solve this, we need to divide both sides by \$30,000 ... which gives us the same $\frac{\$3,383}{\$30,000}$ we got using the effective tax rate formula previously.

$$\frac{P \times \cancel{\$30,000}}{\cancel{\$30,000}} = \frac{\$3,383}{\$30,000}$$

$$P = \frac{\$3,383}{\$30,000} \approx 0.1128 = 11.28\%$$

A lot of consumer math formulas apply the same basic relationships to specific situations.

Marginal Tax Rate

While the effective tax rate can be helpful, **that is *not* the rate you are going to pay on additional income you might get**.

Marginal tax rate is the name for the tax rate you would pay on the next \$1 of income earned. It is the rate of the tax bracket your total income falls inside.

The marginal tax rate is what you should look at when seeing how additional income will affect your taxes. But keep in mind that if you make enough more, you may bump into another bracket.

Look back at the tax brackets on the previous page. In Jamie's case, the taxable income falls in the 12% bracket. The next \$1 he earns will be charged, not at 11.28% (his effective tax rate) but at 12%, up until the taxable income bumps into the next tax bracket.

What if Jamie were to suddenly win a \$5,000 prize? He is going to owe 12% tax on that \$5,000, as he is in the 12% tax bracket. His marginal tax rate — the amount he owes on the next \$1 of income — is 12%.

What if Jamie expected to earn \$20,000 more unexpectedly instead? How much in taxes would he owe on that extra?

Note that this is just estimating the extra taxes from the \$20,000 increase. If Jamie wanted to estimate his total taxes, he would need to find the taxes on all \$50,000, not just the final \$20,000 of it.

Example: Estimate the extra Jamie would pay in taxes if he earns \$20,000 extra on top of his original \$30,000 taxable income, bringing his new taxable income to \$50,000.

Subtract \$30,000 from the greatest value in bracket:
\$44,725 – \$30,000 = \$14,725

Tax on amount in 12% bracket: 0.12 × \$14,725 = **\$1,767**

Calculate amount in 22% bracket by subtracting the least value in bracket from what his new taxable income would be:
\$50,000 – \$44,725 = \$5,275

Tax on amount in 22% bracket: 0.22 × \$5,275 = **\$1,160**

Total tax on extra \$20,000: **\$1,767** + **\$1,160** = \$2,927

>>> To estimate taxes on extra income when it is enough to bump you into another bracket, calculate your taxes for the new taxable income, subtracting out your old taxable income.

KEEPING PERSPECTIVE

Both the effective rate and the marginal rate help with planning for taxes, as they give you an idea of:

- how much of your taxable income you are spending on taxes (effective)
- how much additional you will need to plan for if you make more (marginal)

Compiling Income

As we learned in the last lesson (and as makes sense from the name of the taxes), income taxes are based on income. Knowing your income, then, is important to calculating your taxes. Before we end our tour of taxes, we are going to look at two key income-reporting forms you might receive, as well as talk about tracking other income.

W-2

If you work for a company, you will get a W-2 in the mail. It will tell you the income you received from your employer.

- If a box does not apply to that person, it is left blank.
- Check out boxes 1, 3, 5, and 16. All these boxes contain the income received. Only boxes 3 and 5 are higher. Why? Sometimes, there is income that counts for Social Security and Medicare taxes and not for federal and state taxes.
- This form also reports information you need for state taxes. We will explore those more in the next section.
- Boxes 2, 4, 6, and 17 all list taxes that have been withheld throughout the year (i.e., they are taxes that your employer has taken out of your income and prepaid for you).

»» **Keep track throughout the year of all the income you receive outside of an employer, and collect tax-related forms that are sent to you.** Most of these forms will arrive in January or early February, as that is when employers and other agencies are required to send them out.

Tax forms may be available to you electronically instead of by mail. If so, be sure to save or print a copy for your records.

»» When it comes time to do your taxes, you will need to share your W-2 with your accountant or input information from it into tax software.

22222	VOID ☐	a Employee's social security number 111-11-1111	For Official Use Only OMB No. 1545-0008
b Employer identification number (EIN) 11-11111111		1 Wages, tips, other compensation $25,000	2 Federal income tax withheld $1,118
c Employer's name, address, and ZIP code Company 123 Anywhere Drive City, State, Zip		3 Social security wages $30,000	4 Social security tax withheld $1,860
		5 Medicare wages and tips $30,000	6 Medicare tax withheld $435.24
		7 Social security tips	8 Allocated tips
d Control number 1ABC		9	10 Dependent care benefits
e Employee's first name and initial John	Last name Doe Suff.	11 Nonqualified plans	12a See instructions for box 12 Code
123 Employee Drive City, State, Zip		13 Statutory employee ☐ Retirement plan ☐ Third-party sick pay ☐	12b Code
		14 Other	12c Code
f Employee's address and ZIP code			12d Code

15 State	Employer's state ID number	16 State wages, tips, etc.	17 State income tax	18 Local wages, tips, etc.	19 Local income tax	20 Locality name
GA	999999999	$25,000	$954.72			

Form **W-2** **Wage and Tax Statement**

Department of the Treasury—Internal Revenue Service

For Privacy Act and Paperwork Reduction Act Notice, see the separate instructions.

Cat. No. 10134D

Copy A—For Social Security Administration. Send this entire page with Form W-3 to the Social Security Administration; photocopies are **not** acceptable.

Do Not Cut, Fold, or Staple Forms on This Page

1099 Income

1099 forms show other types of income you have earned, such as interest from a bank. You will want to keep all these forms to calculate your total income.

Here are a few notes about this form:

- **Boxes that do not apply are left blank.** For example, in this case, no income was withheld for taxes, so all those boxes are blank.
- **This is showing an interest income in box 1 of $100.** So that $100 counts as income.
- **A TIN stands for Tax Identification Number.** For individuals, this number is your Social Security number. For companies not operating under an individual's social, the IRS gives out an Employer Identification Number (EIN).

This 1099 is a 1099-INT; the INT stands for interest. **There are other 1099 forms for different types of income.**

9292 ☐ VOID ☐ CORRECTED

PAYER'S name, street address, city or town, state or province, country, ZIP or foreign postal code, and telephone no. Any Bank 123 Any Road City, State, Zip	Payer's RTN (optional) 1 Interest income $ 100	OMB No. 1545-0112 Form **1099-INT** For calendar year SAMPLE TAX YEAR	**Interest Income**
PAYER'S TIN: 11-1111111 RECIPIENT'S TIN: 123-11-1234	2 Early withdrawal penalty $		**Copy A**
	3 Interest on U.S. Savings Bonds and Treasury obligations $		**For Internal Revenue Service Center**
RECIPIENT'S name John Doe	4 Federal income tax withheld $	5 Investment expenses $	**File with Form 1096.**
Street address (including apt. no.) 123 My Street	6 Foreign tax paid $	7 Foreign country or U.S. territory	For Privacy Act and Paperwork Reduction Act Notice, see the **current General Instructions for Certain Information Returns.**
	8 Tax-exempt interest $	9 Specified private activity bond interest $	
City or town, state or province, country, and ZIP or foreign postal code City, State, Zip	10 Market discount $	11 Bond premium $	
FATCA filing requirement ☐	12 Bond premium on Treasury obligations $	13 Bond premium on tax-exempt bond $	
Account number (see instructions) 1111111111 2nd TIN not. ☐	14 Tax-exempt and tax credit bond CUSIP no.	15 State: NH 16 State identification no.: 1111111-AB	17 State tax withheld $ $

Form **1099-INT** Cat. No. 14410K www.irs.gov/Form1099INT Department of the Treasury - Internal Revenue Service

Do Not Cut or Separate Forms on This Page — Do Not Cut or Separate Forms on This Page

Other Income

Some income may not come to you on a form. This includes income such as:

- tips
- babysitting revenue
- other money you earn on your own

This income typically does not come reported to you. Make sure you keep track of this income as it comes in, as you are still required to pay taxes on it if your total income is over the minimum threshold.

A Note About Gifts & Garage Sales

- If you receive gifts from others, that does not count as income for federal income tax purposes, although the *giver* may owe gift tax if they give over the limit ($17,000 in the sample tax year) to an individual.
- If you sell items for *less* than you paid for them (such as at a garage sale or used online), that does not count as income. But if you buy items and sell them for *more* than you bought them for, that would.

While it may be tempting to not declare income that does not come on a form since the government does not know about that income, God says we are to be truthful and pay the taxes we owe. Remember, God sees everything, and He knows if you are doing your best to fill out your taxes honestly.

> *The eyes of the Lord are in every place, keeping watch on the evil and the good* (Proverbs 15:3).

About Other Forms

In addition to forms about income, you may receive forms about amounts you can deduct from your taxes. For example:

- Mortgage companies have to report interest you *paid* (and might be able to deduct).
- Nonprofits (including your church) may send notices about your charitable donations to them either throughout the year as you make them and/or in January.

KEEPING PERSPECTIVE

The key is to keep track of *all* the income you receive throughout the year and to collect any tax forms sent to you in a safe place. This will make tax time much easier.

Intentionally left blank

State Income Tax

One of the beauties of the United States is that each state can set its own laws on certain areas, including state income tax. Thus, state taxes will look different depending on where you live.

- **Some states charge a progressive tax**, like the federal government does.
- **Others charge a flat tax**, taking a certain percent of income regardless of the income level.
- **A few states do not have income tax.**

For the ones that do have income tax, filing state income tax is not nearly as complicated as filing federal income tax. Here are some general principles.

1. Use Your Federal as a Basis.

States use the numbers you have already found on your federal income taxes as a basis. You will be asked for your adjusted gross income, taxable income, or other values from your federal form.

>>> Because you will need numbers from your federal tax form to file with your state, wait to file your state until your federal return is completed.

2. Adjust Based on the State.

Each state is different, and there are variations each year, but you will likely need to add or subtract to what you copied over from your federal income taxes. Standard deductions, for example, vary by state.

3. Calculate Tax Due.

If your state has tax brackets, you will need to look at their chart and figure out the tax owed, much like you did for your federal income taxes. State charts vary from the federal layout but are the same basic idea. Just read them carefully to make sure you are pulling the correct number.

4. Deduct What You Have Already Paid.

As with federal income taxes:

- Employers withhold state income taxes throughout the year. These also get reported on the W-2. Some 1099s may also show taxes withheld. The portion of the W-2 shown shows what was paid to the state (box 17) and potentially to local areas, if applicable.
- Self-employed individuals may also be required to pay state income tax throughout the year.

15 State	Employer's state ID number	16 State wages, tips, etc.	17 State income tax	18 Local wages, tips, etc.	19 Local income tax	20 Locality name
GA	999999999	$25,000	$954.72			

Just as with federal, you will need to figure out the difference between what you have already paid in taxes to the state and what you owe to see if you get a refund or not.

Additional Taxes

States can have additional taxes besides income tax that you file along with your income tax. These additional taxes get added to the income tax you calculate.

Consumer's use tax is an example. This tax is charged on items you purchase out of state but then use in your state on which sales tax was never paid (such as some online purchases).

>>> If your state charges use tax, you will want to keep track of purchases upon which use tax needs to be paid (unless your state allows for estimation). Then, calculating the tax will not be difficult.

If you have to track purchases without sales tax, **be careful to find out what is taxable and what is not so you do not pay extra.** For example, you do not want to pay it on shipping or food items if those are not taxable in your state.

Example: In a sample year, a state charged a consumer's use tax that varies by area. If you bought $405 worth of items on which you owe 6% tax, how much consumer's use tax do you owe? Round to the nearest dollar.

6% *of* $405 *is* $24.30

6% × $405 = $24.30, which rounds to $24

KEEPING PERSPECTIVE

It is a good idea to familiarize yourself with what is taxed in your state so that you know what you will be taxed on *before* tax time, as it is easier to keep track of income or spending throughout the year than to go back and figure it out.

Review

As you finish reviewing the quarter, pause and look again at the big picture. God blesses us with income to steward. We can either:

- give
- save
- spend that income.

Basic math skills help us in that process.

We have looked at:

- how God is a giving God and how our giving is an opportunity to grow in trusting Him and to invest in eternity, as well as to guard our hearts against the love of money
- how spending, especially living expenses, can grow or shrink based on the lifestyle we choose
- how interest works and the importance of avoiding debt and saving where we can
- how taxes work.

In the next quarter, we will continue looking at income, giving, saving, and spending as we discuss planning, or budgeting.

Always remember that we are stewards of what God has given us.

Page intentionally blank

Quarter 2: Budgeting

This quarter, we are going to continue exploring income, giving, saving, and spending with a focus on making a budget, or plan. Along the way, we will also explore topics like planning for medical expenses and saving for big expenses.

A budget is something you can begin as soon as you have any sort of income, even if small, and continue to expand as you earn more. It is a useful, lifelong skill.

Sam's Sample Budget

	A	B	C	D	E	F
1	**Monthly Budget for Sam**	**Budgeted Amount**	**Actual Amount**	**Difference**	**Percent of Total Net Income**	**Percent of Total Gross Income**
2	Gross Income from Job/Salary	$ 6,000	$ 6,000	$ –		92.31%
3						
4	**Paycheck Deductions**					
5	Savings: 401k	$ 500	$ 500	$ –	8.93%	7.69%
6	Spending: Taxes Federal	$ 159	$ 159	$ –	2.83%	2.44%
7	Spending: Taxes Other	$ 41	$ 41	$ –	0.74%	0.64%
8	Spending: Health Insurance	$ 200	$ 200	$ –	3.57%	3.08%
9	Total Paycheck Deductions	$ *900*	$ *900*	$ –	16.07%	13.85%
10						
11	*APPS AND TOOLS WILL START HERE AND LOOK ONLY AT NET INCOME*					
12						
13	Net Income from Job (B2-B9)	$ 5,100	$ 5,100	$ –	91.07%	78.46%
14	Self-Employment Income	$ 500	$ 100	$ 400	8.93%	7.69%
15	*Total Net Income* (B13+B14)	$ *5,600*	$ *5,200*	$ 400		
16	*Total Gross Income* (B2+B14)	$ *6,500*	$ *6,100*	$ 400		
17						
18	***Giving, Saving, & Spending from Net Income***					
19	***Giving***					
20	Church	$ 800	$ 800	$ –	14.29%	12.31%
21	Ministry A	$ 40	$ 40	$ –	0.71%	0.62%
22	*Total Giving*	$ *840*	$ *840*	$ –	15.00%	12.92%
23						
24	***Saving***					
25	Bank	$ 2,030	$ 2,043	$ (13)	36.25%	31.23%
26	*Total Saving*	$ *2,030*	$ *2,043*	$ (13)	36.25%	31.23%
27						
28	***Spending***					
29	Rent	$ 1,200	$ 1,200	$ –	21.43%	18.46%
30	Internet/Cell Phone	$ 80	$ 80	$ –	1.43%	1.23%
31	Transportation	$ 500	$ 100	$ 400	8.93%	7.69%
32	Food	$ 450	$ 437	$ 13	8.04%	6.92%
33	Clothing	$ 200	$ 150	$ 50	3.57%	3.08%
34	Household	$ 100	$ 100	$ –	1.79%	1.54%
35	Medical	$ 200	$ 250	$ (50)	3.57%	3.08%
36	*Total Spending*	$ *2,730*	$ *2,317*	$ 413	48.75%	42.00%

NOTE: You will be referring to this sample budget in the following lessons.

06

Getting Started with a Budget

Whether you make much or little, a general plan — or budget — will help you steward the income God gives you. Budgets deal with income, giving, saving, *and* spending. Now we will explore the big picture of why we need a budget and the basics of how to put one together.

income = giving + saving + spending

Introducing Budgets

Kayla got her first job. She could start saving for that new car — or college — or she could buy that new dress she really wanted. Does she have enough money for saving and the dress?

Kayla knew what she was making. But then she:

- would also have taxes
- wanted to give 10% to the Lord
- will need to pay an auto insurance bill

Suddenly Kayla was not sure if she really would have money for that new dress after all.

Kayla quickly discovered that she needed a budget. Budgets help us make wiser money decisions, living within the provision God has given us and putting our money toward what He is calling us to. Without one, we might find ourselves having spent money we needed to give, save, or pay our bills.

What Is a Budget?

A **budget** is an estimate of income, giving, spending, and saving for a period of time. In this course, we will focus on monthly budgets, as a month is typically a good timeframe.

Most budgets break down spending (and sometimes income, giving, and saving) into smaller categories, showing what is being spent on housing, cars, food, and other categories.

Notice that this hypothetical budget for Sam on the previous page has a lot of saving each month. Notice also that adding the total giving, spending, and saving together equals the income. We are applying the Income Distribution Equation.

income = giving + saving + spending

***income* = $840 + $2,030 + $2,730 = $5,600**

A Simplified Monthly Budget for Sam	
Income	$ 5,600
Giving	$ 840
Saving	$ 2,030
Spending	$ 2,730

These total $5,600.

>> Make sure your budget accounts for all of your income by checking if your giving, saving, and spending equal your income.

Why so much saving on the budget? Well, Sam is trying to save up for a home, so he has prioritized saving as much as he can. A budget helps him see how much he can spend in other areas to set aside the amount he wants to for a house. Without a budget, he might spend a lot more and significantly delay saving enough to buy a home.

How to Budget

You can budget on paper, in a spreadsheet, or with an app/program. If you are using an app or program to budget, they will probably not have giving separated out as its own section. It can instead be thought of as a part of spending. Sam's budget is shown again with giving viewed as a part of spending.

A Simplified Monthly Budget for Sam	
Income	$ 5,600
Saving	$ 2,030
Spending	$ 3,570

These total $5,600.

income* = *saving* + *spending
(modified equation where giving is part of spending)

***income* = $2,030 + $3,570 = $5,600**

However your system tracks it, keep in mind that when you get money (i.e., income), you have three choices: give, spend, or save. A choice to put money in one category is a choice *not* to put it in another.

A Tool, Not a Master

Budgets are useful tools but horrible masters. Imagine if Sam were suddenly presented with an urgent need of a ministry. The need was for $800, but he had already spent his giving for the month. Strictly following the budget would lead Sam to say no, but Sam really should pray and ask God if He wants him to give some of his home savings. The budget is not meant to run his life or be a sneaky way for the love of money to take over. Instead, it is meant to free Sam up to invest in the goals God gives him at different times, be they buying a home for his future family or giving to help others. **Our budgets need to be constantly under the direction of the Holy Spirit.**

KEEPING PERSPECTIVE

- Budgets can help us:
  - guard against squandering the money God allows us to earn.
  - plan wisely so we can be generous and save.
- Budgets can be complex, with numerous categories, or as simple as deciding how much income goes toward giving, saving, and spending.
- You can think of a budget as an overall plan subject to continual revision as you surrender to God's bigger plan.

Skill Introduction: Spreadsheets

Even if you use an app to budget, spreadsheets will still come in handy.

Spreadsheets:

- make it easy to track and compare values
- have built-in calculator-type functions, doing math automatically for you (and updating that math as numbers change)

There are many different spreadsheet programs, some of which are free. If you do not have access to a spreadsheet program, that is okay — you can still learn the basic principles in case you do one day.

Different spreadsheet programs (and different versions of the same program) have slight differences but follow some general principles. Try using the help feature for what you are trying to do. Be sure to get a parent's permission before using spreadsheets or apps.

The Basics

- Spreadsheets have **columns** labeled with letters and **rows** labeled with numbers.
- Each of the little boxes, or **cells**, can be referred to by its column and row. B2 is shown.
- The cells can be formatted to contain text or numbers. Search in the program's help feature for "format cells."
- Multiple cells can be selected at once. An entire column or row can be selected by clicking on the row number or the column letter. In the second spreadsheet image, column B has been selected by clicking on the B.

Just remember that columns on buildings are vertical, so **columns** in spreadsheets are vertical.

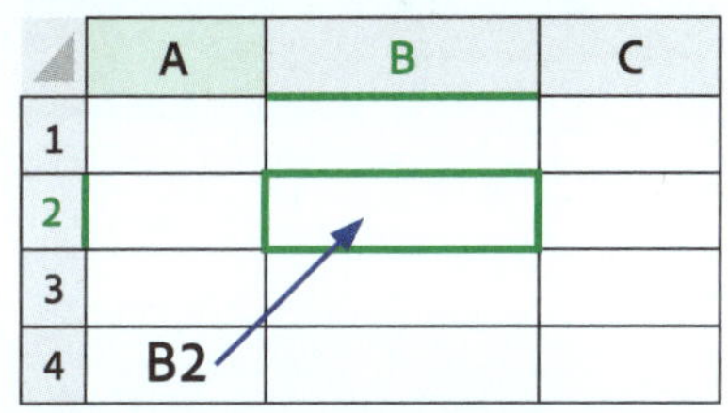

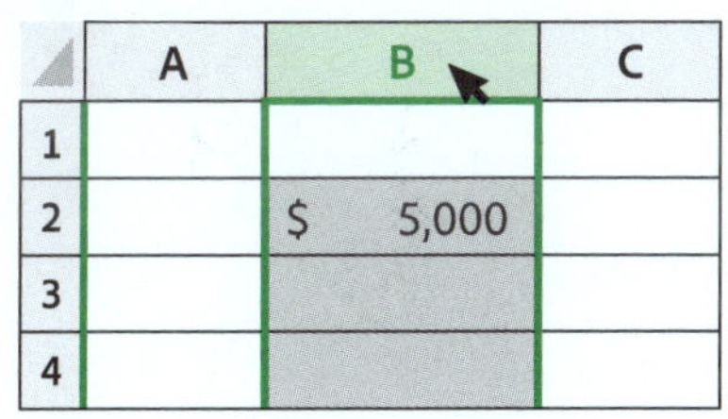

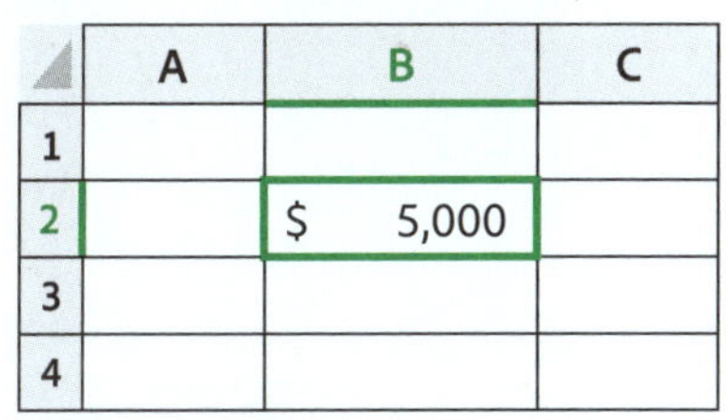

Operations in Spreadsheets

What makes spreadsheets super powerful is that they serve as a dynamic calculator. To use the calculation feature, press the = sign. That tells the program to show in the cell the result of whatever math you type in next.

You can then input math just like you would on a calculator.

- Addition: Use + (50 + 321 is **=50+321**, as shown)
- Subtraction: Use – (50 – 12 is **=50-12**)
- Multiplication: Use * (5 × 3 is **=5*3**)
- Division: Use / (20 ÷ 5 is **=20/5**)
- Parentheses: Use ( ), just as on a calculator
- Exponents: Use ^ (5^2 is **=5^2**)

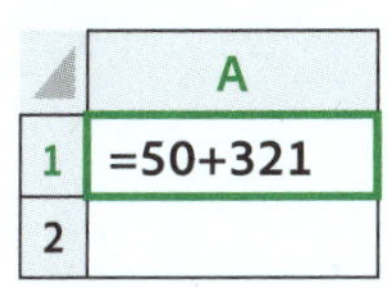

What is typed into cell.

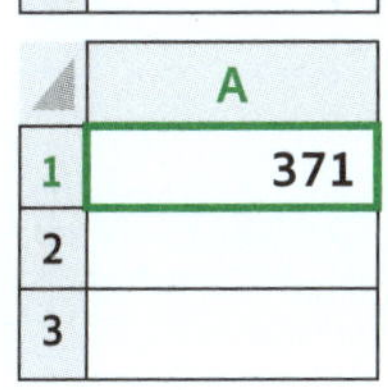

What cell displays when you press enter or click out of cell.

>>> The advantage of spreadsheets over a calculator is that instead of just inputting values, the spreadsheet can also perform math on cells, meaning they will auto-update with changes.

	C	D
15	Yearly Taxes	$12,000.00
16	Monthly Taxes	=D15/12

Division entered.

	C	D
15	Yearly Taxes	$12,000.00
16	Monthly Taxes	$ 1,000.00

Cell displays result.

For example, you have your total yearly expected taxes in cell D15, and you want to divide it by 12 to calculate a monthly amount. You could type **=D15/12** in another cell. That way, as you change the yearly tax value in D15, the program will automatically adjust the monthly tax amount.

Rounding in a Spreadsheet

You can control to what place the cells in a spreadsheet round values. You might find rounding buttons that look like this: ←.0 .00 | .00 →.0 . If not, search in the help feature for "format cells" or "round."

Before rounding	After clicking .00 →.0 twice
$ 417.67	$ 418

When working with budgets in this course (whether in a spreadsheet or on paper), round to the nearest dollar. After all, budgets are just estimates, so getting them to the penny does not make sense.

Example: If your giving is $425.98 a month, what amount should you list in your budget?

Round to the nearest dollar: $426

KEEPING PERSPECTIVE

Whether or not you use a spreadsheet program one day to budget, becoming familiar with one is helpful, as spreadsheets lend themselves well to many financial uses.

By doing math in a spreadsheet, you can go back and check your math easily by seeing what operations you put in the cell ... or adjust it if something changes without having to manually redo all the calculations.

Understanding a Budget Plus Gross & Net Income

Look back at Sam's detailed budget, which is located at the introduction of this quarter. Notice that it breaks down his income, giving, saving, and spending into categories.

Income

Notice that Sam's budget lists two different incomes: gross and net income.

- **Gross income**, or **gross pay**, is what you are actually earning before the employer takes out **paycheck deductions** — what the employer automatically takes out of your pay, such as 401k contributions (we will look at 401ks later; they are savings accounts some employers offer), tax withholdings, and health insurance. These paycheck deductions reduce the amount of money you actually receive.
- **Net income** or **net pay** is the income left after paycheck deductions. Unless you have self-employment income, it is your *actual paycheck amount.*

> *Example:* Calculate the net income if your gross pay is $5,000 a month and you have $1,200 of paycheck deductions each month.
>
> $5,000 – $1,200 = $3,800

Note that self-employment income also counts as income. Since there are not paycheck deductions, gross and net income will be the same for self-employment income. In Sam's budget, his $500 *self-employment income* was added to his *net income from job* to calculate his *total net income*, and to his *gross income from job/salary* to calculate his *total gross income.*

Remember total income, or ***gross*** income? It is the income ***before*** deductions are taken. Notice we are using the same word here (gross) to mean the income ***before*** any paycheck deductions are taken.

If you use an app or computer program to budget (with a parent's permission):

- It will automatically use net pay.
- It will not know anything about paycheck deductions.

Many spreadsheet templates will only list net income. While what you do with net income is the main focus of a budget, still keep track of your gross income and paycheck deductions to have a fuller picture of where all your money is going.

Calculating Gross Income

It is easy to calculate your net pay — just look at your paycheck. For gross income, you will usually be told your yearly salary, or gross income, up-front, and it will likely be stated on your paycheck stub. You can use that information to calculate monthly or weekly gross pay.

>> **To calculate monthly gross income or pay**, divide your yearly salary by 12.

>> **To calculate weekly gross income or pay**, divide your yearly salary by 52, as there are 52 weeks in a year.

The categories you use in your budget might be different than Sam's. Think of where you earn or pay out money and make sure that amount is accounted for in one of your categories.

Example: If you make $35,000 a year, what is your monthly and weekly gross pay?

$$\textit{monthly gross pay} = \frac{\$35{,}000}{12} \approx \$2{,}917$$

$$\textit{weekly gross pay} = \frac{\$35{,}000}{52} \approx \$673$$

Looking at the Rest of the Budget/Categories Based on Percents

- The amounts listed for each category on the budget is what Sam **budgeted**, or planned, to earn, give, save, or spend in that category.
- Sam's budget also has a column for **actual** spending. It is important to track actual amounts and check how they measure up to what was planned. The **difference** column shows where they differed from the plan.
- The **percent** columns give Sam an idea what portion of either his total net or total gross income he is spending where. Focusing on the percent column can help develop and evaluate a budget.
- **Taxes** – Looking at Sam's budget and seeing his employer is only withholding about 2.5% of his gross income should raise a concern, as income tax rates are much higher than that.
- **Giving** – Budget giving as a percent of your *gross* income. After all, that is the real amount you are making. Many use 10% of their gross income as a minimum threshold, but God might have you aim at a different mark. Sam is giving about 13%.
- **Saving** – Many suggest saving at least 20% of your net income. Saving more is helpful when you are young. There is a lot to save for — a house, a car, college, emergency funds — the list goes on. Plus, every dollar you save can grow into many more future dollars if you invest it wisely. Sam can see from his *Percent of Total Net Income* column that he is saving about 36%.

KEEPING PERSPECTIVE

Budgeting does not require fancy math. You just need to think through what income, giving, saving, and spending are for a certain period of time.

Estimating Fixed vs. Variable Budget Categories

How do you figure out how much you are making each month when you are paid weekly? How do you estimate various expenses and bills that are not monthly? We could divide money you receive or distribute into two basic buckets:

Fixed, which earns or costs the same (i.e., a *fixed* amount) each time.

- Garbage pickup (most companies charge a *fixed* monthly fee)
- A salaried job (has a *fixed* amount you receive in a paycheck each pay period)

Variable, which *vary* in amount.

- Groceries (costs *vary* each time)
- Hourly retail position (where hours, and thus pay, *vary*)
- Electric bill (comes at fixed times throughout the year, but the amount *varies*, so variable— it is variable if the *amount* varies)

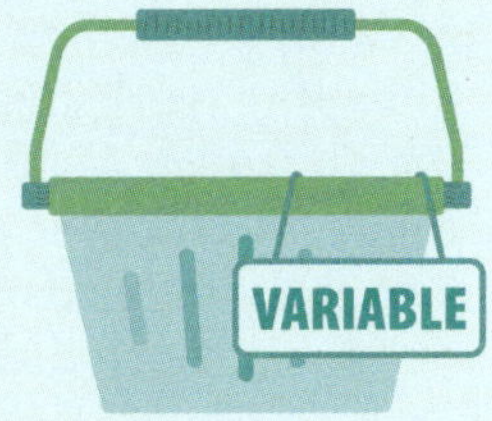

Amounts for fixed or variable budget categories can be calculated:

- based on the bills or income you receive
- based on prior amounts or averages

After looking at your budget as a whole, you might decide to make changes in what you spend in a category. But figuring out the current/historical amount gives you a place to start.

Estimating Fixed Budget Items

It is easy to budget for fixed budget items in your monthly budget.

>>> **If you receive or pay out money once a month**, then that is the amount to plan for in your monthly budget.

>>> **If you receive or pay out money less frequently than each month**, simply divide the amount by the number of months it covers just like you did with yearly salaries.

- Cable – If you pay \$50 a month for cable, budget \$50 for cable.
- Car insurance – A 6-month bill of \$450 would mean we would need to budget $\frac{\$450}{6 \text{ mos.}} = \$75$ per month.
- Property taxes – A yearly bill of \$1,000 would mean we would need to budget $\frac{\$1{,}000}{12 \text{ mos.}} \approx \$83$ per month.
- Income – A 6-month royalty check of \$30,000 would mean we would need to budget $\frac{\$30{,}000}{12 \text{ mos.}} = \$2{,}500$ per month.

>>> Notice that we are rounding to the nearest dollar.

For large but infrequent bills, you need to set aside the budgeted amount each month ***so that you will have the amount needed when the bill is due.***

>>> If you receive or pay out more frequently than once a month, determine monthly cost by:

- calculating the yearly amount and then dividing by 12 (the months in a year)
- calculating the weekly amount and multiplying by 4.33 (the approximate weeks in a month)

Notice the answers were slightly different. Multiplying by 4.33 is not as accurate, since 4.33 is an approximation. Although either is accurate enough for a budget, when you are doing the worksheets in the *Consumer Math Teacher Guide*, the Answer Key will find the yearly cost and divide on these problems.

Example: Your paycheck is $300 weekly. What is your monthly net pay?

Calculate Yearly Amount & Divide

multiplying the number of weeks in a year times weekly amount

$\frac{52 \times \$300}{12} \approx \$1,300$

dividing by 12, the number of months in a year

Take Weekly Amount & Multiply by 4.33

approximate weeks per month × **weekly amount**

4.33 × **$300** = $1,299

Estimating Variable Budget Items

Some expenses, such as groceries, clothing, and gas, get paid in smaller payments throughout the month. You can look at an average and then use that in your budget.

>>> Calculate an average by adding up some prior values and dividing that sum by the number of prior values you used.

Make sure you add up all the prior values before dividing by the # of prior values. Use ( ) if inputting into a calculator, or press enter after you add up all the values.

Example: David is putting together his first budget. He spent the following on groceries the first six months of the year: $101, $120, $89, $200, $114, $99. If he bases his budget on his average, what should he budget?

$\frac{\textit{sum of prior values}}{\textit{\# of prior values}}$

6 values, so put 6 below

$= \frac{(\$101 + \$120 + \$89 + \$200 + \$114 + \$99)}{6}$

$= \frac{\$723}{6} \approx \$121$

KEEPING PERSPECTIVE

We have just focused on how to estimate budget items, including different spending categories. First, decide if you really want to spend that money. With fixed expenses especially, it is easy to forget you are paying for something. As you get recurring expenses (such as cell phone bills, car insurance, or entertainment services), reevaluate them every now and then to make sure they are still the best option for you. Otherwise, you could be spending more than you need to.

More Spreadsheets & Using a Budget

Once you have a budget, how do you check if you are following it? We will look at that as well as at other spreadsheet tools that will help.

Be sure to get a parent's permission before using spreadsheets.

Living Within a Budget

In the budget in the introduction to this quarter, Sam has budgeted \$450 per month for groceries. There is an average of 4.33 weeks per month. That means that he should spend exactly \$103.93 per week (as $\frac{\$450}{4.33} \approx \$103.93$), right?

Not exactly. It means he needs to *average* \$103.93 or less. (It is not a problem to run under budget.) There might be weeks where he spends more than this and other weeks where he spends less.

Now, if Sam has nothing in savings, then he is going to have a hard time when spending in a category is higher one time, even if it goes down the next. Having a "bill buffer" savings (see page 48) makes it much less stressful to live within a budget.

>>> Your budget amount represents your average, not an exact figure.

Tracking a Budget & Summing Cells in a Spreadsheet

A budget tells you where you plan to spend money. You then need to check your actual amounts against that plan.

The sample shows Sam's food expenses for the month of November. The spreadsheet has the date, the amount, and a note about what the amount was for.

	A	B	C
1	**Food**		
2	6-Nov	\$ 80.69	Groceries
3	13-Nov	\$ 101.56	Groceries
4	14-Nov	\$ 20.14	Restaurant
5	20-Nov	\$ 120.56	Groceries
6	27-Nov	\$ 98.45	Groceries
7	30-Nov	\$ 15.16	Restaurant
8	*Total*	\$ *436.56*	

=SUM(B2:B7)

The SUM Feature

Notice that we needed to add a lot of cells together to calculate the total food cost. Rather than adding each individual cell, you can use the sum feature to total them all easily.

Cell B8 has =SUM(B2:B7) in it. This tells the program to sum up B2 *through* B7. SUM lets you add up multiple cells without having to click on each one — you only need to specify the starting and ending cell.

>>> To use SUM, you can type it out (as in =SUM[starting cell:ending cell]) or press a button (which will likely look something like Σ AutoSum) and then click on the starting and ending cells, pressing ENTER at the end.

Notice that each time Sam went to the grocery store, his total varied a little. That is okay. His monthly total is \$436.56, which is still less than the \$450 he budgeted.

>>> It can be hard to track how much you are actually saving. You can find the value to put on your budget using *income = giving + saving + spending*. Just plug in the values for all but *saving* (use your *net income* as *income*), and solve for it, as we did back in Lesson 1.

Many apps or online tools connect to bank and credit cards to automatically list all transactions. While this simplifies tracking actual amounts, it also gives access to sensitive information, requiring great caution ... and a parent's permission for now.

Check out Sam's budget in the introduction to this quarter again. Notice there is a column for the budgeted amount, one for the actual amount, and one for the difference.

- **The budgeted amount** column shows the general plan.
- **The actual amounts** are what was actually done. If using a spreadsheet, actual amounts can be tracked in a separate worksheet in the same file. For example, cell C32 on Sam's budget could be found by typing **=Food B8** in it, provided the food breakdown from the previous page was a different worksheet named Food in the same file (spreadsheets allow multiple pages, or **worksheets**, in the same file).
- The **difference column** is found by subtracting the actual amount from the budgeted amount. (Some resources might subtract budgeted from actual instead — make sure you check if looking at a budget outside of this course.) A spreadsheet can automatically calculate this. For example, in Sam's budget, D14 could be found by typing **=B14-C14** in it. Once you specify to calculate the difference in one cell, you can copy and paste that to the other cells.

>>> ***difference = budgeted – actual***

Skill Sharpening: Negative Numbers

The parentheses in the budget's difference column are just how this spreadsheet program is showing negative values. Parentheses are often used in finance to show negative values, and negative values are used to show the *opposite of* something. In the case of budgets, they mean the *opposite of* having extra money left over; instead, you used more than your budgeted amount. Hopefully you know how to work with negative numbers; if not, in this course, you can use your calculator. Just make sure you include the negative sign (–).

Example: Calculate the difference between a budget of \$60 and an actual amount of \$100.

Subtract actual amount from the budgeted:
\$60 – \$100 = –\$40 (which might be shown as (\$40) on a spreadsheet)

KEEPING PERSPECTIVE

While there are a lot of tools with budgeting, **there are three main steps to having a budget**, all of which should be done with prayer and a heart that recognizes everything belongs to God:

1. Estimate your income, giving, saving, and spending.
2. Track what you actually earn, give, save, and spend.
3. Compare the two and use that knowledge to help guide your daily financial decisions.

Evaluating a Budget

How much of your money is really going where? Should you adjust any categories? Here is a tool to help you evaluate your initial budget, as well as how to reevaluate it based on your actual amounts.

Evaluate Your Initial Budget: Determining the Percent

Look at the monthly budget for Sam in the introduction to this quarter. Notice that columns E and F show the percent of the total net income and gross income each category represents. These columns are based on finding the percent, which we covered back on page 23.

Example: Calculate the percent of the *total gross income* ($6,500) Sam's *net income from job* ($5,100) represents.

We could word this problem like this:

% *of* base *is* percentage — % *of* $6,500 *is* $5,100

Put in equation form: $P \times \$6{,}500 = \$5{,}100$

Divide both sides by $6,500: $P = \frac{\$5{,}100}{\$6{,}500} \approx 0.7846 = 78.46\%$

To get columns E and F to show up as a %, we have formatted the cells as percents.

Rather than doing all this math, the spreadsheet can do it automatically. If you look at row 13 of Sam's budget, the last column is displaying 78.46%; it has been told to do the division of the *net income from the job* divided by the *total gross income* automatically.

	A	B	C	D	E	F
1	Monthly Budget for Sam	Budgeted Amount	Actual Amount	Difference	Percent of Total Net Income	Percent of Total Gross Income
13	Net Income from Job (B2-B9)	$ 5,100	$ 5,100	$ -	91.07%	78.46%
14	Self-Employment Income	$ 500	$ 100	$ 400	8.93%	7.69%
15	*Total Net Income* (B13+B14)	$ *5,600*	$ *5,200*	$ 400		
16	*Total Gross Income* (B2+B14)	$ *6,500*	$ *6,100*	$ 400		

F13 has **=B13/B16** typed into it: it displays the results.

Looking at these percent columns, we can better tell where the money is both coming from and going. Which column we look at depends on what we are evaluating.

Checking out what percent is going where can help you get a better picture of where your priorities are and where your money is going.

Software Tools – If one day you use software or an app to track your progress, it will likely show values in percent notation for you automatically based on your *net income*. You might also make your own chart showing percent of gross pay, at least for certain categories such as giving.

Evaluating Your Budget Based on Actual Amounts

Periodically, you will want to look at your actual spending and see how it lined up, then adjust as needed.

- **Overrunning** refers to when the actual amount is higher than the budgeted amount.
- **Underrunning** refers to when the actual amount is lower than the budgeted amount.

Refer again to Sam's actual and budgeted amounts. He *underran* some categories, such as transportation, food, clothing, and his self-employment income, but he *overran* medical. He underran his spending altogether by $413.

What does this show Sam about how he is doing in terms of following his budget? To answer that question, Sam needs to think through each of those amounts.

- Sam's transportation budget includes car insurance and maintenance costs. These expenses do not get paid every month, but money needs to be set aside for them each month. Sam should *expect* to underrun it many months; doing so does not really mean he is spending less than he planned, but rather that he has set aside that money for a future expense.
- Sam should think through why his self-employment income was so low. Was that expected too — a seasonal low — or is his side business not doing as well as planned? If the latter, he may want to adjust his budget as needed.
- Sam's clothing was also low, but that could just be because he does not buy clothing evenly throughout the year. He should view that $50 as saved for when he needs a new item. His medical, likewise, probably does not happen evenly. He should look at how he is doing over a longer period of time. If he continues to overrun, he needs to plan for more medical.

» Underrunning and overrunning are expected on bills that come less than once a month; for these categories, look at a longer period of time, such as a year, to check how you are doing.

» **Adjust your budget as needed.**

Key Takeaways

- If you have no savings, if you overrun somewhere you will have to cut somewhere else to make up for it.
- If you have enough of a bill buffer, you can look at categories over time and adjust only when you see a trend or know something has changed.
- Check out the big picture. How is your total giving, saving, and spending over time?

KEEPING PERSPECTIVE

As we study budgeting, remember that the point is to learn how to consider what God has provided (your income) and make wise and prayerful choices about where that income goes rather than being subjected to whims or the pull of marketing. A good budget should free you up to keep focused on the financial priorities you have prayerfully set.

Evaluating a Budget: Tools

Spreadsheets or apps contain numerous tools to help you evaluate a budget. Now we will look at creating and using graphs in spreadsheets.

Ask a parent for permission before using spreadsheets or apps.

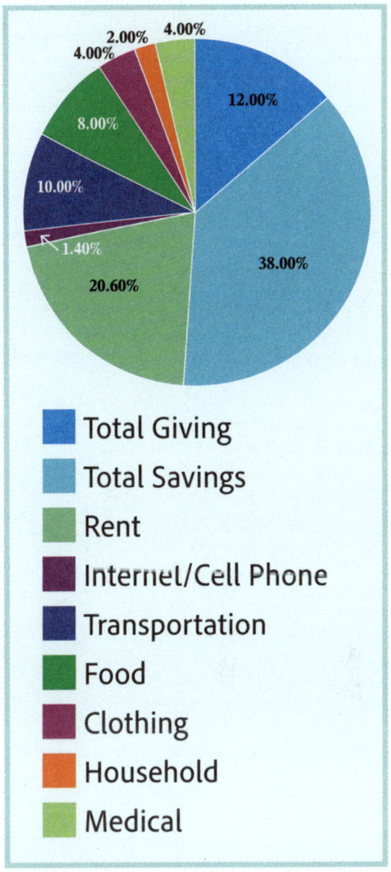

Graphs of the Budget

Here is a pie graph for Sam's budget, made by selecting column A and column B in Sam's budget at the beginning of the quarter, deleting the rows we did not want to show up (we just kept the total for giving and saving, and then the actual spending categories) and inserting a pie graph in a spreadsheet program.

Notice that it gives a visual for instantly seeing where Sam's money is going compared in different places. The pie is divided into pieces, with each piece representing the same percent of the pie as the corresponding value does of the total of all the values.

A pie graph like this one makes it easy to see the portion of a whole being represented. Here, you can tell that saving, rent, and giving account for the vast majority of Sam's income.

If you are trying to make a graph in a spreadsheet and it is not displaying what you need it to, play around with what data you are selecting.

- You may have to make a copy of your budget and delete any rows you do not want displayed in your graph. The graph shown just has the total giving, total saving, and individual spending categories.
- You may also need to adjust what data you are using.
- Note that apps will often create various graphs automatically.

Graphs of Goals

It can be helpful to look at a bar graph of how you are doing toward different goals God has put on your heart. Perhaps you are saving up for something or trying to pay off debt. Tracking your progress toward that goal and seeing how long you have left to achieve it can help you adjust your budget amount accordingly. You can use a spreadsheet (if available, and with a parent's permission) or graph paper.

Total needed	$ 8,000
Total achieved	$ 4,500

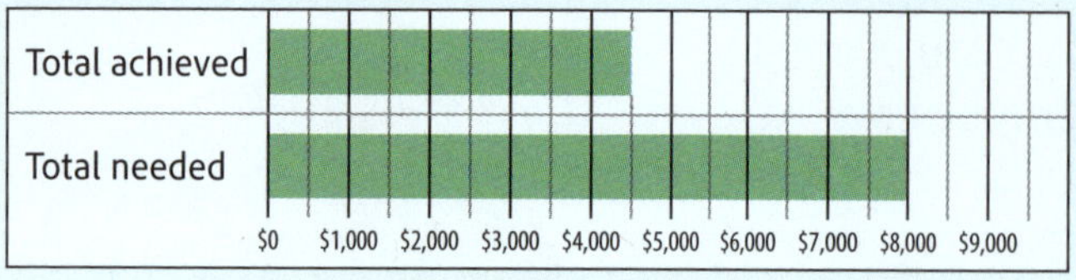

In the bar graph shown, each vertical line is used to represent $500. You can tell what value is represented by seeing at what place the bar stops. If you were drawing a bar graph yourself, you could use graph paper, making each square of the paper represent a certain amount.

Example: You are saving for a trip to Israel for you and a family member. How far do you have left to go?

We see on the chart that the *total needed* is $8,000 and the *total achieved* is about $4,500. Subtraction will show us how much is still needed.

$8,000 – $4,500 = $3,500

There are many other types of graphs out there too that you could use or encounter. Each provides a pictorial look at the data.

Filters, Sorting, Pivot Tables, & More

Spreadsheet programs have a lot of features in addition to graphs that can help you explore financial data.

- **Filters** let you view only transactions with specific attributes (such as just one category).
- **Sorting** lets you arrange data alphabetically by category or by date of transaction.
- **Pivot tables** let you quickly summarize huge volumes of data. For example, they can easily total up all your expenses for each category.

While beyond the scope of this course, it may be extremely helpful in the future to be familiar with spreadsheets, especially if you plan to use them as your main budgeting tool.

KEEPING PERSPECTIVE

Where we spend our money says something about our priorities and what matters to us. Looking at your budget can be a helpful way of honestly assessing what you see as important.

- If you claim to love God and His Kingdom but are not giving toward that Kingdom as your means allow, there is something off.
- If you discover you are spending most of your money on things that do not matter, you need to take a hard look at your heart.

Prayerfully seek God for what His goals are for you financially and budget in light of those and in light of the principles in God's Word.

Your First Budget

When putting together a budget, the big question is this: What would God have you do with what He has entrusted to your care?

Here are a few steps you can take to prayerfully think this through:

1. **Make a list of all your current or historic monthly income, giving, saving, and spending.** If you have never tracked this information, credit card or bank statements can help. You could also ask someone (like a parent) what might be a realistic amount for you. An app or online tool might give you this information quickly (or an online search can turn up averages in your area to use as a starting point), but be sure to get permission from a parent first.
2. **Make a list of financial goals.** Do you have any debt you need to pay off? Are there things you feel you should be saving for, like a car, college, or a future home? How about giving goals — by how much would you like to increase your giving?
3. **Go back and adjust the list** of income, giving, saving, and spending you made to help align with your goals.
4. **Make a budget with the amounts you have adjusted.** Track how you do. Keep adjusting as needed.

Things to Keep in Mind as You Budget

- **Do not miss out on the joy of giving.**

On the first day of every week, each of you is to put something aside and store it up, as he may prosper, so that there will be no collecting when I come (1 Corinthians 16:2).

God wants us to give regularly at every financial level ("as he may prosper"). Develop the discipline of giving to God and trusting God to provide for you. Start budgeting with the goal not of getting wealth for yourself, but of increasing what you can give as God prospers you. If you approach your budget with a heart that views money as a tool to serve God instead of as a tool to serve yourself, you will be amazed at the joyful journey stewarding your finances can be.

The New Testament is filled with examples of people who gave abundantly. We are to live for God's Kingdom, not our own, and that includes with our money.

- **Prioritize getting out of debt.** For every dollar of debt, you are paying interest instead of earning it. It is typically financially prudent to pay off debt as quickly as you can.

- **Save a lot when you are young.** As we learned in Lesson 3, a little set aside when you are young can grow into a lot. Saving lets you increase your wealth little by little, which is the way to lay a good foundation for providing for yourself or your future family and not just getting caught in the here and now.

Wealth gained hastily will dwindle, but whoever gathers little by little will increase it (Proverbs 13:11).

  - Again, it is often recommended to save *at least* 20% of your take-home income (i.e., net pay) if you can.
  - It is also typically recommended to contribute as much as you legally can to retirement accounts (401k/IRA accounts). This will both save you on your taxes and allow your wealth to grow.

There is a limit to the amount of savings you need. But when you are just starting out, you have probably got a long way to go. Saving needs to be a priority.

- **Think about the lifestyle you are called to live considering your income and other goals.** If getting out of debt, giving, and saving are priorities, it may mean making lifestyle cuts — deliberate choices to spend less so you have more money to use elsewhere. If your income starts increasing, do not just let all that increase go toward spending. Make deliberate choices as to what to do with what God has sent your way.
- **Think about the future.** If you are thinking about just what you want now, your budget is going to have you spend money you would really be wise to save. If all you are thinking about is the future here on earth, you are going to miss out on the joy of investing in eternity.

KEEPING PERSPECTIVE

Your budget reflects your priorities, and your priorities reflect your heart.

Keep your heart with all vigilance, for from it flow the springs of life (Proverbs 4:23).

Review

We have talked a lot about the details of budgeting. Now we will review the big picture.

- A budget gives an estimate of average income, giving, saving, and spending.
- You should expect to run slightly over some months and under others, but ideally your average should be close to the budget.
- You can use your budget as a guide, letting you know how you are matching up with what you anticipated and adjusting as needed.

It is easy to lose track of where money is going, spending it on whatever seems important at the moment while forgetting to invest in the future through saving and giving (which invests in eternity). Budgets help us see where our money is going.

Remember when you budget that *all* of your money really belongs to God. While you will assign a specific amount for giving in a budget, it is a good thing if you end up giving out of other spending categories too. For example, if you are giving a certain percentage to your church and specific missions and then you have a family in need over for dinner, that is giving too, even though it may come from your food budget. There is joy in living with the heart that views *everything* we have as belonging to God.

> *The Lord is my strength and my shield; in him my heart trusts, and I am helped; my heart exults, and with my song I give thanks to him* (Psalm 28:7).

Stewarding Time – Just like it helps to have a plan with money to make sure the important things truly get covered, so it helps to have a plan with our time. Time is a gift God gives us to steward too. Making sure time is **planned** for Bible reading, prayer, fellowship, service, and other important priorities can protect us against running out of time for what matters most.

Page intentionally blank

More on Budgeting Income; Limiting Lifestyle (Food)

income = giving + saving + spending

Now that you have the big picture of budgeting down, we will:

- check out some specific situations to make sure you are prepared to budget for whatever type of income or expenses you encounter
- explore in more detail how to make choices when shopping to limit your lifestyle to match your budget. We will focus on food since it is a big area where your daily choices affect how much you will spend, but the same principles apply to other spending

Estimating Income: Biweekly & Bimonthly; Non-Monthly Budgets

Now to make sure you understand how to apply the principles we looked at for income calculation to different scenarios.

Biweekly Paycheck

Biweekly means every other week. With a biweekly paycheck, there are 26 payments a year (there are 52 weeks in a year, and half of that is 26, as $\frac{1}{2} \times 52 = 26$).

Example: Your paycheck is $700 every two weeks (i.e., biweekly). What is your monthly net pay?

calculating yearly amount

$$\frac{26 \times \$700}{12} \approx \$1{,}517$$

dividing that amount by months in a year

Bimonthly Paycheck

Bimonthly means twice a month. **Bimonthly is different than biweekly.** There will be a total of 24 paychecks, not 26, if you are paid twice a month ($2 \times 12 = 24$). However, since there are exactly 2 payments each month, you do not have to know the yearly amount to determine the monthly — you can just multiply the bimonthly amount by 2.

Example: Your bimonthly paycheck is $700. What is your monthly net pay?

$2 \times \$700 = \$1{,}400$

Non-Monthly Budgets

While a month is typically a good period of time for a personal budget, it is possible to budget for other periods of time as well. Businesses, for example, may have quarterly or yearly budgets.

>>> **For periods of time other than a month,** adjust what you multiply or divide by to calculate the amount per the budgeted time instead of per month. **To avoid confusion, in the Answer Key, we will use the yearly method instead.**

We are using *net income* and *net pay* interchangeably.

We could also have **multiplied the weekly amount by 4.33** to find the monthly amount and then multiplied that by 3, since there are 3 months in a quarter.

$3 \times 4.33 \times \$100 = \$1{,}299$

This would have been slightly less accurate, as 4.33 is approximate, but accurate enough for a budget.

Example: What is your average quarterly net income if you have a biweekly paycheck of $1,700?

Calculate the yearly net income and then divide that by 4, as there are 4 quarters in a year. Note that we can do all this in one step.

calculating yearly amount

$$\frac{26 \times \$1{,}700}{4} = \$11{,}050$$

dividing that amount by quarters in a year

Example: What should Sally budget a quarter in tips if she earns an average of $100 a week?

Multiply the weekly amount by 52, as there are 52 weeks in a year, and then divide that by 4 since there are 4 quarters in a year.

calculating yearly amount

$$\frac{52 \times \$100}{4} = \$1{,}300$$

dividing that amount by quarters in a year

Example: What should Jamila budget per quarter for a $50-a-month bill?

There are 3 months in a quarter, so 3 × $50 = $150

Example: What amount would you put in a yearly budget for a weekly expense of $4?

There are 52 weeks in a year, so it would be 52 × $4 = $208.

KEEPING PERSPECTIVE

- Do not try to memorize a formula or rule for every situation.
- Instead, know how different periods of time relate together and multiply or divide as needed to go from the period of time you know to what you are trying to figure out.
- We are really just doing the same thing we discussed in Lesson 6 only adjusting for specific situations.
- Remember that the reason you are learning these skills is so you are prepared to develop a budget to help you better steward what God entrusts to you.

Estimating Income: Hourly Jobs

If you are paid by the hour and your hours vary, you may have to combine a few of the skills we have looked at to determine an amount for your budget.

Hourly Jobs

Example: Calculate your average monthly net income if you had these biweekly paychecks: \$500, \$400, \$450, and \$475.

Start by calculating an average biweekly paycheck, just like we have been calculating other averages.

4 values, so put 4 below

$$\frac{\textit{sum of prior values}}{\textit{\# of prior values}} = \frac{(\$500 + \$400 + \$450 + \$475)}{4} = \$456.25$$

Use the average to calculate a yearly amount and divide it by 12 to calculate an average monthly income. The process is the same as in the last section, except now we are using an average as our basis.

calculating yearly amount

$$\frac{26 \times \$456.25}{12} \approx \$989$$

dividing that amount by months in a year

Example: You worked 20 hours, 25 hours, 30 hours, 35 hours, 40 hours, and 30 hours. You make \$20 an hour. What is your average monthly gross pay?

First calculate the average hours:

6 values, so put 6 below

$$\frac{\textit{sum of prior values}}{\textit{\# of prior values}} = \frac{(20 \text{ hrs.} + 25 \text{ hrs.} + 30 \text{ hrs.} + 35 \text{ hrs.} + 40 \text{ hrs.} + 30 \text{ hrs.})}{6}$$

$$= \frac{180 \text{ hrs.}}{6} = 30 \text{ hrs.}$$

Multiply average hours by hourly rate to calculate average weekly income:
$\$20 \times 30 = \$600$

Use that to calculate a **yearly** income and divide it by 12 to calculate average monthly income:

calculating estimated yearly amount

$$\frac{52 \times \$600}{12} = \$2{,}600$$

dividing that amount by months in a year

>> For hourly jobs:

- Calculate whatever average income you can and then use that to determine a monthly amount for your budget.
- What you have found is just an estimate, not an exact figure.
- Using average paychecks like in the top example reveals a net income or pay, while using average hours and hourly pay as in the second reveals gross income or pay.

Seasonal Hourly Jobs

Some jobs are seasonal, requiring radically different hours at different times of the year. In that case, you might have different average weekly incomes that you add together to calculate the expected yearly income. For seasonal work, it may be helpful to calculate an average for each quarter (3 months) of the year.

Example: You work a job with hours that vary by season, making \$20 an hour. You work an average of 8 hours a week in the first quarter of the year (January through March), an average of 20 hours a week in quarters 2 and 3 (April through September), and an average of 40 hours a week in the fourth quarter (October through December). What is your average gross income?

First quarter

Average weekly: $8 \times \$20 = \mathbf{\$160}$

Average monthly: $\frac{\textit{yearly}}{12} = \frac{52 \times \$160}{12} \approx \mathbf{\$693.33}$

Second and third quarters

Average weekly: $20 \times \$20 = \mathbf{\$400}$

Average monthly: $\frac{\textit{yearly}}{12} = \frac{52 \times \$400}{12} \approx \mathbf{\$1,733.33}$

Fourth quarter

Average weekly: $40 \times \$20 = \mathbf{\$800}$

Average monthly: $\frac{\textit{yearly}}{12} = \frac{52 \times \$800}{12} \approx \mathbf{\$3,466.67}$

We can now calculate the average gross monthly income by adding the gross income for each month to calculate the yearly total. We will use multiplication to calculate the monthly income found by the number of months at that income.

Quarter 1: January–March; Quarters 2 and 3: April–September; Quarter 4: October–December

Total average monthly income: $\frac{(3 \times \$693.33 + 6 \times \$1,733.33 + 3 \times \$3,466.67)}{12} \approx \$1,907$

Multiplied by 6 instead of 3 as the 2nd and 3rd quarters (6 months) were both at this income.

Note that in budgets like this one, you would need to be careful not to spend all your money. If you are hired during the busy fourth quarter, you need to remember that hours will go down after Christmas. You will want to save a lot of your income to cover the first quarter's expenses.

KEEPING PERSPECTIVE

Budgets are very helpful and make your financial decisions easier. The goal is to make an educated estimate and then adjust as needed.

The less certain you are in your income estimate, the more you might need to put some buffer in your budget. Do not plan to spend every dollar if you can help it until you have a better idea what your income actually is. After all, your goal is to live *within* what God has provided for you.

Estimating Income: Self-Employed & Commission

Here are two more types of income you might receive and how to estimate them for your budget. Remember, the point of estimating as accurately as you can is to create a workable plan to steward that income well.

Self-Employed Income

Wesley is starting a landscaping company. He expects to have 5 weekly mowing customers, each of which will pay him \$40 a week during mowing season (which lasts 6 months where he lives). It costs him \$10 a week in fuel. He also expects to have 30 quarterly lawn treatment customers, each of which will pay him \$200 a year, but \$50 of that will go toward the lawn treatment supplies. And lastly, he hopes to have 4 odds-and-ends projects each month, averaging \$300 each after deducting supplies.

Notice that we have used the words *expects* and *hopes*.

- When budgeting for self-employed income, you have to make reasonable guesses of what you think you will be able to make.
- A good rule of thumb is to be conservative in your estimates — extra money can always be donated or go toward savings, but signing up for more expenses than you have income to pay for creates a problem.

>>> Subtract business expenses from what the business makes to calculate the actual income from that business.

Example: How much is Wesley expecting to make from his business given the situation described?

Estimate yearly income from mowing:

weekly mowing receipts – *weekly mowing expenses*

$$26 \times (5 \times \$40) - 26 \times \$10 = \$5{,}200 - \$260 = \$4{,}940$$

Weeks with this receipt or expense

(Since 6 months is half a year, it is 26 weeks $\left(\frac{52}{2} = 26\right)$, so we multiplied both his weekly mowing receipts and expenses by 26.)

Estimate yearly income from lawn treatments:

expected receipts – *expenses per customer*

$$30 \times \$200 - 30 \times 50 = \$6{,}000 - \$1{,}500 = \$4{,}500$$

Expected lawn treatments

example continued on next page

Estimate yearly income from other jobs:

Expected other income per month

$$12 \times (4 \times \$300) = \$14,400$$

Months in year

(We did not have to deduct anything, as we were told this was what Wesley expected to make *after* subtracting supplies.)

Total estimated monthly income:

$$\frac{\textit{total estimated yearly income}}{12} = \frac{(\$4,940 + \$4,500 + \$14,400)}{12} = \frac{\$23,840}{12} \approx \$1,987$$

Since nothing is automatically withheld from self-employment income, \$1,987 is both Wesley's net and his gross income.

Commission

Working on **commission** means that rather than receiving a fixed amount, you get a percentage of the sales you make. Realtors and sales jobs are common jobs paid on commission.

>> Calculate a commission by figuring out a percent of sales you get.

Anna needs to remember this is an estimate. There may be months where she sells *no* houses ... and others where she sells 4 or 5. Anna needs to budget accordingly.

Anna may also have expenses involved in selling homes that she needs to subtract from the value we found when figuring out what she will actually make.

Example: Anna works as a realtor. She earns 3.5% on every home she sells. The average price of a home in her area is \$400,000. She has been told she can expect to sell about 2 homes a month. However, she has to give 1% of the home sales to her agency. What does she expect to make from her commission in one month after she pays the agency?

Anna's projected sales: $2 \times \$400,000 = \$800,000$

Percent Anna keeps: $3.5\% - 1\% = 2.5\%$

Calculate 2.5% of the projected sales:
$0.025 \times \$800,000 = \$20,000$

KEEPING PERSPECTIVE

The point of budgeting is not so you can accumulate wealth. It is a tool to help you steward the resources with which you have been blessed. It also helps you to make wise choices, and is a reminder not to get obsessed with wealth or possessions. King Solomon had all the wealth in the world — and he warned that if you love money, it is never enough. But if we work hard and rejoice in what God *has* given, there can be sweet rest, regardless of how much we have.

He who loves money will not be satisfied with money, nor he who loves wealth with his income; this also is vanity (Ecclesiastes 5:10).

Lifestyle Choice Review: Food Cost

Now we will look at living within the budget you prayerfully set. We are going to focus on food because food is one area where costs can vary drastically depending on the lifestyle choices you make. But the same principles apply in other spending categories.

While your food is probably paid for by your parents right now, it is an expense you will face every day for most of your life.

National or State Averages

Official USDA Food Plans: Cost of Food at Home at Three Levels, U.S. Average for a Sample Year[13]

Age-Sex groups	Thrifty plan		Low-cost plan		Moderate-cost plan		Liberal-cost plan	
Individuals	Weekly cost	Monthly cost	Weekly cost	Monthly cost	Weekly cost	Monthly cost	Weekly cost	Monthly cost
Child:								
1 year	$25.10	$108.90	$36.00	$115.80	$40.70	$176.50	$49.30	$231.70
2–3 years	$37.90	$164.10	$37.80	$163.90	$45.40	$196.80	$55.10	$238.90
4–5 years	$41.30	$179.00	$39.20	$169.70	$48.30	$209.30	$58.40	$253.20
6–8 years	$46.00	$199.40	$55.90	$242.40	$66.40	$287.90	$77.60	$336.40
9–11 years	$53.20	$230.70	$59.40	$256.20	$76.30	$330.40	$88.90	$385.30
Male:								
12–13 years	$56.90	$246.80	$69.20	$300.00	$86.10	$372.90	$100.60	$435.80
14–18 years	$71.80	$311.20	$70.30	$304.50	$88.00	$381.20	$102.40	$443.60
19–50 years	$70.00	$303.20	$69.40	$300.80	$87.00	$377.10	$105.80	$458.30
51–70 years	$61.70	$267.30	$65.30	$283.00	$82.00	$355.40	$97.90	$424.00
71+ years	$58.80	$254.90	$64.70	$280.30	$79.80	$345.70	$97.30	$421.70
Female:								
12–13 years	$49.20	$213.30	$58.80	$254.90	$70.20	$304.20	$87.20	$378.00
14–18 years	$57.10	$247.20	$59.00	$255.50	$70.20	$304.00	$86.90	$376.50
19–50 years	$55.90	$242.20	$60.10	$260.50	$73.40	$318.20	$93.60	$405.80
51–70 years	$51.80	$224.60	$58.60	$254.10	$72.60	$314.40	$87.00	$377.20
71+ years	$57.00	$247.10	$58.50	$253.40	$71.90	$311.50	$85.90	$372.20

The United States Department of Agriculture (USDA) publishes food budget estimates of how much food *at home* costs for four different lifestyles: thrifty, low, moderate, and liberal. Notice how drastically the monthly costs can vary depending on the lifestyle chosen.

- **This chart could give you a starting point for the grocery portion of your food budget.** Keep in mind, though, that food costs vary drastically in different geographic areas.
  - If you live in an expensive state, you will likely need more than this chart indicates.
  - If you live in a cheaper state, you might be able to spend less.
- **This chart does not include eating out**, which is a sizeable portion of many people's food budget.

>>> You can estimate your food expenses by adding what you plan to spend on eating out to the chart's amounts.

Example: How much would a family with a 25-year-old mom and dad and a 6-year-old child budget per month if following the low-cost plan and planning to spend about $40 a week eating out?

dad's food + *mom's food* + *child's food*

at-home food = **$300.80** + **$260.50** + **$242.40** = **$803.70**

$$\textit{eating out} = \frac{52 \times \$40}{12} = \$173.33$$

total = **$803.70** + **$173.33** ≈ $977

Percentage of Income

The USDA tracks the percent spent on food as a percentage of income levels as well, as shown on the chart based on theirs. In a sample year, an average of 11.2% of income was spent on food.

Percent of per capita disposable income spent on food in the U.S.[14]

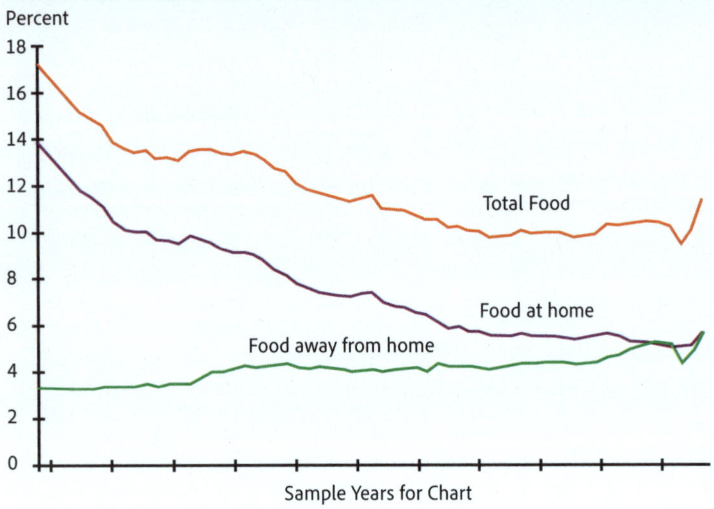

» You can estimate your food expenses by budgeting 11.2% of your income.

When using this guide:

- Know that numbers like this represent an *average*. Some people spend less than 11.2% of their income and others more.
- Know that as your income increases, your food budget does not necessarily need to. Instead, you could put that additional income to saving or giving.

Always think through if the number you have found really does not make sense. This amount is even lower than the thrifty food plan given, so it may not truly be doable.

Example: If you make $400 a week, how much would you spend on food per month if you spend 11.2% of your income?

Calculate your monthly income and take 11.2% of it.

$$\textbf{\% spending} \times \frac{\textit{yearly income}}{\textit{mos. per year}} = 0.112 \times \frac{52 \times \$400}{12} \approx \$194$$

KEEPING PERSPECTIVE

Why might you want to watch your food budget one day? Here are two of many reasons:

- Money saved on food can be used elsewhere. Money saved on food can go toward college, a house, or other goals God puts on your heart.
- We are called to "show hospitality" in Romans 12:13. Knowing how to eat for less can help you stretch your food budget to feed others.

Eating Out vs. In
Plus a Look at Meal Delivery Services

One of the biggest ways to cut food costs is to cook at home instead of eating out or having food brought to you from a restaurant.

The cost of eating in or out varies based on your location. But no matter where you live or what the averages are in your state, the principle holds true: It typically costs significantly more to eat out (or have food delivered from a restaurant) than to cook a meal at home.

Of course, there are less expensive places to eat out. While prices vary by location, year, and item ordered, here are some prices as I am writing this in a suburb of Atlanta, GA:

- A whopper at Burger King® is $5.59, while a meal is $9.89.
- A quarter pounder with cheese at McDonald's® is $4.99, while a meal is $7.79.

In comparison, below are some prices based roughly on those given by USDA[15] on how much it cost to make a burger at home in the same year. Note that these prices assume you will use the rest of the ingredients elsewhere and ignore the cost of condiments.

Note that these savings are for one person. The savings are even higher when you have a family, as you save on each person's food.

Ground beef: $5.16 per lb.; each pound makes 4 burgers, so $\frac{\$5.16}{4} = \$1.29$

Buns: $2.24 per package of 8; $\frac{\$2.24}{8} = \$0.28$

Cheese: $2.80 for 8 slices; $\frac{\$2.80}{8} = \$0.35$

Tomato: $0.92 for tomato that serves 4; $\frac{\$0.92}{4} = \$0.23$

Lettuce: $.70 for 10: $\frac{\$0.70}{10} = \$0.70 \div 10 = \$0.07$

Total: $1.29 + $0.28 + $0.35 + $0.23 + $0.07 = $2.22 per burger

That is a lot less than $5 or more for a takeout burger, and a ton less than what you would pay for a fancy burger at a restaurant.

Example: If you can get each person in a family of 6 a burger at home for $2.22 versus for $5 at a fast-food restaurant, how much have you saved altogether on just the burgers?

Savings per person = $5 – $2.22 = **$2.78**

Savings for 6 people = 6 × $2.78 = **$16.68**

Does it matter? If you save an average of just \$2.78 a day by eating in versus out, that is $\$2.78 \times 365 = \$1,014.70$ a year. If you have a family of 6, that is a savings of $6 \times \$1,014.70 = \$6,088.20$ a year. Food savings can make a significant difference in your budget and in meeting financial goals, such as saving for a home or college, living on one income, or giving.

Meal Delivery Services

Meal delivery may save money over eating out but still typically costs more than making your own meals.

There are now numerous options to get meals delivered to your door from a company in bulk rather than a local restaurant. According to one article, a meal delivery service tends to run \$7–\$14 a meal.[16] If you get a plan on the \$7 side, that would save you $\$13 - \$7 = \$6$ a meal over the average price of eating out. However, you would still be spending a good deal more than making a meal yourself. The difference gets magnified if you are buying meals for a family.

Example: The Thompsons eat out every Monday after their activities. They sign up for a meal delivery service that costs \$8 per person, and they have 5 people. How much does this save them rather than eating out at \$10 per person? How much more does it cost them than making a meal, if they can feed everyone at home for \$15?

Cost of meal delivery: $5 \times \$8 = \$40$

Cost of eating out: $5 \times \$10 = \$50$

Cost of cooking: \$15

Savings of meal delivery rather than eating out:

$\$50 - \$40 = \$10$

Cost of meal delivery rather than cooking:

$\$40 - \$15 = \$25$

KEEPING PERSPECTIVE

Your exercises will help you further explore the cost difference between eating in versus out versus meal delivery — and how that savings can add up.

Does the extra expense mean you should never eat out or have a meal delivery service? Of course not. Economical does not always mean best. Mary poured out perfume on Jesus' feet (John 12:1–8). Food budgeting and choices can look different for different people and seasons. But it is helpful to be aware of the extra cost of various choices should God call you to live on less.

My parents, for example, chose to raise a family on a single income so my mom could homeschool us. To make that work, my mom had to economize, and eating out was a very infrequent treat. While there may be times to eat out or have a meal delivery service, understand the extra costs they bring.

Meal Planning & Estimating

How will you actually stick to your budget when shopping? It is not easy, and it takes practice. Here are a few ideas. Note that these apply whether shopping online or in the store.

- **Have a plan.** If you have a list of what to buy (and maybe a list of what you plan to cook) and avoid extra stuff you do not really need (or maybe plan for just one "extra" item), you will do better at living within the budget you have set.
- **Adjust the plan as needed.** Suppose you find a great deal on meat. You might decide to change the meal you had planned for the night to include that meat.
- **Periodically estimate how much what you have chosen so far costs, especially if you are on a tight budget.** (See "Rounding Practice" on the next page.)
- **Pay attention to unit prices or price per serving.** Do not assume that the item that costs the least amount of dollars is really the cheapest; on the other hand, do not assume that buying the bigger item will save you money.

Pay special attention to unit prices when shopping online, as you can not visually see the size.

Serving Size & Unit Price

A **serving size** is the amount that is eaten at one time. Manufacturers list the serving size based on what they think is reasonable to eat. What really matters for cost pricing is the amount you think you will *actually* eat.

If you start practicing some of these skills whenever you help shop for your family, you will be better prepared when it comes time to run your own household.

>>> To calculate the cost per serving, divide the total cost by the number of servings. You are basically calculating a unit price (a skill we covered in Lesson 2). You can often round and solve mentally for an approximate price.

Example: What is the approximate cost per serving of a slice of garlic toast if you get 6 slices for $2.99?

$\frac{\$2.99}{6 \text{ slices}} \approx \frac{\$3}{6} = \$0.50$ per slice

Sometimes it makes sense to compare items based on the serving sizes and other times based on unit price.

Compare servings, not price per ounce.

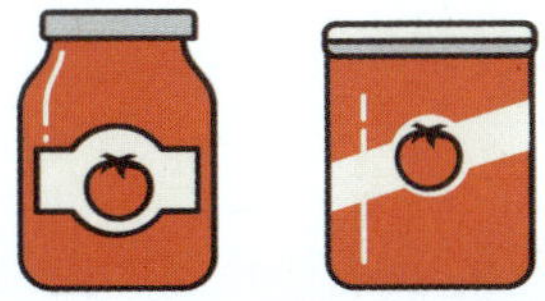

Compare price per ounce, not servings.

You will use the same amount of spaghetti sauce either way, even if manufacturers list different serving sizes.

< means **less than;**
> means **greater than;**
≥ means **greater than or equal to.**

Rounding Practice

To keep a mental estimate of what you are paying, you will want to round. Rounding can also help you mentally compare costs when shopping online. You can round either to the nearest dollar or to the nearest half dollar (i.e., $0.50).

Rounding to the Nearest $1

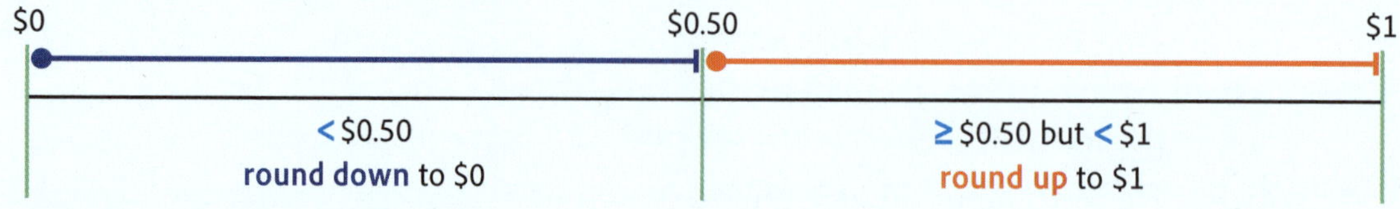

Rounding to the Nearest $0.50

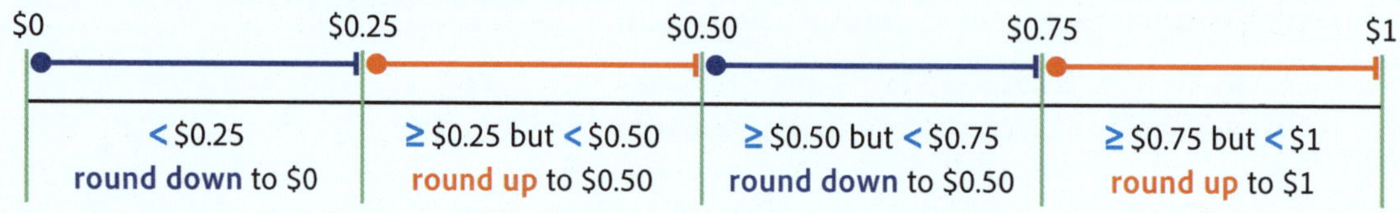

>>> To round to the nearest half dollar, look at if the value is closer to $0, 50 cents, or $1 — if it is halfway or more, round up; otherwise, round down.

Example: Round $1.56 to the nearest $0.50.

$1.56 rounds down to $1.50.

Example: Round $1.75 to the nearest $0.50.

$1.75 rounds up to $2, as $0.75 is exactly halfway between $0.50 and $1.00.

Knowing Your Budget When You Shop

While you might budget by the month, it is helpful to know how much you have to spend each week at the store.

>>> Calculate your weekly budget by dividing your monthly budget by 4.33, the number of weeks in a month.

Example: Calculate your weekly grocery budget if your monthly grocery budget is $433.

$\frac{\$433}{4.33} = \$100$

Keep in mind that the weekly amount you have to spend is an average. So long as you end the month at or under your budget, you have kept your budget, assuming you have a little saved (a "bill buffer") to handle slightly higher weeks.

KEEPING PERSPECTIVE

Sticking to a budget is not easy, but it is doable. Remember, the goal is not to hit the budget amount to the dollar — it is to have your average at or under it. A little planning and some mental math estimations can help you achieve that.

Dish, Brand, & Effort Choices

We learned earlier how eating at home saves money rather than eating out or ordering meal delivery services. We will now talk about some things that affect the cost of eating in.

The Dish You Cook

Meat choice #1:
$11 per lb.

Meat choice #2:
$7 per lb.

Meat choice #3:
$4 per lb.

While the prices above are hypothetical, steak typically costs more than ground beef, which typically costs more than chicken. Plus, you may need more steak to feed the same number of people. Different meals cost different amounts to make.

- Pay attention to the cost of meats in a recipe.
- Check for the cost of other ingredients too.
- Note that sometimes you can rework the recipe to use less expensive ingredients.
- Depending on your grocery budget, you might have a general plan about the cost of meals you will make each week, interspersing more expensive ones on a schedule or for special occasions.

Example: Your recipe calls for fresh basil, which would cost $3.99. If instead you use a few sprinkles of dried basil, which is $3.20 for a jar, but you are only using about 10% of the jar, approximately how much would you save?

We will round since we were asked for an approximate answer.

Calculate portion of basil jar used: $0.10 \times \$3 = \$0.30$

Calculate savings over fresh basil: $\$4 - \$0.30 = \$3.70$

In the example, we are assuming we will use the rest of that dried basil in other recipes. If we had no intention of using the rest of the dried basil, we should have accounted for the whole jar.

>>> Some meals may cost more to make but may stretch further. Look at the cost per serving.

Example: A meal costs $6 to make, but it feeds you for 3 meals. What is the cost per serving?

$\frac{\$6}{3} = \$2$ per serving

It can be helpful to know how measurements relate when finding a cost per serving. For example, you may need to know that 1 TBSP. = 3 tsp. if the serving size on spices is in teaspoons and you are using a tablespoon. Other measurement relationships can be found in the **Reference Section**.

The Brand You Buy

You may have grown up eating one particular brand of a product — or seen ads of a brand that you would like to try. It is worth looking around to see how its cost compares with the others, as switching brands to generic/store-brand or what is on sale can often save a significant amount of money.

- A **brand name** is a recognizable manufacturer.
- **Generic** or **store-brand** products, on the other hand, are ones that you probably do not recognize. Many stores make their own brand of foods that they sell at a discount from the brand names.

Consumer Reports, a retail consulting company, estimates "at least 20 to 25 percent" savings by buying store-brand instead of name-brand.[17]

For example, if a brand-name spaghetti sauce is $4.25 and a jar of store-brand sauce is $2.25, using the store brand will save you $4.25 − $2.25 = $2 just on one ingredient.

Of course, if you can not stand the store brand's taste, then you will want to hunt for those name-brand deals, as saving will not do any good if you will not actually use the product.

The Degree of Preparation

We live in an instant age with lots of instant choices, including with food. You can buy just about everything already premade to one degree or another. While prepped ingredients or a completely premade meal from the store (or a service, as we looked at a few sections ago) will probably still save you more than eating out in general, the more prep someone has done for you, the more you will pay for the same meal.

KEEPING PERSPECTIVE

The meals you choose to make, the brand of ingredients you pick to go in those meals, and the degree of preparation already done for you all affect the cost of homemade meals. While there may be times to make nice meals or buy premade food, knowing that those things add cost can help you live within whatever budget you have set.

Sales & Stocking Up

Planning your meals around what is on sale or what you have a coupon for can save you money, as a little math shows.

>>> Calculate your cost by subtracting coupons from the price (including the sale price), and your savings by subtracting your cost from the original price.

Example: You have a coupon for \$0.50 off an item that is normally \$1.99. You wait to use the coupon till the item is on sale for \$0.99. How much will it cost you, and how much was your savings?

Cost: \$0.99 – \$0.50 = \$0.49

Savings: \$1.99 – \$0.49 = \$1.50

>>> When sales are expressed as a price for multiple items, divide to determine the price for a single item.

Example: How much would a bakery item cost that is on sale for 2 for \$5?

$\frac{\$5}{2} = \$2.50$

Note that stores differ as to whether you would get this price if you just bought 1 bakery item. Find out how your store handles sales like this.

Here are a few things to keep in mind when shopping sales or couponing:

- **Consider planning meals for the week around sale items to save.**
- **Just because an item is on sale does not mean it will save you money if you buy it.** For example, if you do not normally drink soda but buy a liter just because it is on sale, you would have *spent* more than normal. Sales can help get treats you would not normally buy, but that is *spending* a little extra rather than *saving*.
- **Sometimes you have to buy several items in order to get a certain amount off or a free item.** These types of sales make a lot of sense if you need all (or sometimes most) of the items, but beware of just buying a bunch of items you do not need.
- **Make sure you can use what you buy before it goes bad.** Buying multiple packages of something that perishes quickly when on sale does not save you anything unless you can use them before they go bad.

Sales on Non-Food Items

Stocking up can also help you save on non-food items, like toilet paper. Again, though, make sure you are not just buying the item because it is on sale. A good price on a new laptop is not really a money-saver for you if you do not need a new laptop.

Stocking Up

Some of your grocery budget each week should go toward stocking up on items you like to use a lot that are on sale or for which you have a coupon. That way, you are buying ingredients you use often at a lower price than normal.

More Savings

Here are a few other ways you might be able to save on groceries:

- **Buy a stand-alone freezer.** That way, you can buy meats and vegetables in bulk when on sale and freeze extras. You could even join a store where you can buy in bulk.
- **Grow your own fruits and vegetables.** Fruits and vegetables can be pricey, especially for organic items. Growing your own could save your food budget.
- **Shop the sales at different stores.** Different grocery stores put different items on sale. If you are up for going to more than one store, you might be able to save a little extra.
- **Take advantage of member programs.** Many stores (and restaurants) have some kind of free membership program that gives you coupons or discounts. These can result in significant savings.
- **Be aware of and plan for the "extra" stuff you buy.** "Extra" stuff is food other than your main breakfast, lunch, and dinner item. You might be amazed how much that trail mix or bottle of soda or bag of chips really costs. Be mindful of how much you are spending on in-between meal, side, and beverage items.

KEEPING PERSPECTIVE

There is a lot to think through when it comes to grocery shopping. The more time you invest in food planning, shopping, and preparing, the more you will be able to save on groceries. We call this an **inverse relationship**. However, that is also time that could be spent elsewhere, so you have to figure out what the right balance is for you in your season of life.

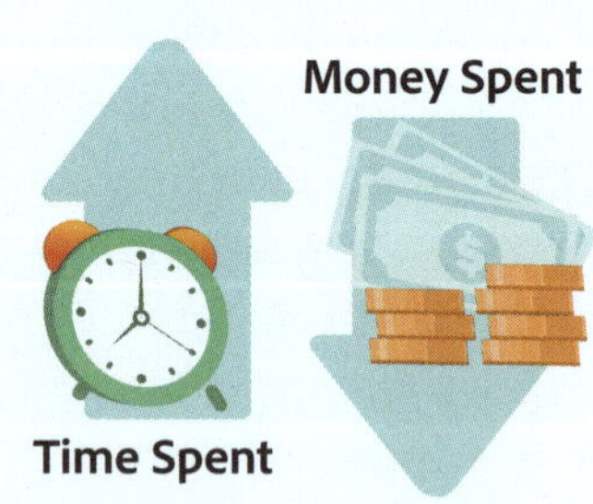

When my mom had a young family and stopped working outside the home in order to raise my brother and me, she was very careful in her budgeting, doing her best to save wherever she could. As life progressed and my dad's income increased, she did not need to invest the same amount of time trying to stretch every penny, although she still tries to follow the basic principles of saving where she can.

Figuring out how to plan meals and grocery shop takes time. Give yourself grace and expect to get better at cutting costs as you go.

Review & Case Study

Hopefully, this lesson has given you an idea of how many expenses, such as groceries, can vary in amount, depending on numerous small choices you make each week. You can change your expenses significantly by the little choices you make.

- The tips we have covered are not designed to be exhaustive but rather to help you understand the basic principles.
- Knowing how to economize does not mean that there are not times to splurge. You need to seek God for what it looks like to live within what He has provided for you in your situation in life.

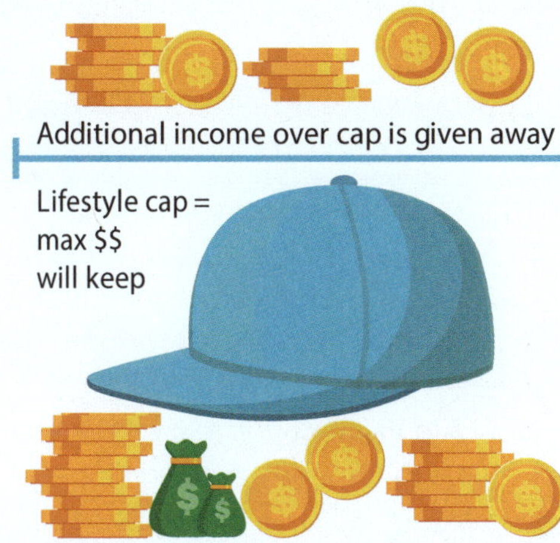

We are going to end this lesson with a story from a couple who, rather than spending to eat out every night and lead an extravagant lifestyle, chose to put a **lifestyle cap**, or limit, on how much money they would allow themselves to keep. The rest was then freed up to give away. This concept is also sometimes referred to as a **finish line** — the amount of spending and/or savings you allow yourself before you say you are finished and give the rest away.

- **There are different types of lifestyle caps/finish lines.** You could have a finish line for how much you spend in a year, and a total finish line for how much you will save for the future (like retirement) that is based on keeping that same lifestyle in retirement.
- **A cap, or finish line, keeps you from working hard just to get more things.** If you know up-front that after you reach a certain amount of money to spend each year and a certain amount saved, you are giving the rest away, then you can have the joy of working hard in order to serve God's Kingdom instead of your own. It is a way to help guard your heart against always wanting more.

Important: A lifestyle cap, or finish line, does *not* mean you do not give until you reach that point. You should give as you are able all along the way — remember the "Handful of Rice" story from Lesson 1. A lifestyle cap, or finish line, is a decision that, should God choose to bless you with wealth, there is a limit to how much of that wealth you will keep for yourself.

Case Study: Choosing a Lifestyle Cap (the Baumers)

This write-up is based on a video from Generous Giving. Used by permission of Generous Giving; all rights reserved.

Greg and Alison Baumer did not start as examples of godly stewardship. When Greg first went to an Ivy League university, he and his wife thoroughly enjoyed spending money. They loved traveling, dining at fancy restaurants, taking fancy vacations, and spending a thousand dollars each month eating out.

One day, however, Greg and a friend began studying what the Bible says about money. Seeing what God said about money changed their lives.

The Baumers set a cap on the amount of money they would spend and save as a family — a cap that allowed for a nice, but not extravagant, lifestyle[18] — choosing to give extra as God blessed them instead of allowing the extra money to make them want more and more. To keep themselves accountable, they got a group of friends they call their "board of directors for life"[19] that holds them accountable financially. The cap they set on how much they would keep for themselves helps keep money's hold on them in check and frees them up to experience the incredible joy and spiritual growth that comes as we give.

The family was tested on their commitment more than once when Greg earned significant unexpected income, but they still embraced gratitude and contentment in obedient generosity. Greg's advice for young people is to set financial caps while you are young, as "it is so hard to unwind a more complex lifestyle later in life."[20]

Quotes from Greg Baumer:

"We ought to be generous, not because God needs our money.... He is going to accomplish His purposes no matter what I do. Rather, generosity is better thought of as such a unique opportunity to stand by God and participate in the fulfillment of His Kingdom."[21]

"I now understand that the point of generosity is not so much the giving in and of itself ... but, rather, it is an opportunity to become more like Him [God] and to grow closer to Him."[22]

Budgeting & Medical Expenses

Medical expenses can be extremely high, especially when facing a major illness or accident. In this lesson, we are going to take a look at different plans to help cover medical costs.

income = *giving* + *saving* + ***spending***

spending = *taxes* + *debt* + ***living expenses***

While you will likely not need to worry about getting your own medical plan for a few more years, understanding medical plans can help you be better prepared for future medical challenges.

Medical Expenses

We live in a fallen world. Sin brought suffering and death into God's perfect creation. Because of the effects of sin on the world, our bodies do not always work the way they should. We sometimes need medicine or medical help. And that can cost a lot of money.

Medical expenses have risen steeply since the mid-1900s.[23] Rising expenses means that medical expenses (or insurance to cover them) is a big part of many people's budgets.

How do you budget for these expenses? Not only is it hard to save the amount of money you might need for major illnesses or injuries, but it is impossible to predict how much you really need to save. You might be healthy today and in the hospital tomorrow from an unexpected accident. I remember showing up to work one day at a bakery as a young person in my 20s, never dreaming that I would fall, hit my head, and start a year-and-a-half traumatic brain injury recovery.

> *Do not boast about tomorrow, for you do not know what a day may bring* (Proverbs 27:1).

- We ultimately have to trust God to provide for us in whatever He allows to impact us. We can rest knowing He works all things "for good" for those who love Him (Romans 8:28).
- At the same time, just like an ant "stores its provisions in summer and gathers its food at harvest" (Proverbs 6:6–8), some families take steps to help cover medical expenses. The steps God would have you take may be very different than the ones I take.

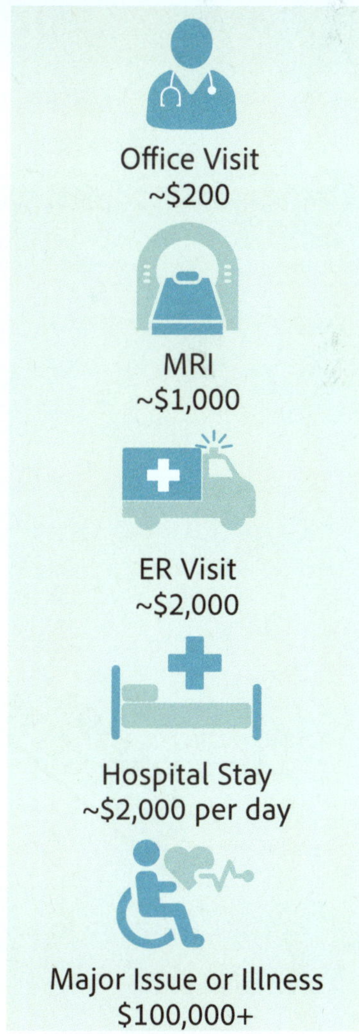

**Prices vary greatly by location and issue; these are hypothetical.*

We will look at some choices and considerations families face when deciding how to approach the topic of covering health care and how math can help you evaluate those choices.

Medical Plans

- Insurance,
- cost-sharing, or
- discount plans

are ways many families help cover and plan for medical expenses. These plans are available to help with different aspects of your medical affairs. They are often broken down like this:

	What They Cover
Health Plans	Your general health but not your teeth or vision unless the issue is considered a health issue (such as TMJ for dental and cataract surgery for vision)
Dental Plans	Teeth
Vision Plans	Vision

Good to Know

- If you have a medical plan, you are referred to as a **member** of or **subscriber** to that plan.
- An insurance plan's agreement is referred to as the **insurance policy**.

We will be focusing on health plans, as health plans cost the most by far due to the potential for significantly higher outlays of money. But the principles we look at could help you select dental or vision plans as well. Also know that some companies only provide limited plans in their health policies.

KEEPING PERSPECTIVE

The prudent sees danger and hides himself, but the simple go on and suffer for it (Proverbs 22:3).

We live in a fallen world. We get sick. Accidents happen. You can trust the Lord to help you through whatever He allows. At the same time, He may lead you to make preparations, knowing you will likely face a medical challenge of some sort.

Insurance, Sharing, & Discount Plans

Now take a look at some different types of plans that can help plan for and cover health, dental, and vision costs. Understanding the different types can help you sort through plans one day.

Insurance Plans

Medical insurance is based off the idea that you pay a fixed payment (called a **premium**) at fixed intervals of time to a medical insurance provider. The insurance provider then agrees to provide certain benefits and coverage. The insurance company makes money by charging enough in premiums and having enough people on the insurance that they are paying out less than they are taking in.

Strengths

- Plans have standardized premiums that are often tax-deductible.
- By law, they can not exclude **preexisting conditions** (i.e., medical issues you have before getting the insurance).
- They offer discounted rates with certain providers.

Weaknesses

- Plans are required by law to support unbiblical procedures.
- They tend to be more expensive than other options unless your employer helps pay for them.

Sharing Plans

Sharing plans are based on pooling together with others to share major expenses. Your monthly or periodic costs for the plan often go directly to support another person. The sharing group is nonprofit, based on people's fees covering the expenses, plus administrative costs. Payment is not guaranteed, although some plans have great testimonies of how large needs were, indeed, covered. The more members that are in the sharing plan, the more likely it is that they will be able to cover your bills.

Insurance and sharing plans exist at different levels, from catastrophic coverage to coverage for most expenses. Their prices also vary drastically. Sharing plans are often faith-based groups. These plans expect you to live a moral and healthy lifestyle. Conditions resulting from unbiblical practices may not be covered.

Strengths

- Plans are exempt from laws to cover unbiblical procedures; many are faith based, and money may go directly to help another believer.
- The cost is typically much less than insurance, unless your employer helps pay for insurance.
- They may offer discounted rates with certain providers.

Weaknesses

- Coverage varies on preexisting conditions.
- Plan can exclude treatments or require waiting periods before covering some expenses. Some are also exempt from the state's insurance regulations.
- Payment is not guaranteed; premiums may vary as member expenses vary.

Discount Plans

In a **discount plan**, the plan organizes discounts but does not provide any coverage of costs. Members pay a small fee to participate in the plan (*much* smaller than for insurance or sharing).

Since a discount plan does not actually cover costs, they tend to work best for dental and vision, as the costs there tend to be more manageable and you might only need a discount. Discount plans do not provide coverage for major health expenses — due to the high costs of health treatments, if you only had a discount plan, even with the discounts, you could still owe a huge amount of money in a bad medical year.

Strengths

- Discount plans are usually much less expensive.
- They are simple — you just pay a discounted price.

Weaknesses

- The plans do not actually cover anything, meaning there is no limit to what you could owe.

Comparing Plans

To compare the costs of plans, calculate the total expected medical costs of each.

total medical costs = plan cost + cost of services with plan

While on years where there are few medical issues you might pay more for insurance or a sharing plan, they guard against astronomical bills.

Example: Calculate your total medical cost if you pay $85.60 in premiums biweekly and then pay $400 for a medical expense. Compare that cost with having no insurance if your costs would have been $1,500 for that same medical expense.

52 weeks in a year; half that (since biweekly) is 26

plan cost = **26** × $85.60 = **$2,225.60**

cost of services with plan = **$400**

total medical costs = **$2,225.60** + **$400** = $2,625.60

Compare plans: $2,625.60 – $1,500 = $1,125.60; the plan costs $1,125.60 more.

Example: Calculate the savings by having a plan if you wound up in the hospital and would owe $127,000 without a plan and $9,000 with one that costs you $2,225.60 a year.

Subtract ***total medical costs with plan*** from ***without***:

$127,000 – (**$9,000 + $2,225.60**) = $115,774.40

KEEPING PERSPECTIVE

Insurance, sharing, and discount plans all have strengths and weaknesses. You will learn more in your exercises how math can help you understand the options out there.

Understanding Terms (Deductibles, Coinsurance, & More)

Now we will check out some terms used in medical insurance that are important to understand if you ever need to use or compare medical insurance plans. Even if you do not, they are still important terms to be aware of, as:

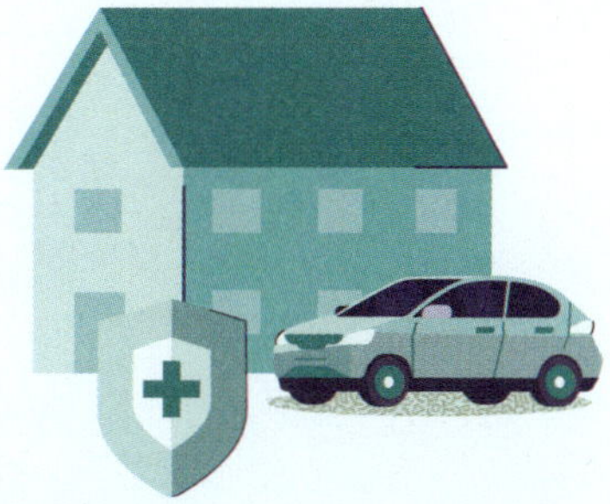

- sharing plans often have the same or similar concepts with different names
- many of these same terms are used in home and car insurance.

Insurance Terms
Deductible – A **deductible** is the amount of money you have to pay to cover your medical bills each year before the plan will start paying. A $1,000 deductible means you will pay the first $1,000 of qualified expenses that year, and *then, once your deductible is met,* the insurance will start paying.
Coinsurance – After meeting your deductible, many plans offer what is called **coinsurance**, where they pay a percent of covered expenses after the deductible and you pay a percent. For example, they may offer 90/10 coinsurance. The first number is the percent your plan pays (so 90%). If only one number is given (such as 10%), that is the percent you pay.
Copay (or **Copayment**) – A **copayment**, or **copay**, is an agreed-upon amount you pay for a particular service, both before and after your deductible is met. Common services that have copays are primary care visits, specialist visits, emergency room visits, and hospital stays.
Out-of-Pocket Maximum – The **out-of-pocket maximum** is the max amount the member will have to pay in a year. After this, the plan will cover 100% of all qualifying expenses. (No more coinsurance or copays!) Deductibles, coinsurance payments, and copays count toward the out-of-pocket maximum.
Limit/Maximum – Plans may have an **annual or lifetime limit on the amount they will cover**. As of the writing of this course, insurances can not limit benefits that are considered essential (by the government's definition, not yours), but they can limit other benefits.

Here is a hypothetical example:

Lillian has a $1,000 deductible, 80/20 (20%) coinsurance, a $50 specialist copay, an $8,000 out-of-pocket maximum, and a $100,000 lifetime maximum for a certain issue.

Lillian has to meet her $1,000 deductible *before* her 80/20 coinsurance starts. This means she pays 100% of the expenses for the first $1,000 of covered medical *before* the plan starts paying 80%. But she will pay those expenses at whatever discounted rate the plan has negotiated.

>>> You pay expenses at whatever discounted rate the plan has negotiated, not at full price. In this case, if Lillian's doctor normally charges $150 but the plan's negotiated rate is $75, she will pay $75.

If she goes to a specialist, she will only pay $50, *even before her deductible is met.*

After she has met her deductible, she will start paying 20% *on the discounted rate.* On a bill of $400, she will pay $0.20 \times \$400 = \$80$, and the plan will pay the rest.

Once Lillian has paid a total of $8,000 between her deductible, copays, and other qualified expenses, the plan will pay the rest. The only exception will be if the plan pays $100,000 on the issue with a lifetime maximum; after that, Lillian will have to pay the rest for that issue, but at the plan's reduced rates.

The chart illustrates these concepts with Lillian's hypothetical values.

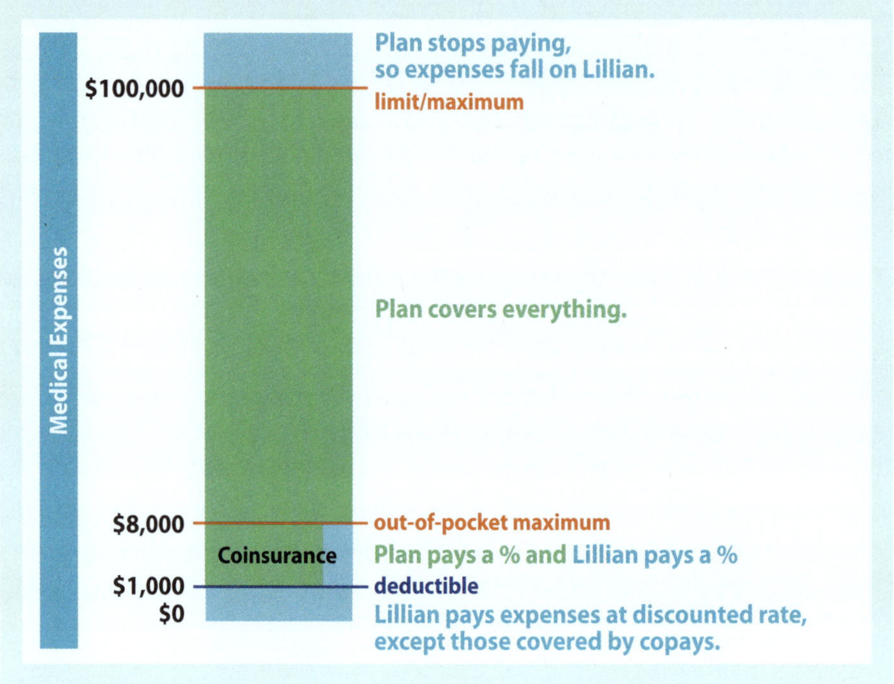

Cost Considerations – Plans that agree to pay out more cost more. Lower deductibles and out-of-pocket maximums typically cost more than high-deductible or maximum plans.

A Few More Details

Plans can get quite complicated — we are summarizing the main aspects here. For example, there are individual and family deductibles and out-of-pocket maximums. If the sum of several individuals reaches the family deductible or out-of-pocket maximum, the plan will start paying even for individuals who have not met their individual deductible or out-of-pocket maximum. The **family deductible** and **family maximum** limit the liability for the family as a whole.

KEEPING PERSPECTIVE

Since we are talking so much about health challenges, remember that whatever medical situation you may face one day, God "works all things for the good" (Romans 8:28). He promises to do just that for those who love Him. The good might not be a good we see this side of eternity, but it is a better good — a forever one.

Accessibility, Coverage, & Types of Insurances

We will now check out some key factors to consider when it comes time to sort through options to help pay for healthcare, as well as examine some different types of insurances.

Accessibility

What good is a plan if you can not use it? None. A primary consideration when looking at a plan — be it insurance, sharing, or discount — should be where you can actually use it.

- **Network** – A plan's network is very important. **In-network** providers have agreed to take discounted prices for services; other providers are considered **out-of-network**. Plans often have different out-of-network deductibles and coverage. Thus, the larger the plan's network, the more options you have for discounted care.

Questions to investigate:

- Are your current providers in network? Search the plan or a provider's website, or give either one a call to see.
- If your providers are not in network, is there a provider in network you are willing to use?
- Are there enough providers in network that should something unexpected come up, you would be able to find a specialist that takes it?
- Is the hospital you would want to go to in network?

- **Access to Specialist** – Some plans require **referrals** to see specialists. In this case, you will have to see your **primary care physician (PCP)** (a doctor who specializes in general care rather than one specialty) before going to, for example, a sleep doctor for sleep issues. The primary care physician refers you to the specialist by specifying in writing that they think you need to go there.

In Network

- Negotiated lower rates that apply to expenses member pays
- Better plan benefits

Out of Network

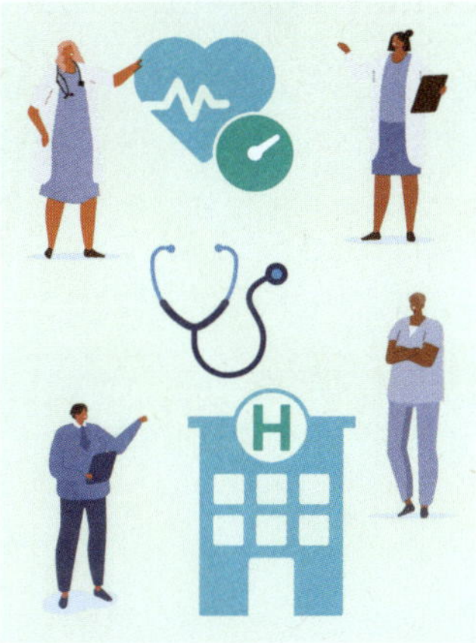

- No negotiated rates
- No or less plan benefits (potentially higher deductible, coinsurance, and out-of-pocket maximum, or no coverage at all)

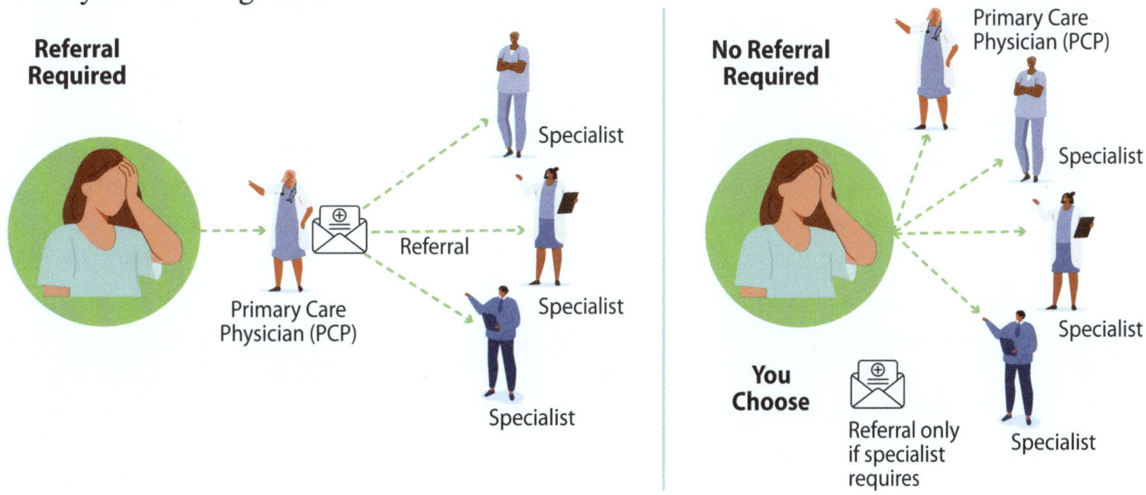

Coverage

Examine what exactly is covered or discounted. For example:

- Pregnancy may be exempted or covered in a limited way.
- The number of times you can go to physical therapy may have different limitations.
- The plan may only cover major medical; emergency room and hospital coverage differs in how much you will pay.
- With dental plans, braces may or may not be covered.
- Natural doctors/treatments may be excluded.
- With sharing plans, there may be a waiting period after you sign up for certain coverage.
- Sharing plans may set different rules as to what they exempt.

Types of Insurance

The chart summarizes different types of insurance. Not surprisingly, **plans with greater flexibility also have higher premium costs**, assuming other factors (such as deductibles) are the same. **Memory Tip:** Notice that the names that start with a *P* have out-of-network coverage.

Category of Insurance	PCP/Referrals TypicallyRequired?	Network	Out of Network	Premium Cost
HMO Health Maintenance Organization	Yes	Limited & number of visits may be limited	⊘	🪙
EPO Exclusive Provider Organization	No	Medium	⊘	🪙🪙
POS Point of Service	Yes	Medium	✓	🪙🪙
PPO Preferred Provider Organization	No	Largest	✓	🪙🪙🪙

⊘ Typically no out-of-network coverage except for emergencies/preapproved.

✓ Out-of-network coverage, although often higher deductible/coinsurance/out-of-pocket maximum & with extra paperwork.

KEEPING PERSPECTIVE

When it comes to medical expenses, there are a *lot* of different options, as within each category of insurance plan we have mentioned, there are many different variations and options. We will look later at comparing policies and how math can help you in that process. When it comes time to pick your own insurance, make the best decision you can and trust the Lord, not your planning, to provide.

Health Savings Account (HSA) & Flexible Spending Account (FSA)

Today we are going to look at a couple of different ways you might be able to someday pay for your medical expenses with tax-free money: HSA and FSA accounts.

HSAs

Think of a **Health Savings Account (HSA)** as a bank account for medical expenses. A certain amount of money each year can be deposited into the account tax free.

- So long as you use the money to pay for qualified medical expenses, you do not have to pay taxes on that money.
- You do not even have to use the money that year — it can stay in the account as savings for when you have medical bills.
- There is a maximum amount you can put in each year, but you can deposit any time of year.

To get an HSA, though, you have to have a **High-Deductible Health Plan**, or **HDHP**. These plans have deductibles that meet federally set limits (which change yearly) and have out-of-pocket maximums.[24] Since these plans do not pay anything until you meet a high deductible, their premiums tend to be less expensive.

A high-deductible health plan (HDHP) can be an EPO, HMO, POS, or PPO.[25] High-deductible health plans tend to save money if you do not use your insurance a lot and are just trying to cover major medical.

You many want to review Types of Insurance chart on page 138.

Higher Deductible

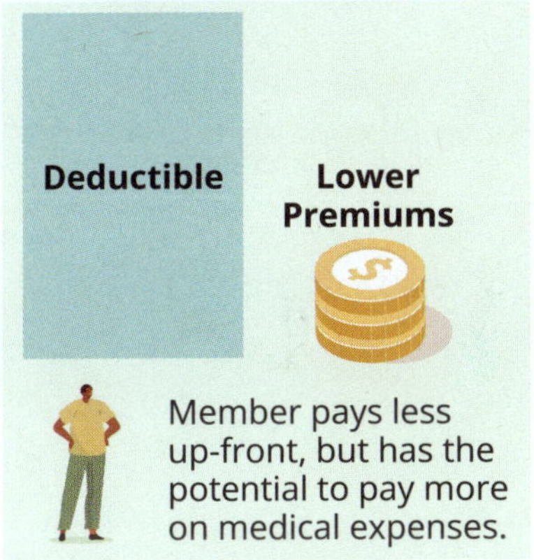

Lower Deductible

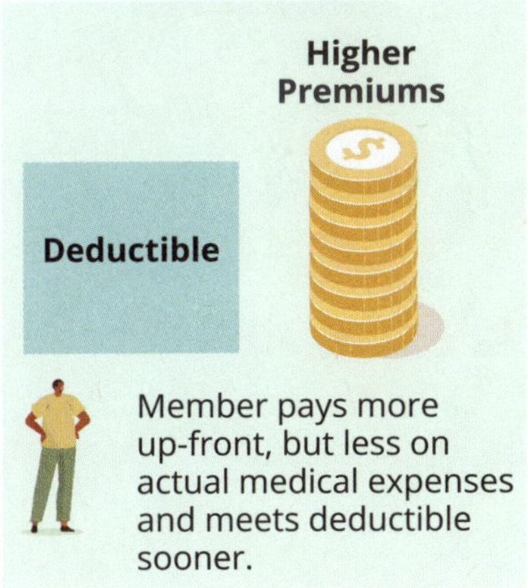

FSAs

Federal **Flexible Spending Accounts (FSAs)** are available through some employers. They work in a similar fashion to an HSA in the fact that they allow you to pay for medical expenses with tax-free money. However, there are three big differences:

- You can use an FSA regardless of if you have insurance or not.
- You have to decide ahead of time how much to put in the FSA for the year (up to a maximum). Whatever you put in and do not use by the end of the calendar year, the government gets to keep.
- You can only get them through some employers.

Calculating the Savings

Paying for medical expenses tax free can save significantly.

>>> To calculate the savings, look at the brackets and see what rate the amount deposited into the HSA or FSA would have been taxed.

Example: In the sample tax year, Suzie deposited the full $3,850 allowed for an individual into an HSA. How much did she save on taxes on that money if her taxable income before the deduction was $70,000, her filing status was single, and she would not have been able to itemize her medical expenses?

Tax bracket on $70,000: 22%

Taxable income with HSA deducted: $70,000 –$3,850 = $66,150

Tax bracket on $66,150: 22%

Since the tax bracket is the same before and after the HSA deduction, all of the deduction would have been taxed at the 22% rate. Multiplying it by that rate will tell us the tax savings.

0.22 × $3,850 = $847

Good to Know:

Tax Considerations – When filling out your taxes, you can not itemize any medical expenses that you paid for from your HSA or FSA account. But since only medical expenses over a percent (7.5% in the sample tax year) of your income can be itemized, HSA or FSA accounts are often more effective at providing tax savings.

Recordkeeping Matters – If you have either an HSA or an FSA, you will need to follow the appropriate rules. Know what the plan requires so you can make sure you get or keep the paperwork you need.

KEEPING PERSPECTIVE

If you ever get an HSA or an FSA, it is worth taking some time to learn what medical-related transactions count and how your HSA and FSA work. Some over-the-counter medicine, for example, may qualify, but if you do not know it, you will not be able to use your tax-free money to pay for it. You might also find that using an FSA debit card simplifies paperwork rather than paying and getting reimbursed.

Supplemental & Disability Insurance

While medical plans cover medical bills, there is also insurance that can provide additional cash in certain situations. We will take a brief look at these, as you will likely be asked someday if you want them.

Supplemental Insurance

There are different types of **supplemental insurance**, insurances that give you extra cash if specific unexpected or out-of-the-ordinary incidences occur.

Here are a few examples:

- **hospital insurance**, which gives you money if you are hospitalized
- **critical illness insurance**, which gives you money if you have a qualifying critical illness
- **accident insurance**, which gives you money if you have a qualifying accident

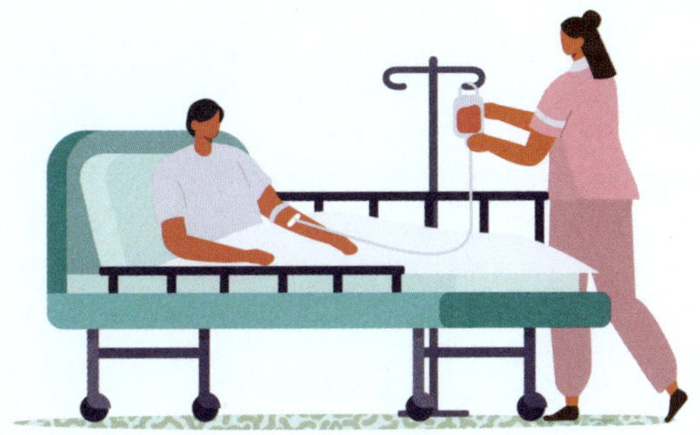

Disability Insurance

Disability insurance only applies if you become disabled and unable to do your job. It provides a percentage of your normal salary while you are unable to work. This payment begins after a waiting period (called an **elimination period**). Some employers will provide disability insurance.

There are different types of disability insurance:

- **Short-term disability** insurance has a short elimination period and provides a percentage of income replacement for a short, pre-agreed-upon time (a few months to up to a year).
- **Long-term disability** insurance has a longer elimination period and provides a percentage of income for a longer pre-agreed-upon time (usually at least several years).

Social Security also offers disability, but there is a lengthy process to obtain this, and you must be unable to work in *any* job (not just your original job).

Example: You have short-term disability insurance that provides 40% of your normal salary for up to 3 months after an elimination period of 14 days. If your normal earnings are $600 a week, how much will you start getting per week after the elimination period?

$0.40 \times \$600 = \$240$

In the above scenario, your income would go from $600 to $240 a week. You would need to reevaluate your budget accordingly. Hopefully, you would have some money saved to help you make up some of the difference.

»» Disability insurance does not take away the need for reasonable savings.

To Buy or Not

There is a different level of risk involved with *not having* each type of insurance or plan.

- **If you do not have a medical plan of some kind**, you run the risk of hundreds of thousands of dollars in medical bills.
- **If you do not have disability insurance**, you could be unable to earn money to pay your bills for a long period of time.
- **If you do not have a hospital supplement**, you are running the risk of not having a little extra cash.

There is also the risk of losing money by *having* a plan. Insurance companies make money because more people do not end up using the insurance than do. A $13 biweekly premium for $10,000 supplemental hospital insurance, if invested at 7% instead of used on supplemental insurance, would grow to about $10,000 after 16 years ... and that $10,000 would be usable whether or not you were hospitalized.

Medicaid supplements are not the same as other supplemental insurances, as they go alongside Medicaid (government insurance) and apply all the time, not just when something like a hospitalization or disability occur.

Only you can prayerfully weigh the risk and the reward of supplemental and disability plans. Here are some things to consider when you have to make these decisions:

- Do you have enough saved up to cover yourself or your family if something happened?
- Are you working in a high-risk industry, or do you have health issues that make you at higher risk of needing supplemental or disability insurance?
- Do you need insurance to provide for a family if you were injured?

KEEPING PERSPECTIVE

Many supplemental insurance advertisements seek to elicit fear, as they want you to buy their insurance. But "God gave us a spirit not of fear but of power and love and self-control" (2 Timothy 1:7). Do not buy any kind of insurance, or make other life decisions, because of fear.

Whether making an insurance decision or a different kind of decision altogether, let your heart rest in knowing God knows the future and has a provision for whatever He allows. Ask Him to direct your steps and help you make wise choices, praising Him that He is able to do so. Then make the best decision you see at the time within the means God has given you to reasonably prepare for life in a fallen world, trusting He will redirect you if needed and will provide no matter what.

Budgeting Medical

When budgeting for medical expenses, you will want to budget both your anticipated *total medical costs* (using an anticipated cost of services, not actuals like we did earlier when calculating the *cost of services with plan*) *and* some extra savings to cover services you unexpectedly end up needing.

> ***budget for medical*** = ***total medical costs*** + ***additional savings***
>
> ***total medical costs*** = ***plan cost*** + ***cost of services with plan***
>
> (If you do not have a plan, then you will have to estimate ***total medical costs*** and plan extra for ***additional savings***.)

Budgeting Saving

- If you have at least your out-of-pocket maximum in a savings account, then, should you have a catastrophic year, you know you can pay your portion of the medical bills. If you stay healthy, the money can sit in savings, earning interest for other uses.
- Of course, having more in savings to cover time off work and other expenses if you were to get sick is also wise. Many suggest always having savings on hand enough to cover 3–6 months of expenses.

The exact amount you save is between you and God; just know that we live in a fallen world, so saving is often very helpful.

Now, getting your out-of-pocket maximum or 3–6 months of your expenses saved may be difficult when you are first starting out.

- Start saving early, while you are still on your parents' health plan and have fewer expenses.
- Begin where you can. The 3–6 months' savings may be a target to aim for, not where you start.

Estimating Cost of Services

Accurate estimates not only help you budget well, but they can also help you decide how much to put in an FSA or HSA if you are using one.

Example: Use the chart to figure out how much you would expect to spend in a year (besides premiums) if you expect to need the bulleted services below.

• 2 primary care visits with a $25 copay	2 × $25 = $50
• 3 specialist visits with a $40 copay	3 × $40 = $120
• 4 generic Rx of $20 copay	4 × $20 = $80

• $300 of bloodwork - With no copay for this, you will have to calculate what you owe based on your $100 deductible and 10% coinsurance.

example continued on next page

Sample Costs	
Primary care visit	$25 copay
Specialists	$40 copay
Generic Rx	$20 copay
Deductible	$100
Coinsurance	10%

Rx is an abbreviation for a prescription.

Subtract deductible, as you owe 100% until that is met	\$300 – \$100 = **\$200**
Calculate coinsurance amount on portion of the bloodwork past the \$100 deductible	0.10 × **\$200** = \$20
Add the deductible with the coinsurance amount to calculate total for bloodwork	\$100 + \$20 = **\$120**

Total: **\$50** + **\$120** + **\$80** + **\$120** = **\$370**

Budgeting for an FSA

When budgeting for an FSA, estimate your *cost of services with plan*. Then use that as a basis for what to put in the FSA. In the above example, you estimated the costs to be \$370, so you might put \$370 a year into your FSA, with a monthly FSA budget of $\frac{\$370}{12} \approx \$31$. Keep in mind that you lose whatever you do not use in an FSA, so make your estimates conservative.

While you have to decide before the year how much goes in your FSA, you will divide that yearly amount for your monthly budget. It is also taken out of your paycheck throughout the year, so would come out of gross, not net, income.

Budgeting for an HSA

If using an HSA, you would want to include at least your *anticipated cost of services with plan*. Since HSA money does not *have* to be used, you could also include additional savings.

In addition to what you put in your FSA or HSA for medical expenses, remember to also budget your plan premiums if needed and have extra savings on hand for the unexpected.

In some cases (such as some unemployment and retirement situations), you may be able to pay your medical insurance premiums from your HSA and would want to include money for them there, but they would typically be budgeted for separately.

Example: Ivy anticipates \$2,000 of medical expenses in a year. If she wants to save an additional \$1,000, how much should she put in her HSA each month?

Yearly amount: \$2,000 + \$1,000 = \$3,000

Monthly amount: $\frac{\$3{,}000}{12} = \$250$

KEEPING PERSPECTIVE

When it comes to budgeting for medical expenses, consider your premiums/membership fees, your anticipated cost of services with a plan (which you would put in an FSA), and savings for the unexpected (in an HSA or a savings account).

Remember Joseph and the wisdom he showed in Egypt in planning for the lean years (Genesis 41:33–36). Lean years come. Plan for them when you are able.

Comparing Policies

When it comes time to choosing policies, the choices can be overwhelming. There are some steps you can take to help narrow down your options. Note that these steps can be adapted to help you with other decisions you might need to make even sooner.

- **Think through how you will get coverage**, as that will narrow your options. Will you be going through an employer or on your own?
  - If through an employer, check out their options.
  - If on your own, note that sharing plans or group insurance coverage (such as through a special-interest group) tend to be less expensive than individual insurance plans.
- **Think about your needs and preferences.**
  - Do you need really good coverage for routine care or just enough to cover you if something unexpected took place?
  - How important is a network to you?
  - Is there a specific type of plan that seems to be a good fit?
- **Think about what you can afford.** You might want an expensive plan, but you might have to settle for something less.
- **Do some research on the company.**
  - Is it well known? How big is its network?
  - Do the providers in your area take it?
  - What do reviews look like online?
- **Compare the plans** after you have narrowed them down to a few. A sample comparison spreadsheet is shown. Notice that it starts by listing some facts about the insurance plans. The facts you pick to compare might differ.

	A	B	C	D
1		**Plan A**	**Plan B**	**Plan C**
2	Deductible	$ 100.00	$ 1,000.00	$ 6,000.00
3	Coinsurance	10%	20%	20%
4	Out-of-Pocket Maximum	$ 5,000.00	$ 10,000.00	$ 7,000.00
5	Primary Care Copay	$ 25.00	$ 50.00	$ 25.00
6	Specialist Copay	$ 50.00	$ 50.00	$ 30.00
7	ER Copay	$ 300.00	$ 150.00	$ 100.00
8	Rx Copay (generic)	$ 15.00	$ 5.00	$ 10.00
9	Bloodwork	10% after deductible	30% after deductible	0% after deductible

Think through the cost of services on each plan in different scenarios. How you define those scenarios may differ from what we did — the idea is to compare healthy and non-healthy years.

If you search online for "online calculator to compare health plans," or reviews, you should find some **online tools that can help you compare different plans. Make sure you get a parent's permission first.**

Cost of Services by Plan for Different Scenarios

		Plan A	Plan B	Plan C
Healthy Year	Year with 2 primary care and 1 specialist visit and 2 generic Rx	$ 130.00	$ 160.00	$ 100.00
Moderate Year	Year with 2 primary, 5 specialist, and 4 Rx	$ 360.00	$ 370.00	$ 240.00
Major Health Year	Year with 1 ER copay, major surgery of $10,000, plus everything from year above	$ 1,750.00	$ 3,320.00	$ 7,000.00
Extreme Year	Meeting out-of-pocket maximums	$ 5,000.00	$ 10,000.00	$ 7,000.00

Values on this chart are found by calculating *cost of services with plan* for that type of year using values on previous page.
cost of services with plan* = 2 × *primary care* + *specialist* + 2 × *generic Rx
***cost of services with plan* = 2 × $25 + $50 = 2 × $15 = $130**

The cost of services for each plan were found by looking at how much things would cost on that plan. Right now, Plan A looks the best, right? But we have not accounted for the cost of the plan itself — that is, for the premiums, or plan cost. Calculating it and the total medical costs tells a different story.

	Plan A	Plan B	Plan C
Plan Cost/Premiums Per Year	$ **6,144.00**	$ 5,320.00	$ 3,600.00
Total Medical Costs (*plan cost + cost of services with plan*)			
Healthy Year	$ 6,274.00	$ 5,480.00	$ 3,700.00
Moderate Year	$ 6,504.00	$ 5,690.00	$ 3,840.00
Major Health Year	$ 7,894.00	$ 8,640.00	$ 10,600.00
Extreme Year	$ 11,144.00	$ 15,320.00	$ 10,600.00

$6,274 = $130 + $6,144

Looking at these totals, Plan A no longer seems like the best deal. Except in the case of a major health year, Plan C saves money. So long as we have or will have money saved up or a supplemental plan to cover the high deductible if we have a major health year, Plan C looks like the best choice.

We ignored the fact that Plan C would also qualify us for an HSA, which could result in tax savings.

Insurance comparisons do not always yield a clear winner like this one did. At the end of the day, you have to pick what you think is reasonable, considering your situation.

KEEPING PERSPECTIVE

Wading through choices, whether insurance or something else you have to decide on, is an opportunity to make sure you get what you need at a cost you can afford. And the best option for you might not be the same as the best option for me — or the best option for you next year.

You may later find yourself wishing you had made a different choice — and that is okay. Trials are what God uses to help us to grow (James 1:2–4). Whatever happens, whether seemingly good or bad, God has promised to be with you always (Matthew 28:20b) if you are His child.

When I was injured, I remember being awed at how God had a provision all figured out. My employers had workers' compensation medical coverage that gave me some income and covered my medical expenses, which proved quite extensive. At the time, I would have struggled if I had had to use my personal insurance instead of workers' compensation. God knew what I needed, and He knows what you need too.

Review

There is a lot to think about and pray through when it comes to budgeting for medical expenses. I hope this lesson has given you some thoughts to ponder and that you have been able to sort through some of your own thoughts and convictions in this area. It is helpful to think through these things ahead of time. Whatever route you take, ultimately trust God for His provision.

Remember that the most important preparation for hard times is spiritual preparation. Learn to trust God now, growing in Him so you are prepared for other trials that will come. Above all, make sure you are part of God's family through Jesus so that no matter what happens, you can have the peace and joy of knowing Heaven is yours and God is walking through the challenge with you.

There is no need to let fear grip us when we think about medical challenges (or any other kind of challenge or decision). God promises that He works all things together for good to those who love Him.

> *And we know that for those who love God all things work together for good, for those who are called according to his purpose* (Romans 8:28).

Page intentionally blank

Budgeting for Big Expenses

Some expenses require saving in order to afford them. In this lesson, we will take a look at saving and budgeting for life's big expenses. We will also look at some measuring skills. These skills are not only needed to budget for home improvement projects (a big expense you might save for one day) but also for projects you might be doing now, such as arts and crafts, sewing, or building projects.

income = giving + ***saving*** *+ spending*

Saving for Big Expenses & Skill Sharpening: Units of Distance, Unit Conversion, & Months to Years

As a child, you might have saved up to buy a toy you really wanted. As we get older, we end up saving for bigger-ticket items, such as mission trips or computers or cars or houses or college. You will find that there are:

- **long-term big expenses** (ones that cost a lot and require a long time to save for, like a home)
- **short-term big expenses** (ones that can be saved for relatively quickly, like a typical vacation)

While the focus of this lesson is on saving for big expenses, remember that there are many other reasons to save, including to have a buffer for bills, an emergency fund, and money in retirement.

Skill Sharpening: Converting Months to Years and Months

As you figure out how long you will need to save for a goal, you may find yourself needing to convert months to years. For example, you see something that will take 50 months. How many years is that?

Since there are 12 months in a year, we will divide by 12 to see.	$\frac{50}{12} \approx 4.17$
It would be easier to have the decimal portion listed in months. Calculate how many months are in 4 years.	$4 \times 12 = 48$
Now we know that 48 months is 4 years. We have 50 months altogether. How many months are left?	$50 - 48 = 2$
50 months is 4 years and 2 months.	

>>> To convert a fraction of a year into months, first determine the number of months accounted for by the years. Then subtract that from the total months to calculate the remaining months. Much of this can be done mentally.

Skill Sharpening: Units of Distance & Unit Conversion

Measurement and unit conversion skills will serve you well both now and in the future.

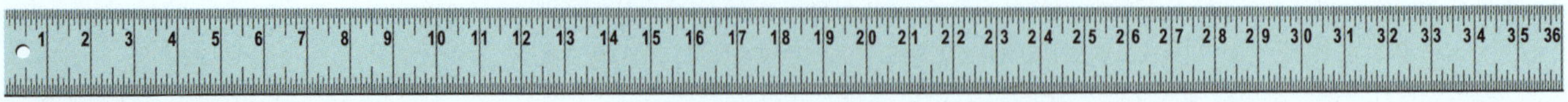

1 yd. = 3 ft. = 36 in.

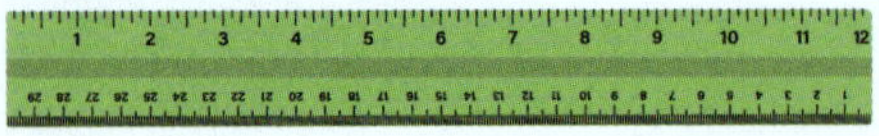

1 ft. = 12 in.

1 in.

»» To convert between units of measure, use a **conversion ratio**, a fraction where both the numerator and denominator represent the same amount, only in different units, making the fraction worth 1. Multiply by the conversion ratio needed to cancel out the unit.

1 yd. and **36 in.** represent the **same distance**, just in different units.

Example: Convert 24 inches (in.) to yards (yd.), given that 36 in. = 1 yd.

Conversion ratios: $\frac{36 \text{ in.}}{1 \text{ yd.}}$ or $\frac{1 \text{ yd.}}{36 \text{ in.}}$

Multiplying by the conversion ratio needed to cancel inches:

$24 \cancel{\text{in.}} \times \frac{1 \text{ yd.}}{36 \cancel{\text{in.}}} \approx 0.66 \text{ yd.}$

For a review of multiplying fractions, see the introduction to Quarter 4.

Remember, a fraction line means to divide. If we are multiplying by inches (as we did with *24 in.*) and dividing by inches (as we did with *36 in.*), they are going to cancel each other out, leaving us with an answer in yds. **Choose the conversion ratio needed for the units to cancel.**

Sanity check yourself. Does your answer make sense? 1,800 is more than 50, which makes sense, since we are going from a larger unit to a smaller one. It will take more of the smaller unit to measure the same distance. If we had gotten an answer of 10 in., we should know we did something wrong.

»» Since it is worth 1, multiplying by a conversion ratio will not change the value. This holds true because of the consistent way God governs creation.

Example: Convert 50 yards to inches.

This time, we will put yd. in the bottom of the conversion ratio so the yds. cancel out.

$50 \cancel{\text{yd.}} \times \frac{36 \text{ in.}}{1 \cancel{\text{yd.}}} = 1{,}800 \text{ in.}$

KEEPING PERSPECTIVE

The thought of a financial goal might seem materialistic — and it certainly can be if your heart is just to amass treasures here on Earth or to replace your daily dependence on God for your provision. But financial goals with the right heart can be a way of pursuing God's plan for your life. For example, owning a home one day in which to raise a family and show hospitality to others lines up with God's desire for mankind to "be fruitful and multiply" and "show hospitality" (Genesis 1:28; Romans 12:13). Saving for godly goals can be a way of stewarding your finances for God's Kingdom.

The Saving Process

How do you start saving to reach financial goals?

1. Seek Wisdom.

Seek God's wisdom (and the wisdom of godly people) as to whether you should pursue the goal. Ask for wise advice. Keep seeking throughout the process.

> *Plans are established by counsel; by wise guidance wage war* (Proverbs 20:18).

2. Calculate the Total.

It may sound obvious, but you need to figure out how much you need to save. This might involve a little bit of research. You do not need exact amounts — just fairly decent estimates.

- If you are trying to save up for a home, you might look online or talk with a realtor about the price of homes in your area and about what kind of a down payment (i.e., money required to put down on a mortgage) you need.
- If you are looking to buy a car or a computer, you might visit a few stores or websites and get a sense of the costs.
- If you want to go on a trip, you would want to ascertain the total cost of the trip (including airfare, food, activities, and lodgings).

Once you have gotten a better idea of what is reasonable to spend, evaluate how much you can spend (or will be able to spend after saving a bit). Can you afford what you thought? If not, is there a way to modify your plans to afford it by the timeframe you hoped?

Looking at costs and at what you can afford can help you settle on a goal of how much to save.

3. Add It to the Budget.

Now that you know your saving goal, you need to build saving into your budget to reach that goal. You can do this in different ways.

Way 1: Figure out how much you can save each month and then figure out how long it will take you to reach your goal. For example, if you need to save \$50,000 for a home down payment and you can save \$1,000 a month, it is going to take you 50 months $\left(\frac{\$50{,}000}{\$1{,}000} = 50\right)$ to meet that goal. Since there are 12 months in a year, that is 4 years and 2 months.

Way 2: Divide the total amount you need to save by the time you have to save for it. For example, if you want to go on the trip in 10 months, divide the total by 10 months. That is the amount you need to save per month.

4. Adjust as Needed.

As you figure out how much you would have to save each month to meet your goal, you might decide you really need to reach it faster. To do this, you could:

- **revisit your expenses** to see if you can lower anything else to save more each month. We have seen how food budgets especially can often shrink based on choices made.
- **get some side jobs** or find additional ways to earn extra income.

Just remember the general income equation:

income = giving + saving + spending

Increasing income or decreasing spending can result in saving increasing. Alternatively, you could:

- **reexamine your goal and find a way to get what you need for less**, thereby decreasing the amount of savings you need. For example, you could decide to save for a less expensive car or cut down the number of days you are staying on a trip.
- **repurpose some general savings** you already have on hand to pay for this expense.

Example: You want to spend $800 on a trip but can only save $100 a month. You have $300 in general savings already. How many months will it take you to reach your goal if you use your general savings?

Calculate amount left to save: $800 – $300 = $500

Calculate months left to save enough: $\frac{\$500}{\$100} = 5$

≫ Having general savings can prove very useful in helping cover big expenses as they come up.

KEEPING PERSPECTIVE

For which of you, desiring to build a tower, does not first sit down and count the cost, whether he has enough to complete it? Otherwise, when he has laid a foundation and is not able to finish, all who see it begin to mock him, saying, "This man began to build and was not able to finish" (Luke 14:8–30).

If building a tower, you would think through if you can complete it. You would not want to end up with half a tower. While in context Jesus' main point in this passage was to count the cost spiritually, this parable also illustrates the importance of counting the cost financially and having a plan to complete the different goals we embark on in life.

Budgeting for Events & Trips

As was the case with food, the amount people spend on events and vacations varies greatly.

You can vacation:

- on a shoestring (staying at a campground and eating food cooked over the fire)
- on a medium budget (staying at a hotel and eating at low- to medium-priced restaurants)
- on a large budget (staying near the ocean in a five-star hotel and eating caviar).

Obviously, the option you choose drastically affects how much you need to save.

Tradeoffs

Determining and sticking to a budget is all about tradeoffs. What is most important to you? That is going to determine how much of a budget you set and then how you end up spending that budget.

When I was getting married, my parents gave me a certain amount of money for the wedding. I had some additional savings set aside that I could have added to it if I chose, but I opted to economize instead. I decided it was more important to keep the savings to help buy a home after I got married.

Throughout the wedding process, I had to constantly make tradeoffs to keep within the budget. The tradeoffs I made might be different than the tradeoffs another person would have made. **The Holy Spirit and wise counsel from others can help you navigate the tradeoffs you face.**

Making & Tracking the Budget

- It can be helpful to set up a way of listing your anticipated expenses for a big event or trip. A simple spreadsheet or lined paper will work. An expense budget can be similar to a full budget, only streamlined for an event. You will want to list all the expenses you anticipate by category.
- As you make purchases, record your actual amounts and compare the two. If you find yourself going over budget in one place, see if you can cut expenses somewhere else. If you can not, at least you will know the amount you are over budget so you can figure out how to cover that.

- Notice in this trip budget that there is a column for the *budgeted amount*, one for the *actual amount*, and one for the *difference*. Since the actual amounts were $55 higher than the budgeted amount (shown by the values being in parentheses — that is how this spreadsheet is showing negative values), we would say this person has **gone over budget** by $55, or they **overran** the budget by $55.

	A	B	C	D
1	**Trip Budget**	**Budgeted Amount**	**Actual Amount**	**Difference**
2	Lodging	$ 400.00	$ 450.00	$ (50.00)
3	Airfare	$ 300.00	$ 300.00	$ 0.00
4	Food	$ 150.00	$ 200.00	$ (50.00)
5	Activities	$ 175.00	$ 150.00	$ 25.00
6	Souvenirs	$ 30.00	$ 10.00	$ 20.00
7	Total	$ 1,055.00	$ 1,110.00	$ (55.00)

- Spreadsheets make the math easy. Notice that each cell was automatically calculated. The formula in D2 was copied and pasted into the remaining rows of D.

=5*80 =5*90

	A	B	C	D
1	**Trip Budget**	**Budgeted Amount**	**Actual Amount**	**Difference**
2	Lodging	$ 400.00	$ 450.00	$ (50.00)
3	Airfare	$ 300.00	$ 300.00	$ 0.00
4	Food	$ 150.00	$ 200.00	$ (50.00)
5	Activities	$ 175.00	$ 150.00	$ 25.00
6	Souvenirs	$ 30.00	$ 10.00	$ 20.00
7	Total	$ 1,055.00	$ 1,110.00	$ (55.00)

=SUM(B2:B6) =B2-C2

KEEPING PERSPECTIVE

Setting and keeping a budget for an event or trip involves an element of *planning*.

- We know from the Bible that God plans — "For I know the plans I have for you," declares the LORD, "plans to prosper you and not to harm you, plans to give you hope and a future" (Jeremiah 29:11).
- We are told in Proverbs 16:3 to "Commit your work to the LORD, and your plans will be established." Notice the verse talks about plans — and work that is committed to the Lord.

James 4:13–15 reminds us that we really do not know what the future holds. All our plans are subject to the Lord's leading. We can plan, but we should do so in a surrendered way, trusting Him to redirect.

Come now, you who say, "Today or tomorrow we will go into such and such a town and spend a year there and trade and make a profit"— yet you do not know what tomorrow will bring. What is your life? For you are a mist that appears for a little time and then vanishes. Instead you ought to say, "If the Lord wills, we will live and do this or that" (James 4:13–15).

Skill Sharpening: Square & Cubic Units, Plus More Unit Conversion

Now we will switch gears and focus on some measurement skills. If you ever need to figure out how much to save for a home improvement project, you will likely have to start by measuring and determining the supplies needed. These skills might also come in handy in helping your parents with their projects or in doing other projects that involve measuring.

There is a chart in the **Reference Section** on page 345 explaining how to find the perimeter, area, and volume of some common shapes; consult it as needed on exercises.

Square and Cubic Units

Area, or the space enclosed, is measured in square units. For example, you are trying to price out the paint you need to paint a wall. The wall is 10 ft. wide and 8 ft. tall. We can measure the area of the wall itself in square feet.

A **square foot** is a square with 1-foot sides.

Think about how you calculate the area of a rectangle. You multiply the length by the width. That counts up the number of square units that could fit in that space.

$$10 \text{ ft.} \times 8 \text{ ft.} = 80 \text{ ft.}^2$$

Notice we wrote the unit not as *ft.*, but as *square feet*, or *ft.*^{2} After all, we have 80 squares with 1-foot sides, not just 80 feet. An exponent is a way of showing repeated multiplication, and we just multiplied ft. by ft., resulting in ft.^{2} **We work with units of measure just like we would with numbers.**

$$10 \times 10 = 10^2 \qquad \text{ft.} \times \text{ft.} = \text{ft.}^2$$

Suppose you needed to calculate the space of an entire room, or what we would call its **volume**, or space contained. You might need to do this if figuring out how big of an air conditioner or heater you need. You would use cubic units.

A **cubic foot** is a cube with 1-foot sides.

Once again, multiplication can help add up all the cubes that would fit inside of a space, such as the room shown.

$$11 \text{ ft.} \times 10 \text{ ft.} \times 8 \text{ ft.} = 880 \text{ ft.}^3$$

Note that ft.^{3} means cubic feet. It is standing for the number of 1 ft. by 1 ft. by 1 ft. cubes that would fit in a volume ... and we can count them up by multiplying ft. a total of 3 times.

$$\text{ft.} \times \text{ft.} \times \text{ft.} = \text{ft.}^3$$

We really just calculated the area of the floor and then multiplied it by the height of the ceiling to calculate the total volume. We could write out how to calculate the area of a rectangular prism, which is what the room really is, like this:

$$\textit{volume of prism} = \textit{area of base} \times \textit{height}$$

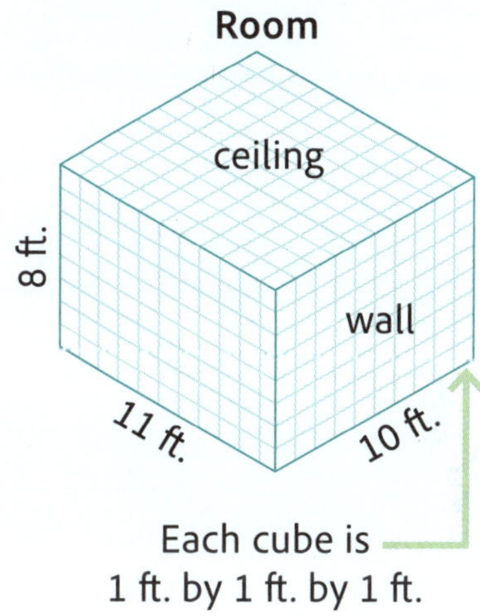

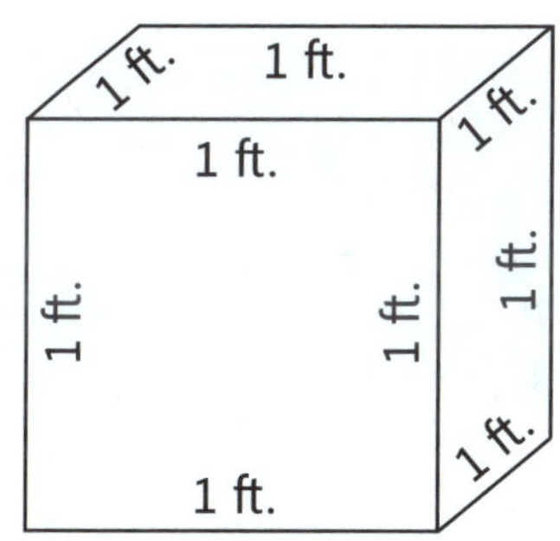

1 cubic foot

ALGEBRA Connection

Rather than using words, letters are often used to represent formulas for shapes. The volume of a prism can be written like this: $V = Bh$. When using letters, there is no need to put multiplication signs — just putting the letters next to each other (like Bh) means to multiply.

More Unit Conversion

We saw a few sections ago that to convert between units, you multiply by a conversion ratio.

Example: Convert 4 ft. to in. (1 ft. = 12 in.).
$4 \cancel{\text{ft.}} \times \frac{12 \text{ in.}}{1 \cancel{\text{ft.}}} = 48 \text{ in.}$

As you get familiar with the relationship between different units, you may not always need to write out the conversion ratio. For example, you might know the relationship between feet and inches well enough that you automatically know that you need to multiply 4 ft. by 12 in. to solve the problem.

You may also only need an approximation, in which case you can round as needed and approximate a value to the nearest value that will evenly divide.

- Going from a larger to smaller unit, you will need to multiply.
- Going from a smaller to larger unit, you will need to divide.

Going from smaller unit to larger unit.

Example: Approximately how many inches will 5.75 feet be?
Round to 6 ft. and then multiply by 12: $5.75 \times 12 \approx 6 \times 12 = 72$; answer: 72 in.

Going from larger unit to smaller unit.

Example: Approximately how many yards will 13 feet be?
Approximate 13 as 12 so it will evenly divide by 3, as 3 ft. in a yd.: $\frac{13}{3} \approx \frac{12}{3} = 4$; answer: 4 yd.

KEEPING PERSPECTIVE

You might be surprised to learn how much of what you know about shapes and units of measure helps with home improvement projects. Circles, rectangles, and squares can all be used to help figure out supplies needed.

- While we are looking at inches, feet, and yards, there are other units of measure that we could use instead. For example, God used a cubit when telling Noah the ark's dimensions.
- God made an orderly universe, which makes it possible for us to use math to measure areas and volumes — and to record formulas for how to do so for various shapes. Be sure to praise the great Creator of all as you refresh your measuring skills today.

Skill Sharpening: Converting Square & Cubic Units, Problem Solving & Circles

In the last section, we looked at calculating areas and volumes using cubic units. Sometimes, figuring out how much money you need to budget for home improvement projects (and other projects you might be doing) requires converting from one unit to another. Doing so is not hard, but we will review that skill in this section.

Converting Square Units

You measured the area in square inches of a table you are trying to paint. But the paint can only lists the square *feet* that it covers. How do you convert the square inches of the table to square feet?

>>> To convert square units, multiply by the appropriate unit ratio twice.

Example: Convert 300 square inches to square feet.

$$300\ \cancel{\text{in.}}^2 \times \frac{1\ \text{ft.}}{12\ \cancel{\text{in.}}} \times \frac{1\ \text{ft.}}{12\ \cancel{\text{in.}}} \approx 2.08\ \text{ft.}^2$$

Get the Units to Cancel Out

Note that in.^{2} means in. × in. When we multiply by the conversion ratio twice, we are really canceling out both the inches. We multiply by a unit ratio the number of times needed to cancel out the units.

2 inches in the numerator

$$300\ \text{in.}^2 = 300\ \text{in.} \times \text{in.} =$$

$$300\ \cancel{\text{in.}} \times \cancel{\text{in.}} \times \frac{1\ \text{ft.}}{12\ \cancel{\text{in.}}} \times \frac{1\ \text{ft.}}{12\ \cancel{\text{in.}}} \approx 2.08\ \text{ft.}^2$$

2 inches in the denominator

Cubic Units

You see that the volume of one swimming pool you are considering buying is 120 cubic feet. Another one is 50 cubic yards. To compare the volumes, you need to convert them both to the same unit.

>>> To convert cubic units, multiply by the appropriate unit ratio three times.

Example: Convert 120 cubic feet to cubic yards.

shown writing cubic feet as ft.^{3}:

$$120\ \cancel{\text{ft.}}^3 \times \frac{1\ \text{yd.}}{3\ \cancel{\text{ft.}}} \times \frac{1\ \text{yd.}}{3\ \cancel{\text{ft.}}} \times \frac{1\ \text{yd.}}{3\ \cancel{\text{ft.}}} = 10\ \text{yd.}^3$$

shown writing cubic feet as ft. × ft. × ft.:

$$120\ \cancel{\text{ft.}} \times \cancel{\text{ft.}} \times \cancel{\text{ft.}} \times \frac{1\ \text{yd.}}{3\ \cancel{\text{ft.}}} \times \frac{1\ \text{yd.}}{3\ \cancel{\text{ft.}}} \times \frac{1\ \text{yd.}}{3\ \cancel{\text{ft.}}} = 10\ \text{yd.}^3$$

In comparing the two pools, one is 5 times larger than the other.

X ALGEBRA y Connection

We multiply cubic units by the appropriate conversion ratio 3 times because that is what is needed to cancel out the units.

The rule about converting cubic units is based on the consistent way God governs multiplication and division and the methods (like fractions and exponents) we use to describe those consistencies.

Problem Solving

When working with real-life measurement problems, you will often need to think through what you are trying to calculate. For example, if you are adding trim around the floor of the 10 ft. × 10 ft. room shown, you might think you need to know the perimeter, or distance around, the room. However, there is a door. You will not need trim across the doorway. To calculate the amount of trim needed, you would need to calculate the perimeter and then subtract the doorway.

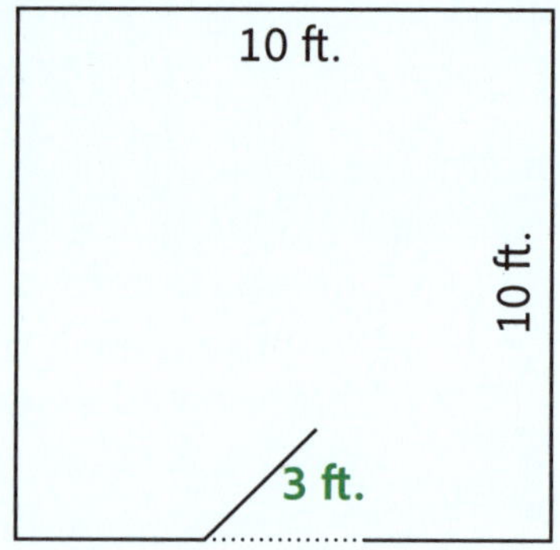

Trim needed: 10 ft. + 10 ft. + 10 ft. + 10 ft. – **3 ft.** = 37 ft.

Always think about what you are really trying to calculate.

Circles

In home improvement projects, you might encounter circular objects, such as lamps, whirlpools, and bowls. Use the formulas given (which you have hopefully seen before in previous math courses) when working with them, being careful to think through what you are really trying to calculate.

Cylinder
(circle with height)

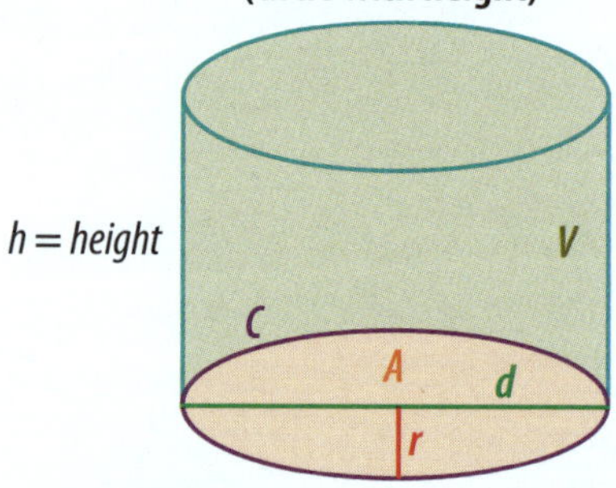

Formulas

d = *diameter* = $2 \times r$

r = *radius* = $\frac{d}{2}$

C = *circumference* = $2 \times 3.14 \times r$

V = *volume of a cylinder* = $3.14 \times r^2 \times h$

A = *area of circle* = $3.14 \times r^2$

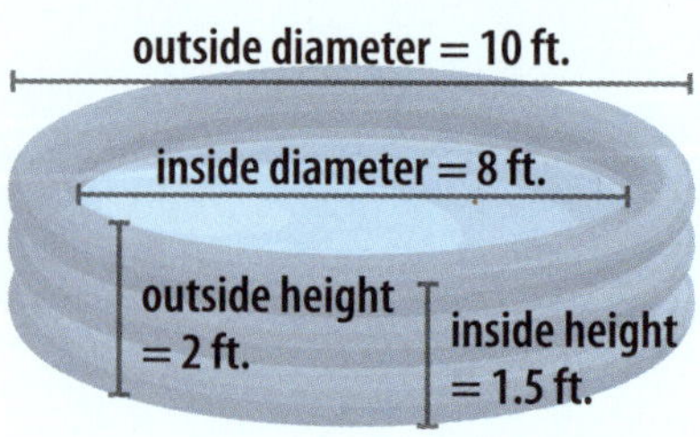

Example: Calculate the circumference around the outside of this inflatable pool and the volume of water it could hold.

Outside radius: $\frac{d}{2} = \frac{10\text{ ft.}}{2} = 5\text{ ft.}$

(using outside diameter, as want the circumference around the outside)

Outside circumference: $2 \times 3.14 \times r = 2 \times 3.14 \times 5\text{ ft.} = 31.4\text{ ft.}$

Inside radius: $\frac{d}{2} = \frac{8\text{ ft.}}{2} = 4\text{ ft.}$

Volume could hold: $3.14 \times r^2 \times h = 3.14 \times (4\text{ ft.})^2 \times 1.5\text{ ft.} = 3.14 \times 4\text{ ft.} \times 4\text{ ft.} \times 1.5\text{ ft} = 75.36\text{ ft.}^3$

$(4\text{ ft.})^2$ means $4\text{ ft.} \times 4\text{ ft.}$

(used inside radius and height, as looking at what inside space can hold; put the () to show we have to multiply both the 4 and the ft., not just the ft.)

KEEPING PERSPECTIVE

Unit conversion can seem challenging at first, but stick with it. It is a useful tool in a lot of areas, including home projects.

Skill Sharpening: Scale Drawings

It has been said that a picture is worth a thousand words — and that applies when thinking through many home improvement and other projects. It often helps to make a drawing of the project — and making those drawings to scale can sometimes help even more. Scale drawings can help you picture what you plan on doing and spot errors *before* you invest a lot of time and money.

You might have learned about **similar shapes** in geometry — a scale drawing shows a similar shape to the actual.

A home or apartment blueprint is an example of a scale drawing. Notice that the ratio (i.e., comparison using division) of the length over the width of the actual apartment and the ratio of the length over the width of the scale drawing are both equal.

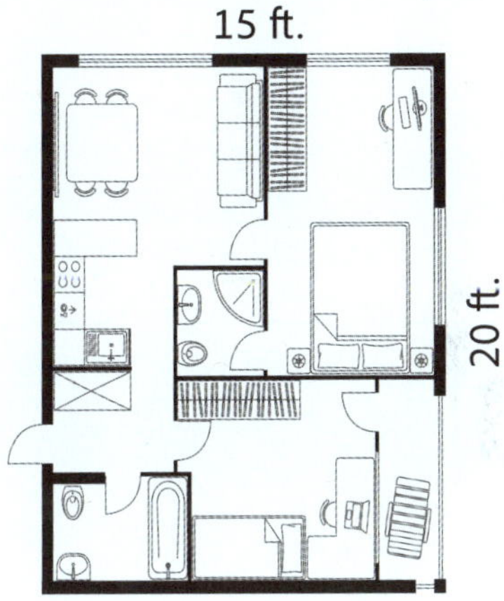

$$\frac{\text{actual length}}{\text{actual width}} = \frac{20 \text{ ft.}}{15 \text{ ft.}} = 20 \text{ ft.} \div 15 \text{ ft.} \approx 1.33$$

$$\frac{\text{drawing length}}{\text{drawing width}} = \frac{2 \text{ in.}}{1.5 \text{ in.}} = 2 \text{ in.} \div 1.5 \text{ in.} \approx 1.33$$

Since both ratios equal 1.33, they equal each other, forming what we call a proportion: $\frac{20 \text{ ft.}}{15 \text{ ft.}} = \frac{2 \text{ in.}}{1.5 \text{ in.}}$. A **proportion** is simply a name we give to two equal ratios.

>>> In a **scale drawing**, the ratios between dimensions equal the ratios in the actual. The angles and proportions are the same, making the drawing a mini version of the actual.

An easy way to make a scale drawing is to use graph paper. That way, you can make each square of the paper a square inch, foot, yard, etc. Drawing a garden area that is 6 ft. by 5 ft.? Just make it 6 squares by 5 squares on the graph paper. Use a ruler or straight-edge to draw straight lines.

$$\frac{6 \cancel{\text{ft.}}}{5 \cancel{\text{ft.}}} = \frac{6 \cancel{\text{squares}}}{5 \cancel{\text{squares}}}$$

We showed the units canceling out (fraction means division, and when you divide the same unit by itself, such as ft. by ft., the units cancel out) so you could see how these really do form a proportion. Since each square is representing 1 foot, the relationship between the length and width is the same.

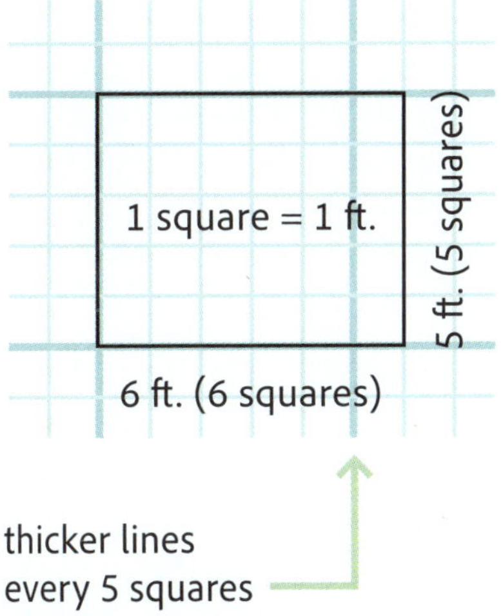

- Some graph paper has thicker or colored lines every 5 or 10 squares. This type of graph paper is helpful when drawing projects. It is easy to tell the dimensions if you know the thicker lines go every five squares.

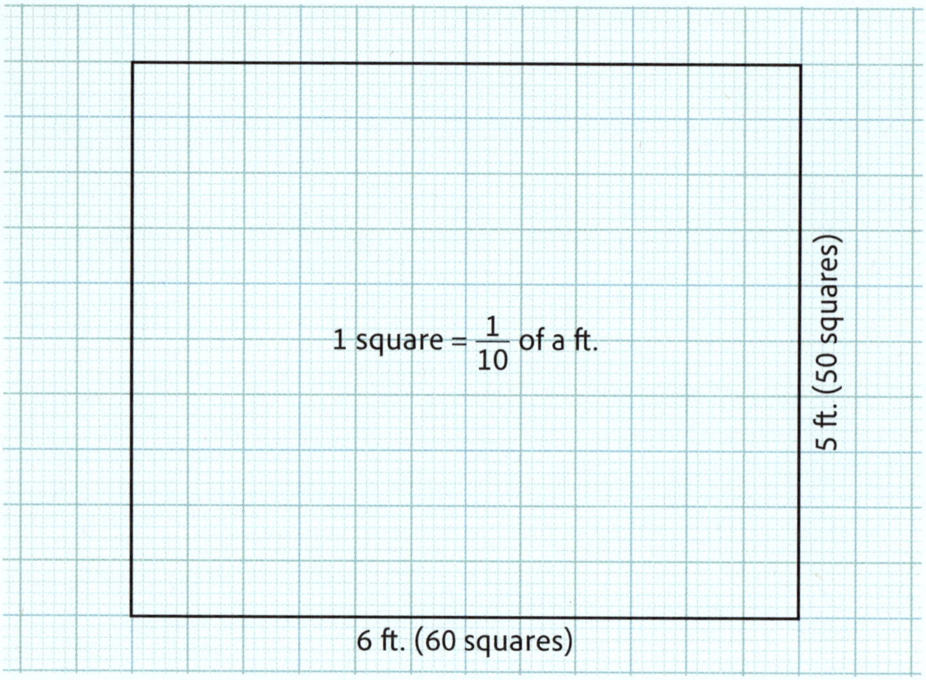

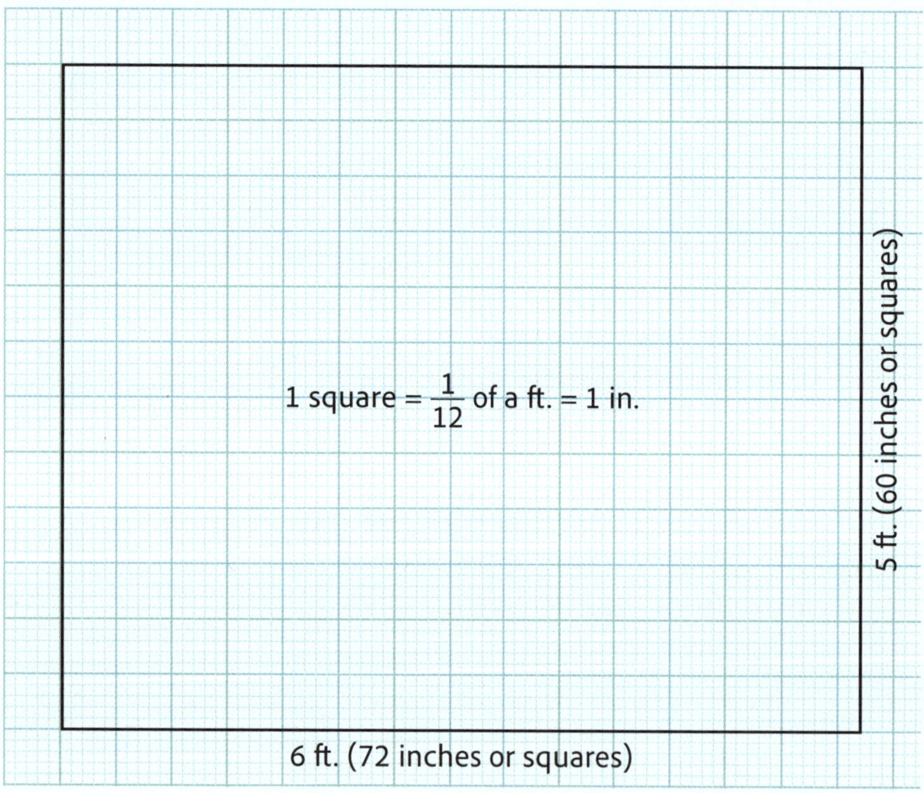

- Sometimes, you will want to use more than 1 square to represent a unit in order to give you space to draw. For example, you could make 10 squares represent 1 foot instead, drawing the garden like this.

$$\frac{6\,\cancel{\text{ft.}}}{5\,\cancel{\text{ft.}}} = \frac{60\,\cancel{\text{squares}}}{50\,\cancel{\text{squares}}}$$

>>> To make an equal ratio, multiply the original values by a fraction with the same amount in the numerator and the denominator. This is multiplying by a fraction worth 1, which does not change the value.

$$\frac{6}{5} \times \frac{10}{10} = \frac{60}{50}$$

- Or you could draw the rectangle where 1 square represents 1 inch. Six feet would then be $6\,\cancel{\text{ft.}} \times \frac{12\text{ in.}}{1\,\cancel{\text{ft.}}} = 72$ in., or squares, and 5 ft. would be $5\,\cancel{\text{ft.}} \times \frac{12\text{ in.}}{1\,\cancel{\text{ft.}}} = 60$ in., or squares.

$$\frac{6\,\cancel{\text{ft.}}}{5\,\cancel{\text{ft.}}} = \frac{72\,\cancel{\text{squares}}}{60\,\cancel{\text{squares}}}$$

Note that $\frac{6}{5} \times \frac{12}{12} = \frac{72}{60}$.

KEEPING PERSPECTIVE

Our ability to design and draw reflects God's creativity and engineering. We are made in His image, and He commanded mankind to "fill the earth and subdue it" (Genesis 1:28; NIV). Working with measurements is a tool we can use in that process.

Savings vs. Debt, Budgeting Savings, & Skill Sharpening: Drawing Circles

We are going to look at saving again today, but we will still keep reviewing some measurement and home improvement skills.

You might be wondering why you do not just go ahead and buy that big-ticket item on credit instead of saving up money to buy it. Then you can have what you want now, without having to wait.

While that certainly is an option, keep in mind all we learned about debt; remember that "the borrower is the slave of the lender" (Proverbs 22:7b). When you borrow money, you have to pay it back with interest. That car is going to cost you a *lot* more over the long term if you have to finance it than if you can buy it outright. You are also making assumptions that you will be able to meet those obligations as you go forward.

- What happens if you lose your job or the car gets wrecked? Do you really want that debt over your head?
- Is it wise to presume on the Lord's future provision to get what you want today?

In general, it is best to avoid debt. Be very, very careful and seek wise counsel before going into debt. Waiting and saving can be hard. It requires bucking the culture that says get what you want now. Waiting and saving can help teach you perseverance and contentment if you let it.

With that in mind, we will keep reviewing some skills that can help you estimate home improvement projects.

Budgeting Savings

We have talked about a lot of different types of saving. Now we will talk for a minute about budgeting it.

- **You might budget some savings from your paycheck as general savings, for which you have not assigned a purpose yet.** For example, you might set aside 20%.
- If you are saving up for a one-time expense (like a big vacation), you might **create a separate category under savings and create a separate worksheet** (or section on your budget worksheet) to see how much you have saved over time. (Apps/online programs may have features to let you save for big goals too. Ask permission from a parent to search on the Internet for apps or online programs.)
- **Other savings might be built into your budget as expenses.** For example, the $100 a month you set aside for your $600-every-six-months car insurance bill is savings you are building in as an expense. You *expect* to underrun that expense most months, with the extra going to savings for a short time until the month that $600 is due.

Separate category to save $100 each month for a vacation

Savings	
General Savings	$ 1,000
Vacation Savings	$ 100

Separate worksheet showing how you are doing toward vacation goal

Total Needed	$ 800
Saved	$ 500
Amount Left to Save	$ 300

Savings built in as expense (most months).			
Spending	Budgeted	Actual	Difference
Car Insurance	$ 100	$ 0	$ 100

Savings built in as expense (on months pay bill).			
Spending	Budgeted	Actual	Difference
Car Insurance	$ 100	$ 600	$ (500)

>> In looking at spending that happens less frequently than once a month, **look at spending over time.**

The car insurance category in this example will look like you are underrunning the budget most months and then will look like you overran when you pay the $600 bill. The key is that over time, you are staying on the budget. If that $600 bill you planned for ends up being $700 instead, then you *did* go over budget and you will need to adjust your budget for next time.

App Trick – Most apps let you look at expenses both monthly and over longer periods of time. Be prepared to switch back and forth depending on the expense you are looking at to get an accurate picture of your finances. If you want to use an app, ask a parent for permission first. Apps are not always safe. Some gather your personal information and then sell it.

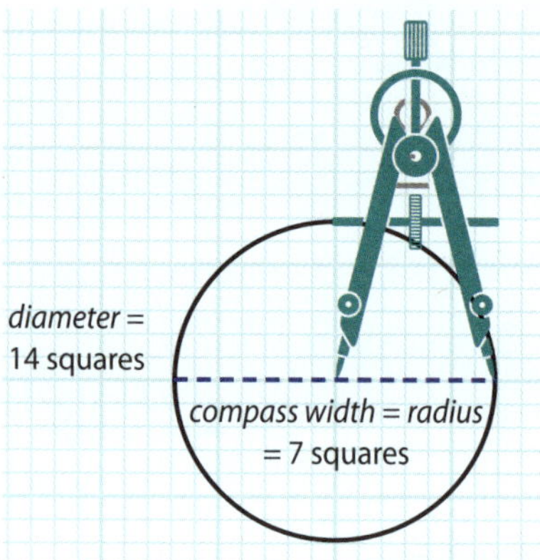

Drawing Circles

Circles can help with drawing a variety of things around the house and yard, which, in turn, can help in planning a project.

To draw circles to scale, you will need a compass. If drawing with graph paper, simply figure out how many squares the radius of the circle needs to be and set the compass that many squares wide.

KEEPING PERSPECTIVE

Just like we reflect God's image in our ability to design and plan, God wants us to reflect His image in how we live in holiness and righteousness too. Through Jesus, God is in the business of restoring fallen image bearers to more fully reflect His image to the world.

And walk in love, as Christ loved us and gave himself up for us, a fragrant offering and sacrifice to God (Ephesians 5:2).

Hiring Home Improvement & a Note About Scraps

Instead of doing a home improvement project yourself one day, you might want to hire someone to do it for you. When you do, you will likely find yourself using math to evaluate the options.

In general, it is wise to get several bids on a project. That way, it is easier to tell which is really a good deal. Here are a few things to keep in mind when evaluating those bids:

- **Some contractors include labor in their price; others break it out.** The bids you receive might look very different from one another in layout. Make sure you understand what is included and what is not.
- **If they underestimate their materials or time, some contractors will cover that additional cost; others will charge you.** It is helpful to know up-front who is covering underestimations of materials or time, as home improvement projects are known for involving more than initially thought.
- **If something breaks, some will warranty their work; others will not.** You are having a fence built, and a month later, the gates will not shut. Will the workers come to fix their work? Some warranties include parts only, while others cover labor too, so ask what the warranties cover.
- **Look at examples of work done and references.** I was once having a wall built in my home. One contractor bid tremendously lower than the other two. I asked to see a picture of other walls he had built, and I understood why. His wall did not look professional.
- **Look at experience and insurance.** Before you let someone touch your home, you want to make sure you trust them to do a good job. Looking at their experience with your type of project can be helpful. If they are doing something where they could really mess up your home (such as building an addition), you might also check to see if they are insured. For any project, if they are not insured, you run a risk should things go wrong.
- **Discuss the timeliness of the project and penalties if it takes longer.** This applies more to larger projects, such as a bathroom or kitchen remodel. It can be hard living without a bathroom or kitchen, so you want to make sure you have discussed the timeframe.
- **Get everything in writing.** Should something go wrong, verbal agreements create a mess. Make sure everything you are agreeing to is written down … and carefully read what you sign.

You do not always want the cheapest bid. When evaluating contractors, you will want to look not only at cost but also at a variety of other factors.

Tools or equipment for a project can be pricey if you are only doing one job and are not able to rent what you need.

- In general, unless you need to buy expensive tools or equipment, hiring someone costs more than doing it yourself. After all, you owe another person for their time and experience.
- On the flip side, your time gets freed up and you do not have to figure out something you may not know how to do.

As we saw back with food, there is an inverse relationship between time and cost. When one increases, the other decreases. You have to decide if it is worth the extra time to save.

>>> Think through what you are doing when you determine supplies needed. There is no formula to follow, as each situation is different, but a drawing and basic math can help you think it through.

A Note About Scraps

When doing home improvement projects, sometimes you can use portions of a whole and sometimes you can not.

For example, you are tiling a bathtub wall. You can use square footage to figure out the approximate number of tiles that will fit, but then you also have to think through how many will go across each row in your pattern. You may need more tiles than the square footage alone will tell you in order to have them lay out the way you want ... or you may not.

In the case of tile, you will want to figure out how many you need per row and then how many rows you have. If you need part of a tile, you will have to buy an entire additional tile (as you can not buy a part of a tile) ... unless you have a part left over from another row that you can use.

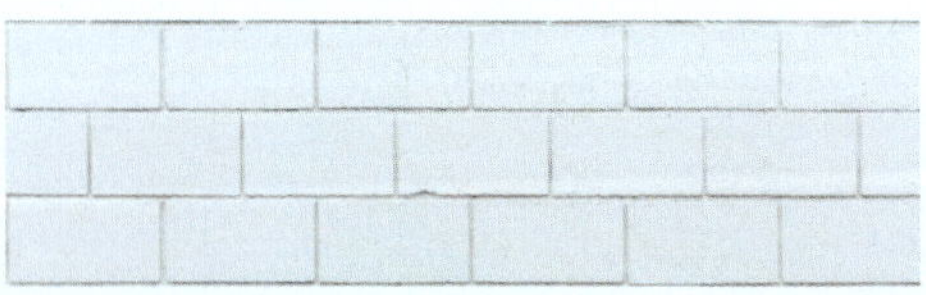

Example: Calculate the bricks needed for the middle row if the beginning and end are each half of a brick (so you only need 1 brick to make both).

Counting the bricks, we see we need 5, plus another 1 we can cut in half for the beginning and end. So 6 total.

Example: Calculate the bricks needed for the middle row if you need a separate brick for the beginning and end.

This time we will count the beginning and end, each as a separate brick, getting 7 total.

KEEPING PERSPECTIVE

As with any decision, hiring someone should be a matter of prayer and trusting the Lord to lead you. God might direct you to hire someone you know needs a job even though they are not necessarily the most qualified. The points we looked at in this section are simply things to be aware of if you ever hire help.

Review & Case Study

In this lesson, we looked at planning for big expenses, whether they be big purchases, events, trips, projects, or something else. Planning involves forethought and effort.

Some people naturally like to plan; others hate it.

- If you like it, you might find yourself with elaborate plans for big expenses.
- If you hate planning, your plans might be more of a guesstimate.

Either way, know that a little **planning can help you save up for the goals God puts on your heart.**

As you plan, whether it be financially or in other ways, keep in mind that God is over every plan. Our plans may need to be adjusted as life happens, but God's do not. They stand firm. And remember, it is His will to sanctify you — to mold you into His image. And He uses all things, even plan changes, for the eternal good of those who love Him and are called according to His purpose.

> *Many are the plans in the mind of a man, but it is the purpose of the Lord that will stand* (Proverbs 19:21).

> *For this is the will of God, your sanctification* (1 Thessalonians 4:3a).

> *And we know that for those who love God all things work together for good, for those who are called according to his purpose* (Romans 8:28).

Throughout the lesson, we also reviewed some measurement skills that come in handy with various projects, including home improvement ones. Remember that we are able to design and build because God gave us that ability. He is the Master Designer, and we are made in His image.

Case Study: God's Plan & Ours (Joseph)

I think it is safe to say that Joseph's life did not go according to his plans. God gave him dreams of greatness, then he found himself sold by his own brothers into slavery. On top of that, he landed in prison for a crime he had actually refused to commit.

Yet all along the way, Joseph faithfully served where he had been placed. Listen to what Genesis tells us about his time in Potiphar's house, where he arrived as a common slave.

> *His master saw that the Lord was with him and that the Lord caused all that he did to succeed in his hands. So Joseph found favor in his sight and attended him, and he made him overseer of his house and put him in charge of all that he had. From the time that he made him overseer in his house and over all that he had, the Lord blessed the Egyptian's house for Joseph's sake; the blessing of the Lord was on all that he had, in house and field. So he left all that he had in Joseph's charge, and because of him he had no concern about anything but the food he ate* (Genesis 39:3–6).

Joseph was put in charge of Potiphar's entire estate — of both his house and his field. Even though his plans had not worked out, we see Joseph serving faithfully. When he wound up in prison, he was put in charge of the prisoners. He again served faithfully.

Later, He was put as second-in-command over all of Egypt. Once again, Joseph served faithfully.

Joseph's life reminds us to trust God's plans and faithfully steward whatever God entrusts to us — whether a pit, prison, or kingdom. Looking back years later after he had been reunited with his brothers, Joseph could say to his brothers who had sold him into slavery, "So it was not you who sent me here, but God" (Genesis 45:8a). He recognized God's plan over his life, including things that at the moment were horrible and a complete overturning of Joseph's own plans.

Joseph's life also reminds us that it is God who makes plans work or fail. As manager of Potiphar's estate, Joseph likely would have overseen various projects that would have used math and required planning. Later, he would lead the whole nation in saving up for the years of famine. No doubt Joseph utilized many skills as he managed first his master's and then Egypt's wealth. Yet the Bible makes it clear that God was the one who gave him success, not Joseph's own abilities.

> *The keeper of the prison paid no attention to anything that was in Joseph's charge, because the Lord was with him. And whatever he did, the Lord made it succeed* (Genesis 39:23).

Never forget that it is God who ultimately makes plans succeed or fail.

> *The heart of man plans his way, but the Lord establishes his steps* (Proverbs 16:9).

Review

As we wrap up another quarter, we are going to:

- do review exercises
- talk more about our heart in stewarding what God has given us
- learn one additional skill that comes in handy as you shop.

As we review budgets, remember that budgets are ways to help us plan for the goals and priorities God puts on our hearts. They are working templates, not cast-iron musts. They require periodic reevaluations based on the new situations God allows into our lives.

Prudent Saving vs. Mere Wealth Accumulation

I hope you have been seeing the importance of saving.

- Saving helps us to be able to afford big expenses.
- Saving helps plan for the challenges we can expect in a fallen world (think medical issues, car breakdowns, and more).
- Saving makes budgeting easier by taking away stress if we run slightly over or under, or if bills come at different times than income.
- Saving helps plan for the distant future when we are unable to earn money.

However, we need to be careful to watch our heart in saving.

- We can save with a heart of serving the King of kings, wisely providing for our family and funding the goals God puts on our heart.
- We can save from a heart of greed or selfishness or fear.

The Bible warns about a man who saved and saved and saved some more, building bigger barns to hold his savings. Yet his soul was not right with God.

> *And he said to them, "Take care, and be on your guard against all covetousness, for one's life does not consist in the abundance of his possessions." And he told them a parable, saying, "The land of a rich man produced plentifully, and he thought to himself, 'What shall I do, for I have nowhere to store my crops?' And he said, 'I will do this: I will tear down my barns and build larger ones, and there I will store all my grain and my goods. And I will say to my soul, "Soul, you have ample goods laid up for many years; relax, eat, drink, be merry." ' But God said to him, 'Fool! This night your soul is required of you, and the things you have prepared, whose will they be?' So is the one who lays up treasure for himself and is not rich toward God"* (Luke 12:15–21).

We do not want to save out of a spirit of greed, desiring more and more and more. When money is the goal, it is never enough.

> *I have seen everything that is done under the sun, and behold, all is vanity and a striving after wind* (Ecclesiastes 1:14).

KEEPING PERSPECTIVE

As you think through financial decisions, remember to **check your heart**. Know that saving is prudent and helpful, but living to amass money is not.

Total Abandonment & Mental Sales Tax

We have talked some this quarter about saving money and living frugally. The skills we are learning can help you grow in wealth, assisting you in meeting long-term goals like buying a car or home one day. But be sure to not make wealth accumulation the goal.

A rich ruler once came to Jesus, asking how he could inherit eternal life. Jesus challenged him to keep the commandments; the man thought he had. Now, there is no way he truly had — no one but Jesus has. Then Jesus told him to sell everything he owned.

> *When Jesus heard this, he said to him, "One thing you still lack. Sell all that you have and distribute to the poor, and you will have treasure in heaven; and come, follow me"* (Luke 18:22).

Does this mean we are all called to sell everything we have? No. Jesus was trying to get this man to see he was not really loving God and others with all his heart (which is what the Law is really about [Matthew 22:38–40]) — he was really loving wealth. Jesus was trying to humble him so he would realize he needed a Savior.

Elsewhere in the New Testament, we see Christians whom God entrusted with wealth. 1 Timothy gives this instruction:

> *As for the rich in this present age, charge them not to be haughty, nor to set their hopes on the uncertainty of riches, but on God, who richly provides us with everything to enjoy* (1 Timothy 6:17).

- These instructions to the rich did not say they needed to sell everything and lead a poor life.
- Instead, the rich were warned not to be proud about or set their hope in wealth.

God is after our hearts being entirely His. He wants us totally abandoned.

It is okay to enjoy gifts God gives us — notice these instructions to the rich mention that God "provides us with everything to enjoy." It is okay to enjoy spending some of the money God has given you on things you want. What is not okay is to set your heart on riches.

- Wealth and enjoying the provision God has given is not wrong.
- Living for money/this present age — or being proud about what you have — is wrong.

Think of Job in the Bible. He was blessed with riches. He trusted God. They all disappeared in a moment. He still trusted God. In the end, he wound up with riches once more.

It was not wrong for Job to have wealth. He is an example of someone who served God both with and without it — his hope was not set upon wealth.

Mental Sales Tax

Back in Lesson 2, we looked at some mental math related to percentages. Now we will revisit this skill, applying it to sales tax.

>>> **When mentally calculating sales tax**, round the tax amount to either 5% or 10% (whichever is closer).

Example: Calculate the total with 6% sales tax of an item that costs $48.69.

6% is closer to 5% than to 10%, so we will estimate by calculating the total with 5% more.

1. Round the item cost to $49.
2. 10% is $4.90, which rounds to $5.
3. 5% is half that, or $2.50. (Or, for a more accurate answer, we could have found half of $4.90 by taking half of $4, which is $2, and half of $0.90, which is $0.45.)
4. If we add $2.50 to $49, we get $51.50.

The actual cost is $51.61, so that was not a bad estimate.

KEEPING PERSPECTIVE

We are called to live in total abandonment for God's Kingdom.

- For some, that may mean selling all and leaving the comforts of home behind.
- For others, it may mean wisely stewarding what God gives, enjoying God's gifts while living for an eternal Kingdom.

The skills you have been learning can help you, whether God entrusts you with a lot materially or a little.

God's Heart for the Poor

As you think about wise stewardship, pause and look at God's heart for the poor.

Throughout Scripture, we see that God cares about the poor, and that we are supposed to as well. Here are just a few of many Scriptures that make this point.

> *Only, they asked us to remember the poor, the very thing I was eager to do* (Galatians 2:10).

> *Whoever oppresses a poor man insults his Maker, but he who is generous to the needy honors him* (Proverbs 14:31).

> *He executes justice for the fatherless and the widow, and loves the sojourner, giving him food and clothing* (Deuteronomy 10:18).

> *Behold, this was the guilt of your sister Sodom: she and her daughters had pride, excess of food, and prosperous ease, but did not aid the poor and needy* (Ezekiel 16:49).

> *Behold, the wages of the laborers who mowed your fields, which you kept back by fraud, are crying out against you, and the cries of the harvesters have reached the ears of the Lord of hosts. You have lived on the earth in luxury and in self-indulgence. You have fattened your hearts in a day of slaughter. You have condemned and murdered the righteous person. He does not resist you* (James 5:4–6).

In the final quote James does not say it is wrong to be rich — only that, if rich, you should not unfairly mistreat those under you.

Bottom line: We are called to love God and love others. If we truly have that heart, we will want to use the resources God has given us to help others around us. How? That is a question each one of us must seek God for the answer to.

Note that we are not called to eradicate poverty — Jesus warned that the poor would always be with us (Matthew 26:11) — but to do what we can to love those we can.

As Christians, we understand that helping others includes caring for their souls as well as their bodies. The Great Commission to preach the gospel should guide our care.

Quick Heart Check

What do you think about when you think about being rich?

- More money than you can spend?
- Huge homes and expensive cars?
- Money to bless others?

You might not be rich in money, but think about what you are rich in:

- Time?
- Talent?
- Technology ability?
- Knowledge?

Think about how you can use what you are rich in to help others. As you use what God has given you to reach out to the poor and needy around you, you will discover that it truly is more blessed to give than to receive.

> *"In all things I have shown you that by working hard in this way we must help the weak and remember the words of the Lord Jesus, how he himself said, 'It is more blessed to give than to receive' "* (Acts 20:35).

KEEPING PERSPECTIVE

As you continue reviewing the quarter and learning money-management skills, ask God to give you His heart for others, including the poor around you.

Review & Case Study

You have learned a lot this quarter about budgeting. We will continue to touch on it and use what we have learned already as we head into the next semester, only our focus will be more on specific consumer math decisions: choosing a career and buying a car or a home.

Here are a few key things to remember from this quarter:

- **A budget is a guide, not a law.** It should help you keep your finances focused on what matters in life, avoiding living for the moment and not being able to meet your obligations. However, a budget makes a horrible master — do not use it as an excuse to be greedy or miserly or to not trust God when He asks you to do something.
- **Your heart matters.** Math can help you make and evaluate your budget, but your heart will determine what you actually put in that budget and how you approach it. Beware of the love of money.
- **Planning, of which a budget is a tool to help, is useful and important.** We see examples of people planning in the Bible, such as:
  - Joseph
  - the Proverbs 31 woman
  - God Himself, who had plans for His people

  But we need to hold our plans loosely, surrendering them to God's greater plan.

Here is another story about a real-life couple who had their plans changed and chose to do things with their money (and budget) that to the world sounds strange. But they have found tremendous spiritual blessings and joy in the process.

Case Study: In the Background (the Barnharts)

This write-up is based on a video from Generous Giving. Used by permission of Generous Giving; all rights reserved.

Alan and Katherine Barnhart had a heart to serve the Lord. Shortly after their marriage, they were on their way to a closed country, using Alan's civil engineering ability as a way to get in with the gospel, when Alan's parents offered him their business.

Running a business did not seem nearly as important to Katherine, but she reluctantly let Alan accept the business, thinking he was making the wrong decision. Little did she know that God would use that business to give away millions to missions.

Alan studied what God said about money and came away realizing that everything he had belonged to God. He became keenly aware of the dangers of greed, and he feared wealth would destroy him if the business did well.

Both he and his brother (his business partner), together with their wives, decided to view God as the owner of the business and to set a lifestyle cap, meaning that even if the business prospered, their lifestyle would not increase beyond that cap (which, for Alan and Katherine, meant a salary between $40,000 and $160,000 at different seasons of life, and $3 million for retirement). They also decided to get accountability by telling employees that if the business did well, the increase would not go into Alan's or his brother's salaries.

Alan discovered that he had a gift and love for business, and his business did incredibly well. They put half of the profit back into the business, and the other half funds ministry through a charitable trust. As of the interview, they were giving away $21 million a year. Eventually, the Barnharts gave away the business itself to trusts, enabling them to still steward the business but not legally own it.

Katherine still has not accomplished her dream of going to a foreign country to share the gospel, but through godly stewardship, she and Alan have been able to fund much gospel work. She has lived surrendered to the path God chose for her.

Their story reminds me of how we are all different pieces of God's body, with different gifts and roles to play. God decides what those are, not us. Wherever He puts us, though, let us live all-out for Him.

Quote from Katherine Barnhart:

> *"For so long, I thought that ... the stewarding of these funds was second best. And the Lord had to grow me through that ... to show me that ... it is not about me serving Him. It is about Him — it is about the Lord glorifying Himself in and through me, in and through you. And the paradigm shift there is that I'm not the one in control; He is the one in control. I'm not the one with the influence; He is the One with the influence. I'm not the one with the power; He is the One with the power. I'm not even the one with the resources; He is the One with the resources."*[26]

Quarter 3: Financial Decisions

Throughout life, we face numerous financial decisions. We have already talked about little decisions, like what we buy at the grocery store and how they add up. In this quarter, we are going to move on to bigger decisions, such as buying a car or a house and choosing a career.

As we examine these bigger decisions, we will also be looking at what God's Word has to say about making decisions. Many of the principles examined apply to non-financial decisions as well.

In looking at decisions, we will see that the basic operations and what we learned about percents applies over and over again. Rather than trying to memorize a bunch of formulas, keep in mind what operation does what, as that will help you figure out how to solve a problem.

Addition – Combines values; helps calculate a *total* or a *sum*.

Subtraction – Takes away; calculates the *difference*; a way of *comparing* values.

Multiplication – Adds repeatedly. For example, we multiply weekly amounts to calculate monthly or yearly amounts for a budget.

Multiplication by a value between 0 and 1 is finding a portion *of* another value. In a **percent** problem, for example, you look at the percent (portion *of* 100) of a value something is. Remember that *of* and *is* can help you figure out where to put what.

% *of* base *is* percentage

% × base = percentage

Division separates out a quantity; shows us an amount per some other amount.

- Keep in mind that *per* means division.
  - When needing a *monthly cost*, we determine the cost *per* month by dividing the total cost by the number of months.
  - When needing a *unit price*, we calculate the cost *per* unit by dividing the total cost by number of units.
- Division can also show us how long it will take to earn or pay back a value. We saw this in Lesson 2 and applied it in Lesson 9 with savings to see how many months you would have to save for an item.

Example: If you save $45 a month, how long will it take to save $300?

$\frac{\$300}{\$45} \approx 6.67$ months

We are seeing how many times it takes of saving $45 to get to $300 — that is, we are essentially separating $300 into piles of $45 and seeing it will take 6.67.

Intentionally left blank

11

Cars

income = giving + saving + ***spending***

spending *= taxes + debt +* ***living expenses***

Ask yourself:

- Should I get a car?
- What should I look for when choosing a car?
- Should I buy new or used? Or should I lease?
- How much can I really afford?
- What should I put in my budget for a car?

We will explore these questions and more in this lesson.

Transportation Options & Payment Plan

What is a car? To many, it is a symbol of status. Yet as God's child, your status should come from knowing you belong to the King of kings. A car is simply a vehicle — a way to get from point A to point B. It is something you can use to get to places to serve God and others. These three questions can help you sift through and see what kind of transportation option might be best:

- Where is God calling you to go?
- Whom is God calling you to transport?
- How much has God provided for you to spend?

Here is an overview of some options:

- **Brand-New Car** – While new and hopefully not needing repairs for a while, new cars come with a hefty price tag that might necessitate a loan. New cars also lose a lot of value quickly.
- **Used Car** – Used cars come at every price point. Depending on the age, there is more potential for repair/maintenance, but you can also buy fairly new cars that run well but cost less than a brand-new car.
- **Leased Car** – Always a newer car, but with a continual high monthly fee and no ownership of anything at the end (like rent).
- **Public Transportation** – A way to get around a city. *Examples:* city bus, subway, taxis, etc.
- **Ride-Sharing Services** – These let you rent a seat in another person's car.

Loan Amount:		$33,000		Interest Rate:	5%
Month	**Payment Amount**	**Interest**	**Principal**	**Loan Balance**	**Cumulative Interest**
1	$622.75	$137.50	$485.25	$32,514.75	$137.50
2	$622.75	$135.48	$487.27	$32,027.48	$272.98
3	$622.75	$133.45	$489.30	$31,538.17	$406.23
4	$622.75	$131.41	$491.34	$31,046.83	$537.83
5	$622.75	$129.36	$493.39	$30,553.44	$667.20
6	$622.75	$127.31	$495.44	$30,058.00	$794.50
7	$622.75	$125.24	$497.51	$29,560.49	$919.74
8	$622.75	$123.17	$499.58	$29,060.91	$1,042.91
9	$622.75	$121.09	$501.66	$28,559.24	$1,164.00
10	$622.75	$119.00	$503.75	$28,055.49	$1,283.00
11	$622.75	$116.90	$505.85	$27,549.64	$1,399.90
12	$622.75	$114.79	$507.96	$27,041.68	$1,514.69

Payment Plan

Car commercials are always advertising their low monthly payments. But before you decide to take on a car loan, we will review loans, as these "low" payments really add up.

Just as we did with house loans in Lesson 3, we can explore car loans using an amortization schedule, such as the one shown.

Notice that after 12 months (i.e., 1 year), the cumulative interest is more than $1,500. The cumulative interest over the 5-year life of this loan is $4,365.04. That is a lot extra to pay for a car! If you do need to take a loan, look for 0% interest sales to avoid the interest charges.

In addition to the extra interest, **loans represent debts that must be paid back, even if:**

- job situations change
- the car gets wrecked
- you sell the car at a loss.

While you might think the loan will be easy to repay when you take it out, life changes. In the case of an accident, your car insurance will give you money, but not necessarily enough to both pay off your loan and get a new car. Think twice before you take out an auto loan, remembering that "the borrower is the slave of the lender" (Proverbs 22:7).

Cars *lose* value, meaning you are not borrowing for an investment (such as a house or business) that can then hopefully be sold for the same or greater value but rather paying for an item that will *lose* value over time. This means if you need to get out of the loan, you can not just sell the car to cover its cost.

KEEPING PERSPECTIVE

- Take time to define what you really need in regard to transportation and look at the possibility of a car in light of your entire budget and overall goals.
- Make a decision up-front about how much you are willing to pay and whether or not debt is an option.
- Prayerfully looking at your needs and budget and deciding on the maximum you could spend will help keep you from a car decision and payment plan you regret.

Depreciation & Brand-New Cars

Depreciation, the loss of a car's value over time, is key to deciding if and what car you buy.

>>> Calculate car depreciation by subtracting what you could sell a car for from what you paid for it. That is, subtract its ending value from its starting value.

Example: If Stan bought a car for $20,000 and 1 year later it is only worth $16,000, how much did the car depreciate?

$20,000 – $16,000 = **$4,000**

If Stan were to sell his car at this point, he will have lost $4,000. Depreciation is essentially a hidden ownership cost of a car.

>>> Calculate the monthly cost of depreciation by dividing the total depreciation by the months since it was purchased.

Example: How much did it cost Stan per month in depreciation to own that car that year?

$\frac{\$4{,}000}{12} \approx \$333.33$

Example: If Stan kept the car 10 years and then sold it for $8,000, what did it cost Stan per month in depreciation?

Depreciation: $20,000 – $8,000 = **$12,000**

Monthly depreciation: $\frac{\$12{,}000}{10 \times 12} = \$100$

Divided by 10 × 12, as this depreciation was over 10 years, and each year has 12 months in it.

>>> You can also use what you know about percent problems (% *of* base *is* percentage) to explore depreciation.

Example: After 1 year, the depreciation on Sam's car ($4,000) was what percent of the car's original value ($20,000)?

$P \times \$20{,}000 = \$4{,}000$

Divide both sides by $20,000: $P = \frac{\$4{,}000}{\$20{,}000} = 0.20 = 20\%$

Example: If a $50,000 car depreciates 20%, how much did it depreciate?

20% of $50,000 = **0.20** × **$50,000** = **$10,000**

Example: If a $50,000 car depreciates $10,000, how much is it worth now?

$50,000 – **$10,000** = $40,000

Brand-New Cars

Annual depreciation rates by automobile age[27]

Age of automobile, in years	% of Depreciation
0 (brand new)	23.9%
1 year	11.3%
2 years	10.8%
3 years	14.0%
4 years	13.7%
5 years	13.1%
6 years	11.4%
7 years	8.7%
8 years	9.9%
9 years	6.9%
10 years	4.9%
11+ years	10.3%

Now that you know about depreciation, we will look at some pros and cons of buying a brand-new car.

All cars depreciate, or lose value, over time. Cars do this at different rates. But **all cars depreciate at a higher percentage the first several years of their existence, and especially when they are purchased brand new.**

Notice in the chart that the **depreciation of brand-new cars (year 0) is significantly higher** than for used cars, even if the used cars are almost new (like 1 year old). In one year, a brand-new car would depreciate 23.9%, while a one-year-old car would only depreciate 11.3%. This is because brand-new cars lose value the moment you drive them off the lot. You will have to decide if the advantages of a brand-new car (see below) outweigh the extra you will pay in depreciation. Seek advice from your parents or other trusted adults.

Pros:

- **No repair cost** – Warranties typically cover anything that goes wrong, at least in the first 3 years or 36,000 miles. Extended warranties are also sometimes available.
- **Ability to maintain car from beginning** – You are able to make sure maintenance is done on the car as needed.

Cons:

- **Up-front cost** – New cars come with a big initial price tag.
- **Depreciation** – The moment you drive it off the lot, a new car loses significant value. The car continues to lose value rapidly for several years, based on mileage and condition.

KEEPING PERSPECTIVE

Depreciation is a key concept to understand. You might be tempted to buy a really expensive car as an investment, but cars *depreciate* in value, not gain value.View a vehicle as an expense, not an investment.

Cost Per Mile, Used Cars, & Leasing

As a car ages, it has less life to it. But slightly aged cars can still have plenty of miles left. **Many cars last 200,000 miles.**

Just like we used unit prices to compare the cost *per* unit between different packages, **we can look at the cost *per* mile to compare the cost of different cars**. For now, we will compare just the purchase prices (i.e., the up-front cost) and not any maintenance or use costs.

>>> Divide the up-front cost (i.e., the purchase price) by the expected miles left to calculate the up-front cost per mile.

Example: If a car expected to go 200,000 miles costs $20,000 new, what would the up-front cost per mile be if you bought it new?

$$\frac{\$20{,}000}{200{,}000} = \$0.10$$

>>> If needed, first calculate the expected miles left by subtracting the miles the car has gone from its total expected miles.

Example: You instead buy the same car as in the previous example after it has gone 45,000 miles (it is expected to last 200,000). If the purchase price is $12,000, how much per mile is its up-front cost?

Expected miles left: 200,000 − 45,000 = 155,000

Up-front cost *per* mile: $\frac{\$12{,}000}{155{,}000} \approx \$0.08$

Since $0.08 is less than $0.10, this means the used car's price is less per mile than the new car's price. This is because of all the depreciation the new car experiences the first several years.

Buying Used

>>> Used cars vary greatly in age, price, and longevity (some makes and models tend to break down more than others). **Newer used cars** can be good working vehicles.

When buying used, do not assume that the seller will tell you everything. Look at the car yourself, or get someone who knows cars to help you examine it. If you go to sell a car, be sure to share everything that you know about it. God wants us to walk in truth (Colossians 3:9).

Pros:

- **Lower upfront cost**, making it easier to buy without taking on debt.
- **Lower depreciation rates**, meaning a much lower depreciation cost per month.

Cons:

- **Knowledge** – You do not know exactly how the car was maintained.
- **Age** – As a car ages, parts tend to wear out. (Some used cars, though, are almost new, so this would not apply.)

Leasing a Car

Leasing is a way to drive a brand-new car without a big price tag. It is sort of like renting a car. Sounds great … until you use math to explore what you are really paying.

Example: How much will Ann pay altogether if she signs a 3-year lease for just $421 a month on a new $20,000 car?

Yearly amount: 12 × $421 = **$5,052**

For 3 years: 3 × **$5,052** = $15,156

We could have solved this in a single equation: 3 × 12 × $421 = $15,156

At the end of that $15,156, Ann owns nothing. The car is now worth less due to depreciation, but the dealer does not mind. He made sure the amount Ann paid him is much more than 3 years of depreciation on the car.

Pros:

- **Constant new car** – You always drive a new car that is under warranty.
- **No or low up-front cost.**

Cons:

- **No ownership** – You own nothing in the end, as you are basically renting.
- **Higher monthly fee** than even a brand-new car's depreciation.

KEEPING PERSPECTIVE

We are going to explore other costs of car ownership besides the initial price tag that you will want to take into account when you consider what car you can afford. After all, you do not want to buy a car you then can not afford to drive.

While buying a car might still be a few years off for you, keep in mind that it is something you can start saving for now. Cars come with an initial price; while leasing can lower that, it means paying extra over time and owning nothing in the end.

Car Maintenance Costs

Before we continue discussing the car-buying process, we will pause for a few sections to talk about some costs of owning a car, as they are important to think about when selecting a car too. We will start with maintenance costs.

Scheduled Maintenance

The table shows a few common car maintenance items.

Service	Frequency	General Cost	
Oil and Filter Change	regularly		
Tire Rotation	regularly		Is included in tire purchase at some places.
Alignment	regularly		Lifetime alignments available at some places, meaning you pay once and get alignments there as long as you own the car.
Brake Pads	periodically		
Brake Rotors & Pads	periodically		
Tires	periodically		
Radiator Hose	periodically		
Timing Belt/ Chain	once around 100,000 mi.		

Legend	
regularly = often yearly (or more)	
periodically = only after several years	
	~$100 or less
	~a few hundred dollars
	~several hundred to $1,000+

Many car manufacturers suggest a car inspection at 30,000, 60,000, and 90,000 miles. Following whatever your recommended inspection schedule is should help you catch wearing parts *before* they stop working. Note that as a car gets older, you will need to plan for some unscheduled maintenance too.

Consider Maintenance When Buying a Car

While all cars need maintained, cars cost different amounts to maintain. When buying a car, look into the car's general maintenance costs. Some cars tend to cost more to maintain than others. For example, one car might cost $6,000 over 10 years, while another might cost $16,000. That is a $10,000 difference!

If buying a used car, find out if regular maintenance, like oil changes, has been done. Has the car been well cared for? It is important that you understand how a used car has been (or has not been) used and maintained when you buy it.

Also see what periodic maintenance has already been done and budget for that. How close is the car to needing more work? If buying used, consider taking it to a mechanic to see what major work might need to be done soon.

»» When buying a used car, include any immediate maintenance the car needs as part of its up-front cost.

Example: Daniella is considering a $6,000 used car. Her mechanic says it will likely need $1,000 worth of work in the near future. After researching, Daniella anticipates that it will cost about $2,000 a year to maintain. How much does Daniella need to plan to spend up-front, and how much in maintenance does she need to budget each month?

Total up-front cost:	$\$6{,}000 + \$1{,}000 = \$7{,}000$
Monthly maintenance:	$\frac{\$2{,}000}{12} \approx \$166.67$

"For which of you, desiring to build a tower, does not first sit down and count the cost, whether he has enough to complete it? Otherwise, when he has laid a foundation and is not able to finish, all who see it begin to mock him" (Luke 14:28-29).

KEEPING PERSPECTIVE

Part of the cost of a car includes the cost of any work it needs up-front and ongoing costs of maintaining it, so those costs are important to factor into your car-buying decision.

Gas Mileage

Not every car will run the same distance on the same amount of gas. When buying a car, it is helpful to look at how many miles it gets per gallon of gas, as that affects your ongoing gas cost.

The **combined gas mileage** listed on stickers is based on driving 55% in the city and 45% on the highway. Notice it is in between the city and highway mileage listed below it, and that highway is higher than city (this is typically the case for gas-only cars).

⋙ **Gas mileage** is the distance a car can go *per* gallon of gas (the **mpg**, or **m**iles **p**er **g**allon). It is found by *dividing* the miles gone by the gallons of gas used. (*Per* indicates division.) The higher the gas mileage, the less you will pay in gas, as you will travel farther per gallon.

Example: Calculate the gas mileage if you travel 200 mi. and use 10 gal. of gas.

$\frac{200 \text{ mi.}}{10 \text{ gal.}} = 20 \frac{\text{mi.}}{\text{gal.}} = 20 \text{ mpg}$

Tracking Gas Mileage

Cars do not always get exactly the gas mileage advertised.

⋙ Calculate the actual gas mileage by dividing the miles you have driven between fill ups by the number of gallons the car used.

⋙ Calculate the miles you have driven (your mileage) by resetting a trip odometer each time you fill up, or by writing down your car's total mileage at each fill up, subtracting to find the change between fill ups.

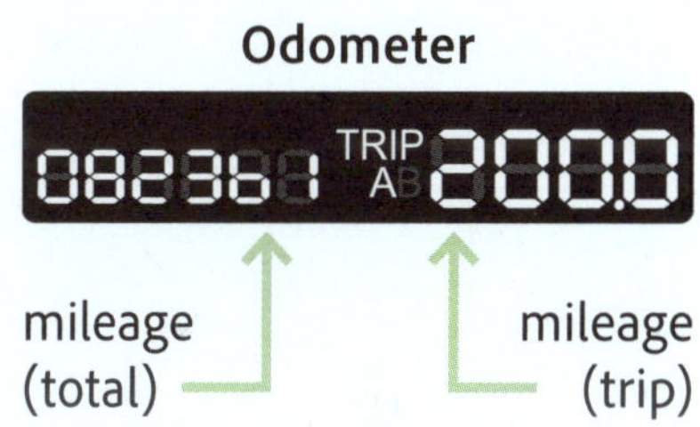

Example: Your previous odometer read 103,401 and it now reads 103,620. If you used 10 gallons, what gas mileage did you get?

Mileage: 103,620 – 103,401 = 219

Gas mileage (i.e., miles *per* gallon): $\frac{219 \text{ mi.}}{10 \text{ gal.}} = 21.9 \frac{\text{mi.}}{\text{gal.}} = 21.9 \text{ mpg}$

Budgeting for Gas

You can use your gas mileage and basic math to estimate how much you should budget for gas.

Larger vehicles tend to need more gas than smaller ones, but you can still look for one with better gas mileage than similar vehicles.

Example: If you expect to drive 15,000 miles, your car gets 30 mpg, and gas costs $5 per gal. in your area, how much should you budget for gas per month?

Gallons of gas needed: $\frac{15{,}000 \text{ mi.}}{30 \text{ mpg.}} = $ **500 gal.**

Expected cost per year: $5 × **500 gal.** = **$2,500**

Monthly cost to budget: $\frac{\$2{,}500}{12} \approx \$208$

Electric & Hybrid Vehicles

With rising gas prices, electric and **hybrid vehicles** (gas with some electric) have become more popular. A little math can help you compare them with other options.

There are often tax incentives for getting electric cars; include that into your cost comparison. However, there could be higher costs for batteries, service, etc., and you may have difficulty finding charging stations.

While you do not have gas mileage to calculate with electric vehicles, you do have to consider the range, or distance, you can travel on a charge *and* make sure there are stations to charge. Note that range may be different in cold weather.

Example: You are considering a hybrid vehicle that costs $2,500 more than a gas car. You estimate the hybrid will save you $40 a month on gas and about $200 a year on maintenance. How long would it take you to break even?

Monthly maintenance savings: $\frac{\$200}{12} \approx \$16.67$

Total monthly savings: $\$16.67 + \$40 = \$56.67$

Months to recuperate: $\frac{\textit{extra cost}}{\textit{monthly savings}} = \frac{\$2{,}500}{\$56.67} \approx 44.12$ mos.

Convert months to years: $\frac{44.12}{12} \approx 3.68$ yrs.

Example: If a car can go 200 mi. on a charge, how many times would it need charged to go 420 mi.?

$\frac{420}{200} = 2.1$, so 3 times.

We listed 3 instead of rounding to 2, as the car can not go beyond its range, not even a small amount.

KEEPING PERSPECTIVE

The importance of gas mileage depends on how much you drive and the price of gas in your area. Depending on these, vehicles with better gas mileage can result in significant savings.

Car Insurance

In most states, having at least a minimal amount of car insurance is required by law. It is another expense to consider before you buy a car. In fact, when considering a car, talk to a car insurance agent to see how much that car would cost to insure. New cars tend to cost more to insure than old cars, as they would cost more for the insurance company to replace after an accident. Following are a few things that are helpful to know when you consider car insurance and think about car costs:

Car Insurance Terminology

- **Premiums** – As with health insurance, car insurance has payments you make called premiums. You might pay this premium monthly, annually, or somewhere in between. Your premium cost is determined by a variety of factors, including:
  - the coverage you select
  - your driving record
  - the car you drive
  - how many miles you drive per year
  - your age
  - your gender
  - your credit score
- **Deductible** – As with health insurance, car insurance can have deductibles (amounts you will pay before the insurance starts paying). If you have a $500 deductible, then you will pay the first $500, after which the insurance will start paying.
- **Limits or Maximums** – There will usually be limits or maximums that the insurance will cover for different categories. These limits are often per accident or per person.

Be sure to get quotes from a couple of different insurance companies and ask if there is anything you can do to lower your premium. For example, some will offer discounts if you take a safe driving course or get certain grades.

If damage is the other driver's fault, your insurance might not have to pay anything, meaning you will not have a deductible.

Coverage Choices

The table shows some terms that we did not cover with medical insurance.

Type of Insurance:	Covers:
Liability	what you do to other drivers and their cars
Bodily Injury	injuries to other people
Property Damage	damage to others' property (like their car)
Underinsured/Uninsured	you in case the other driver's liability insurance is not enough or the other driver does not have any
Collision	your own car regardless of fault
Comprehensive	other events like theft
Medical Payments	medical expenses regardless of fault
Roadside Coverage	getting towed/roadside assistance
Transportation/Rental Car/Travel Coverage	a rental while your car is being repaired

When you get car insurance, you will have to choose how much coverage you want in each of these categories. Your agent should be able to help give you guidance as to what is reasonable in your area.

Using Your Insurance

If you get in an accident that is not your fault, you may not owe a deductible, provided the other driver's insurance pays for the damage. If it is your fault or your insurance has to pay (such as in a no-fault state, if the other driver did not have proper coverage, or if you opt to use your insurance anyway instead of the other driver's), you will owe your deductible.

If the repairs needed on your car are more than the car's value, the insurance company will likely just give you the car's value. This amount will probably not be what you paid for it, especially with newer (or brand-new cars). Remember, new cars depreciate quickly.

If you have a car loan, consider getting **gap insurance** — insurance that helps you pay the "gap" between what the insurance gives you and what you owe on the loan.

After an accident where your insurance has to pay, some insurances will immediately increase your premiums. Others will allow you one or two accidents first. When you get insurance, be sure to ask what happens in the case of an accident.

Using Math to Explore Car Insurance

Basic math can help you explore various aspects of car insurance.

If you decide to pay for insurance periodically, be sure to still budget a monthly amount so you are not surprised by the bill.

Example: An insurance company will charge you $900 if you pay every 6 months, or $170 monthly. How much will you save if you pay every 6 months?

6-month cost if pay every month: $6 \times \$170 = \$1{,}020$

Savings if pay every 6 months: $\$1{,}020 - \$900 = \$120$

Example: You are currently paying $800 every 6 months for insurance. If you switch to another company, the cost will go down to $755. How much will you save by switching?

$\$800 - \$755 = \$45$

KEEPING PERSPECTIVE

Living in a fallen world means that accidents happen. Property gets damaged. People get hurt.

Yet in the midst of the pain, God promises that "all things work together for good, for those who are called according to his purpose" (Romans 8:28). It is a truth to cling to, no matter what.

Total Cost of a Car (Including Hidden Fees)

Now that we have looked at maintenance, gas mileage, and insurance, we will examine how those costs, as well as others, come into play when you consider buying a car.

Upfront Costs

You might think the initial cost of buying a car is just the purchase price (or the down payment on a loan, if you take out a loan), but there are also some hidden costs.

- **Purchase price or down payment** – This is likely the biggest upfront expense you need to be prepared to pay.
- **Taxes** – Some states charge sales tax on cars bought from dealers. Sales tax on an item that is thousands of dollars can be a hefty amount. For example, a 7% sales tax on a $20,000 car would be $0.07 \times \$20,000 = \$1,400$ — definitely not a cost you want to be surprised by.

  Before you head to a no-sales-tax state to buy your car, check to see if your state will charge you tax anyway. Some states charge the difference between their sales tax and the state's sales tax in which you bought the car.
- **State Fees** – Instead of (or in addition to) taxes, some states charge hefty fees when you register the car or change its title. In Georgia, for example, in a sample year, there was an Ad Valorem Title Tax of 7% of the vehicle's value (not the price you paid but how it is valued). Again, that can be quite hefty.
- **Dealer Fees** – Dealerships want to make money. Be sure to ask about any fees they charge on top of the purchase price of the car.

Ongoing Costs

Before buying a car, make sure you have a budget that can handle the cost of maintaining and using the car, as well as saving for the next. These prices will depend on the car you buy:

- **maintenance price**
- **loan payment** (if applicable)
- **gas**
- **car insurance**
- **yearly inspections and/or emissions** (usually not large expenses)
- **depreciation** (Saving up what your car depreciates will help you have money on hand for your next car when that time comes.)

Comparing Cars

Comparing cars can seem overwhelming. You could **use a spreadsheet or sheet of lined paper**, listing all the buying costs and maintenance costs for each car as best as you can. If a parent gives permission, there are also a variety of online tools.

Ownership Cost Summary		
Total Driving Cost		
Per Year	$ 6,950.00	
Per Mile	$ 0.70	
5-year Total	$ 34,752.00	

Cost Breakdown	Cost per year	5-year Total
Fuel	$ 1,077.40	$ 5,387.00
Maint/Repair	$ 952.60	$ 4,763.00
Depreciation	$ 2,221.20	$11,106.00
Insurance	$ 1,426.80	$ 7,134.00
Fees & Taxes	$ 495.40	$ 2,477.00
Finance Charges	$ 777.00	$ 3,885.00

AAA Car Calculator
2022 Toyota Corolla LE 4d
in Georgia, with 55/45% city/highway driving and $3.32 per gallon estimated for gas cost, as of 2024

In using any tool, **you might need to adjust the values.**

Insurance – An average will likely be shown; if you know you pay more or less than average, you will want to adjust.

Finance charges – You would want to subtract if you buy the vehicle without a loan.

Example: If you are looking at the ownership cost summary shown and know that your insurance costs $800 every 6 months, how much more would you need to account for than listed here?

Yearly insurance cost: 2 × $800 = **$1,600**

Additional from what chart shows:
$1,600 – $1,426.80 = $173.20

When making a budget, whatever you are saving for a new car would go under *savings*, not *spending*, as you do not actually owe anything until you buy your next car. Yet depreciation is really a hidden cost of owning a car.

KEEPING PERSPECTIVE

While you can use math to determine whether or not you can afford a car and to compare different cars, remember that no human knows the future. You could buy a car that typically needs a lot of maintenance and God could keep it running perfectly, or vice versa. However, if you know a particular car tends to be a nightmare, you would be wise to take heed to that unless you are really sure God is leading you forward.

The prudent sees danger and hides himself, but the simple go on and suffer for it (Proverbs 22:3).

Making a Fair Deal

Now that we have talked about costs of car ownership, we will look at the buying process. Buying a car is not like buying a phone. Not only does it probably cost a lot more, but it often involves some negotiation.

Car-buying Hints

→ **Do your research.**

→ **Consider the timing**. Dealers often get incentives for selling a certain number of cars by the end of the month, quarter, or year, so you might get better deals then. Watch for sales, or if you need financing, a 0% interest special.

→ **Be willing to wait.** If you need something right away, you might end up paying for a car with features you do not really need just because it is on the lot — although if you timed it right, the dealers might be willing to give a good deal to move cars off their lot.

→ **Be willing to walk away.** Many car salesmen want to pressure you into not leaving the lot without a car, but do not agree to something under pressure. Make sure you have done your homework and sought wise counsel. Visit several car dealers so you get a better picture as to the market and make sure you really know what you want. You might even leave your wallet (except your driver's license) at home when you first look or bring someone with you who will keep their head and help you walk away. Even once you are sure you want the car, if you are willing to walk away and wait for that car, you may end up getting a better deal. Salesmen may increase their prices if they think you are not willing to walk away.

→ **Beware of loans.** In negotiations, the salesman will probably try to talk you into buying the car via a loan instead of cash. That is because the dealer makes more money if they get your interest too. Look at the total cost of the vehicle, not just the monthly payments. Decide ahead of time if you want a loan. Sometimes, dealers will offer interest-free financing for a time. While that can be a helpful way to build your credit, know that you will save money if you have what you need to pay it off (and follow through on it) before the interest starts.

Whether you buy from a dealer or from another person, there is usually some wiggle room in the price. How do you figure out how to negotiate for a deal that is fair? Understanding the standard for pricing both brand-new and used cars can be very helpful in this process.

Brand-New Cars: MSRP

Car manufacturers set a price that they think is a fair sale price: the Manufacturer's Suggested Retail Price, or **MSRP**. This price should be posted on all new cars. It is a price that includes what the manufacturer thought was a fair markup for the dealer.

Does that mean you have to pay that price? No.

- If cars are not selling, dealers may be willing to sell under MSRP to meet a sales quota that gives them incentives or bonuses.
- If cars are in high demand, dealers may be able to charge more than MSRP on cars.

Used Cars

Be sure to get a parent's permission before using online tools or apps, and always be cautious with what information you share about yourself and your finances on apps or the Internet.

Online or library pricing guides can give you an idea of what is a reasonable range for the car. To use them, ask a parent's permission first. You will need to know the car's:

- make
- model
- condition
- history

Some sites will let you plug in a VIN or license plate number and look up some of that for you. You will also need to list the zip code, as location also affects the price. You will then get to see a range of values as to what that car is worth. You can compare what is being asked to that range and negotiate from there if needed.

For example, if someone is asking a price for their car as if it were in very good condition but it really is a good- or fair-condition vehicle, you might want to talk them down or find a different vehicle to purchase.

Selling or Trading in a Car

When it comes time to sell your car, these same online or library pricing guides used to buy a car can help you set a price. If you sell through a dealership, you will typically get less than if you sell on your own, as the dealer needs a markup. However, sometimes dealers will give good deals if you are also buying a car from them.

Negotiation Tip – While they might not be able to lower the price of the car you are buying, dealers might be able to give you a higher price for your old car. Or they might be able to throw in some free oil changes.

Be prepared to do some math throughout car negotiations to see how much you are really paying and spending overall, and compare that to other options.

KEEPING PERSPECTIVE

Understanding MSRP and used car pricing and doing your homework before you look at a car can help you know what a good, but fair, deal looks like.

And if you make a sale to your neighbor or buy from your neighbor, you shall not wrong one another (Leviticus 25:14).

Review & Case Study: Contentment & God's Heart (Hsieh Family)

Whew. I am not sure about you, but I am glad I do not have to buy a car every day. There is a lot to think through with purchasing one. Hopefully what we looked at in this lesson has equipped you with some things to consider besides just which model car you like the most.

We are going to end with a case study of a couple that had the means to buy a fancy new sports car yet chose to continue driving an older vehicle, thereby freeing up that money to invest in eternity instead. When considering the car you buy (or do not buy), remember that there are more important things than driving the newest and best car. Keep eternity in mind.

Case Study: Contentment & God's Heart (Hsieh Family)

This write-up is based on a video from Generous Giving. Used by permission of Generous Giving; all rights reserved.

Bree realized in college that God had a heart for the poor. Wanting to share His heart and discover His love for the poor, she moved to a poor community. While engaged to Tom Hsieh, they both made a radical decision: they would live at or below the median income level (i.e., they would set the median income as their lifestyle cap). This meant cutting out their elaborate honeymoon plan. Winning a 7-day trip to Italy encouraged them that God would take care of them.

Living in a poor community has opened up many doors for the Hsiehs to share God and His love. Neighborhood boys asked if the Hsiehs could teach them the Bible, something that might not have happened in a more affluent area. The kids were facing so many challenges. The Hsiehs also got to open their home and pour hope into different people in need. Being in the inner city, the couple has had no lack of opportunities to help and make a difference in people's lives.

The Hsieh's choice to live radically did not come from a lack of ability to live a much wealthier lifestyle. Tom worked for Earthlink, and when they went public, his stock options became worth several million dollars. While coworkers began sporting fancy cars, Tom continued driving his Geo Metro. Rather than living a wealthy lifestyle, the Hsiehs have donated the extra they have been blessed with instead.

While the Hsiehs do not believe everyone is called to do what they have done, they challenge others to ask God how He is calling them to steward the resources He gives them. Their lives illustrate that it is possible to be content with less and to choose not to allow money to gain a grip on our hearts but instead use it to experience the joy of giving and being a part of what God is doing.

Quotes from Bree:

"... [W]e have to just believe as people who follow Jesus that our lives can make a difference when we follow Him in what we are called to do."[28]

"... [Y]ou really have to ask Jesus what He has given you, and what He is asking you to give away, and what He is asking you to hold onto. So, there is no formula for what kind of life to live; there is only a listening to Jesus and a conviction, and all you have to do is ask how with what God's given you."[29]

12 Career Preparation

***income** = giving + saving + spending*

In this lesson, we will explore some different ways to prepare for what God has for you next, as well as look at how basic math skills can help you compare different jobs.

The Importance of Thinking Through Your End Goals, Plus a Biblical View of Work

Throughout this lesson, we will check out some different ways to prepare for a career. We will start with college (and how to eliminate or minimize college debt), but we will also explore other options and how to evaluate the costs of each one.

Before jumping into how to prepare for something, though, it is important to step back and think about what you are preparing for. Otherwise, you might end up spending a lot of time and money preparing for something you do not really want to do — like heading down a road that does not lead where you want to end up.

- According to one report based on a sample year, about 45% of college graduates were not working in a job that requires a college degree 10 years after they graduated (that percent is more after one year).[30]
- On the flip side, student loan debt at the time totaled more than 1.7 trillion dollars.[31]

Putting these two stats together means that a lot of people are spending a lot of time and going into a lot of debt for something they do not necessarily use in the end.

As we have mentioned before, debt can be crippling, hampering you from other goals. Student debt interest rates for a sample year are between 5.5% to 8.05% for federal debt. That means every year, graduates are paying about 5.5% to 8.05% (if using federal loans) on their loans.

Example: If we assume no payments are made, how much is Rosa paying that first year in interest on an $80,000 loan at 8.05% compounded daily?

Total debt after 1 year:

$$P = \$80{,}000\left(1 + \frac{0.0805}{365}\right)^{(1 \times 365)} = \$86{,}705.54$$

Total interest for that year: $86,705.54 – $80,000 = $6,705.54

Student loans, like home loans, make sense in some situations. The justification behind them is that a college degree will allow you to earn more your entire career, thereby paying for itself many times over. While this can be true, it is not all the time.

Before jumping into how to get trained, and especially before signing up for lots of debt, **think through *what* you want to do and your goals for your future**. Not everyone is good at doing the same things. God made us all different on purpose.

- Pay attention to what you like and dislike and what you excel at. Ask those who know you best what skills they see in you.
- Learn about different careers or ministries. Consider asking people at church or in your social groups to let you shadow them at work or learn more about what they do.
- Think about how different careers or ministries may affect a family one day.

***Then* look at how to best get trained** for the goals God has put on your heart as well as the gifts and interests He has given you. The answer may be college — or it may be another route.

While thinking through what you want to do first can save time and money, sometimes God redirects our steps as we go. Do what you can to research ahead of time, but know that God may redirect you down the road. Pray for His guidance and direction for your life.

If you do go to college, you will need to select a **major** — an area of focused study. You may also select a **minor** — a secondary, lesser area of focused study.

A Biblical View of Work

Work existed *before* the Fall. God worked when He made all things. As His image bearers, work is a privilege: "The Lord God took the man and put him in the garden of Eden to work it and keep it" (Genesis 2:15). Some might get paid for their work and call it a job, and others might not. But we are all called to work in one way or another and to honor God through that work.

Sin has affected that work — the ground now produces "thorns and thistles" (Genesis 3:18). Even if we are not farmers, the effects of a cursed world affect our work, but that does not change the fact that work itself is a beautiful gift from God.

In fact, if you are in Christ, you were created "for good works" that God has *already* lined up for you — both now and in your future.

> *For we are his workmanship, created in Christ Jesus for good works, which God prepared beforehand, that we should walk in them* (Ephesians 2:10).

KEEPING PERSPECTIVE

As you think about choosing a career or answering a calling, start by thanking God for the privilege of work. Thank Him that, regardless of what field or ministry you pick, God has good works already lined up for you. Ask Him to help you start walking now in those good works. Learning to follow God now is the best preparation you can make for the future.

College: Overview of Costs & Tuition Plus Credits

The price of college varies drastically based on where and how you go. Before you sign on with a college, it is a good idea to make sure you have looked at the total cost to finish the degree and compared that with other options.

- **Tuition** – Tuition is the cost of instruction at a college. It does not include your room, board, or materials.
- **Room and Board** – Room and board is typically a sizeable chunk of money to think through — and a place where you could potentially save money (as in the neighborhood of $40,000, or potentially even much more).
- **Books** – College books can be pricey, costing hundreds or even $1,000 or more each year. However, you may be able to lower/recuperate some of this cost by buying used books and/or reselling your books when finished, although you may find ones in your major useful down the road.
- **Fees** – Most colleges have fees that get added to your total cost. Some might involve you having additional expenses too, such as travel to get home to family for weekends or holidays.

>>> To see how much you will spend on a degree, total all the costs together, subtracting any incentives (such as scholarships and grants) the college offers that would *lower* your cost there.

When seeking a college, it is important to think through the worldview you will be receiving. Private colleges may have higher tuition than others. However, some colleges offer discounts if you live in state. Private colleges have other important advantages, such as the potential for a Christian mission and perspective. Also consider proximity to home and recognition of the degree.

More with Tuition

To help you evaluate costs, we will unpack tuition a little further.

- **College classes give you credits.**
- **It typically takes 120 credits to graduate** (provided all credits count toward your degree).
- **Most courses are worth 3 to 4 credits**, although some are worth more or less.

Colleges will often have the option to pay a specific amount per credit or pay a set amount per semester, regardless of how many credits you take.

» If paying by the credit, the cost for each course you take equals the credits you get for that course times the college's credit cost.

Example: Ethan enrolls in a 3-credit class at a college that charges $600 a credit. How much does that class cost him in tuition?

$3 \times \$600 = \$1,800$

» If paying by the semester, divide the semester fee by the number of credits you actually take that semester to calculate the cost per credit.

Example: If a college charges $5,000 a semester, how much is Sophia paying per credit if she takes 10 credits?

$\frac{\$5,000}{10} = \$500$

» When calculating the total cost or comparing costs between different colleges, look at the cost of the total credits needed to graduate or at the total semester prices.

TOTAL CREDITS

Example: How much will Jason spend on his bachelor's degree in tuition if he gets 120 credits at a school that charges $600 a credit?

$120 \times \$600 = \$72,000$

TOTAL SEMESTERS

Example: How much will Sarah spend on her bachelor's degree if she pays $7,000 per semester for 8 semesters?

$8 \times \$7,000 = \$56,000$

Credits & Time

The number of credits earned from a course is based on the number of hours of class time it should take each week in a normal semester. While it can vary by college, typically:

- You get 1 credit per 1 hour of class time each week (so a 3-credit course has 3 hours of class each week).
- You are expected to study an additional 2 hours per credit.

In summer or half-semester courses, you would have more hours of class time and study time each week to get the same total hours in a shorter number of weeks. The time it takes you may vary from these estimates.

KEEPING PERSPECTIVE

While we are focusing on helping you compare the costs, there are a lot of non-cost factors worth considering as well, such as the worldview of the college. There is an old saying that cheaper is not always better — and it is true. Just because a school costs less does not mean that is the way for you to go. It is always helpful to count the cost before you begin something.

College: Ways to Save

Whole books have been written on how to save money when getting a degree. We will just check out a few possibilities besides picking a less expensive college altogether.

- **Think through your plan as best as you can.** If you change plans partway through college, you may have to get more than 120 credits, which adds cost.
- **Transfer courses from a local, less expensive college.** Courses tend to be less expensive, but many do not offer 4-year degrees. Make sure whatever you take will transfer to your 4-year college. Some also offer free enrollment for high school students, where you can get both college and high school credit at the same time.

>>> Figure out your savings for making some cost-saving changes by calculating the cost with and without those changes and finding the difference.

Example: You realize you can take a 2-credit course at a local 2-year college for $150 a credit that would count toward your degree. How much would that save over paying $600 per credit at your 4-year college?

Price at local 2-year college: 2 × $150 = **$300**

Price at 4-year college: 2 × $600 = **$1,200**

Savings: **$1,200** – **$300** = $900

Savings add up. If you save $400 a credit on 60 credits, that would be 60 × $400 = $24,000.

- **Test out of credits.** Some colleges will accept certain tests or other evaluation tools to allow you to skip courses, particularly courses not in your major. Note that you will need to be good at studying on your own and test taking, and be sure to research what (and how many) test credits your college will take first.
- **Live at home.** Staying at home and doing college either online or at a local college saves room and board costs.
- **Take more credits each semester.** Some schools offer package prices over a certain number of credits. At these schools, you will save more if you are able to take more credits a semester.

Example: A college charged $6,000 per semester for in-state students taking more than 6 credits. What would your per-semester savings be if you took 15 credits a semester instead of 10?

Cost per credit if 10 credits: $\frac{\$6,000}{10}$ = **$600**

Cost per credit if 15 credits: $\frac{\$6,000}{15}$ = **$400**

Per-semester savings: **$600** – **$400** = $200

- **Earn scholarships.**
  - Schools, states, organizations, and businesses sometimes sponsor students, meaning they give money toward a student's education.
  - Scholarships can be based on various factors, such as grades, sports, income, or an attribute such as community service, religious affiliation, or military background.[32]
  - There may also be program-specific scholarships designed to attract students to certain fields.
  - Online scholarship search tools, used with a parent's permission, can help you learn about scholarship offerings.
- **Get military funding.** The government may pay for college education if you commit to a certain number of years of military service.
- **Get help from an employer.** Some employers will help pay for students to get a college education.
- **Work for the college.** Some colleges have positions where students can work to earn part or all of their tuition or room and board.
- **Use financial aid.** We will explore financial aid more in the next section.

There are sometimes options through employers that will help pay for the cost of college courses related to your work.

It does not have to cost a fortune to attend college. There are ways to significantly cut down on the amount. Basic math can help you evaluate the savings.

KEEPING PERSPECTIVE

Whatever job you pick one day and however you train for it, remember that if you are God's child, then He is your true boss. This admonition to bondservants (who worked for a master) applies to us in our work today:

> *Whatever you do, work heartily, as for the Lord and not for men, knowing that from the Lord you will receive the inheritance as your reward. You are serving the Lord Christ* (Colossians 3:23-24).

When the time comes to get a job (or if you already have one), look for ways to be God's hands, feet, and voice. The real reward for a job well done is not just a paycheck — it is eternal. Keep this in mind both when picking a career and in the schoolwork and other assignments you have now. God wants you to serve Him *wherever* you go and *whatever* you do.

College: Financial Aid

You may qualify for financial aid for college from the United States government. To get this aid, you will need to fill out a form called the **FAFSA** (**F**ree **A**pplication for **F**ederal **S**tudent **A**id). This form can be filled out online, with a parent's permission, or you can call their office to get a printed version.

States offer financial aid too. Check to see if your state offers it through the FAFSA form or if they have their own application process.

Filling out the FAFSA may be a bit of a process. It reads a bit like a combination of a tax form and the forms you fill out at a doctor's office. It asks about things like:

- financial information (for both you and your parents)
- your parents' education (as well as your own)
- demographic information.

Read each question carefully as you fill it out.

Federal financial aid can be either need based or non-need based.[33] We will look at each.

There are online EFC calculators you can use to get an idea of how much you might qualify for before filling out the FAFSA. Before you search, get a parent's permission.

Need-Based Aid

Grant: Federal Pell Grant[34]	$7,395 a year in a sample year.
Grant: Federal Supplemental Educational Opportunity Grant (FSEOG)[35]	$100 to $4,000 a year in a sample year.
Federal Work-Study Program	Helps you work your way through college.
Subsidized Loans[36]	A loan where the government will pay your interest for you while you are in college, for 6 months afterward, and sometimes for other periods.

A **grant** is money that is given that does not need to be repaid.

A **subsidized loan** is still a loan, just one where the government pays your interest for a time. You will eventually have to pay the money back and start paying interest.

The maximum amount of need-based aid you can receive is called your **financial need**.

>> ***maximum need-based aid*** = *financial need* = ***COA*** – ***EFC***

- **COA**, or Cost of Attendance, is the expected cost for you to attend a particular school (including things like room and board and textbooks).
- **EFC**, or Expected Family Contribution, is the amount the government expects you/your family to pay toward your COA. This amount is based on what you put on the FAFSA.

Maximum need-based aid is based on your costs minus what the government deems you can pay toward those costs.

> ***Example:*** What is the *maximum* need-based federal aid you could qualify for if your COA is $80,000 and your EFC is $77,000?
>
> *financial need* = ***COA*** – ***EFC*** = **$80,000** – **$77,000** = $3,000

Non-Need-Based Aid

You might also qualify for aid that is not based on your need.

- **Unsubsidized loans** – Loans where your interest is *not* subsidized, meaning the government is not paying the interest for you while you are in college and shortly after. You will not need to start repaying these loans until 6 months after you graduate (or drop below a certain threshold); however, you will be charged interest for the entire time.
- **Specific grants**, such as for those willing to agree to teach low-income students in certain fields.

Again, there is a simple formula for figuring out the *maximum* non-need-based aid you might qualify for.

All this is really saying is that you can not get more non-need-based aid than your costs and the other aid you are receiving.

>>> ***maximum non-need-based aid*** = ***COA*** – ***aid received***

The *aid received* here includes *any* aid, not just federal.

> ***Example:*** What is the maximum non-need-based aid you could qualify for if your COA is $80,000 and you receive $10,000 of other aid or scholarships?
>
> *Maximum non-need-based aid* = ***COA*** – ***aid received***
> = **$80,000** – **$10,000** = $70,000

KEEPING PERSPECTIVE

The numbers we looked at were for calculating *maximums*. That does not mean it is the amount you will get. For example, in a sample year, the maximum Pell Grant[37] was $7,395, but the minimum was $740. That is a big span. Federal aid comes in different forms, including loans.

Apprenticeships, Certificates, Trade School, & Other Non-College Options

You may find there are ways to get trained for your desired occupation besides college. We are going to touch on just a few of these other options here. **Careers where there are a lot of hands-on skills to be learned lend themselves especially well to non-college options.**

Apprenticeship – Many apprenticeships pay a decent salary. You could *receive* money while you learn rather than pay.

- If you search online with a parent's permission for "apprenticeships," you should be able to find listings of some.
- The current most-viewed listings on the government's apprenticeship database[38] are electricians, information security analysts, and software developers.
- Many apprenticeships require you to have some knowledge of the field first.

Certificate

- A certificate is a way of showing that you have learned a specific set of knowledge, be it mastery of a specific software or knowledge of how to do something, like court reporting.
- You can find certificate programs online (with a parent's permission) or at a local college.
- While there is usually a cost for a certificate, it should be significantly less time and money than a college degree.

Trade School – Some fields that can be learned via trade school might surprise you. For example, you can be a dental assistant via trade school.

Portfolio – For some fields, such as art-based fields, employers care more about what you can do than what you know. A portfolio is a compilation of your work that showcases your skills. You could develop one by offering your services *pro bono* (without cost) or at a low cost.

License – Some fields, such as a real estate agent, are based on licensing. Licensing requirements vary. You might need to take a course, or study for a test and then pass that test. Some licenses may also require a degree or a certain number of related college courses.

Self-Employment – Many innovative people have been successful without a degree by running their own business.

Looking at the Salaries

One reason many people choose a college degree over non-college options is the belief that it will lead to a higher salary in the end. This is not always the case.

Basic math and some research can help you explore costs and different training options for careers of interest to you.

Example: Tyler earns $20 an hour as an electrical apprentice while at the same time learning the skill. If he works an average of 40 hours a week but also has $2,000 in training fees each year, at the end of four years, how much will he have made?

Earnings per week: 40 × $20 = **$800**

Earnings per year: **$800** × 52 = **$41,600**

Earnings per 4 years: 4 × **$41,600** = **$166,400**

4-year earnings minus 4-year training fees:
$166,400 – (4 × $2,000) = **$158,400**

Notice that we just calculated the amount Tyler *made*. When exploring college, we have been looking at what it *costs*.

>>> To look at the total difference financially, add together what could have been made with what was spent.

Example: In contrast, Tyler could have *spent* $100,000 on a college degree. How much less well off financially would he be after 4 years with his college degree than if he has an apprentice?

$158,400 + $100,000 = **$258,400**

>>> Compare what can be made longterm by dividing the total difference financially by the extra made each year.

Example: Tyler can earn $70,000 a year as an electrician after his training. In contrast, he can earn $90,000 a year in a different field with a degree. How long will it take for Tyler to recuperate the $258,400 less that he is starting out with?

Amount extra made each year: $90,000 – $70,000 = **$20,000**

Number of years to reach same place financially:

$$\frac{\$258,400}{\$20,000} \approx 12.92 \text{ years}$$

After almost 13 years, Tyler would end up better off financially with his degree *in this hypothetical scenario.* In other scenarios, it might have taken shorter or longer for the degree to pay for itself — or it might never have.

KEEPING PERSPECTIVE

There are many examples of successful people that remind us that a college degree is not always needed. If a college degree is truly the best way to get into the career God has called you to, it can be an investment in your future, but college is not a guaranteed way to a better job one day. **Also remember that making the most money is not the most important consideration in the career you pick.** Seek God for what He wants you to do.

Comparing Jobs: Benefits

When comparing jobs, you might think all you need to do is compare the salaries. But that is not true. You want to compare all aspects (such as benefits), as well as look at things that can not be measured in dollars, such as convenience to family and hours and lifestyle required.

Time Off

Different jobs offer a different amount of paid time off.

>>> To determine the monetary value of the time off offered, divide your salary by 260 (the number of working days in a year) to calculate the daily rate and then multiply that daily rate by the number of days off.

Example: Company A offers an $80,000 salary and 15 days off. Calculate the monetary value of the vacation days.

Daily rate: $\frac{\$80,000}{260} \approx \mathbf{\$307.69}$

Value of days off: $\mathbf{15} \times \mathbf{\$307.69} = \$4,615.35$

Days off may include both holidays and vacation days.

In the example above, we rounded the daily rate. We would get a more accurate answer by **doing all the calculations in one step on the calculator.**

Example: Previous example without rounding the daily rate.

$\frac{\$80,000}{260} \times \mathbf{15} \approx \$4,615.38$

Notice the answer when done in one step is slightly different, as we did not round the daily rate. You will get the most accurate answer if you solve multi-step problems by plugging into your calculator as one step.

>>> If you are paid hourly, use your hourly rate to calculate the daily rate and multiply that by the number of days off. This can all be done in one step, as shown.

Example: Company A offers $25 an hour with 15 paid days off per year. Workdays are 8 hours.

$\mathbf{(8 \times \$25)} \times \mathbf{15} = \$3,000$

Now, there is also a non-monetary value to paid time off. You may decide you want a job with more time off, even if monetarily it pays less overall than a higher-paying job with less time off.

401k and Retirement Plans

Some companies give you money toward your retirement and offer an account called a 401k where you can invest money tax-free. Some employers will also offer to match contributions you put in this 401k account up to a certain percent of your salary — meaning if you put in $1, they will put in $1 too. The amount the company will match varies.

>>> If the employer matches $1 for $1 you contribute, they will give the same amount you contribute; if they match $0.50 per $1 you contribute, that means they are giving $\frac{1}{2}$ of what you contribute. There is usually a maximum they are willing to give that you can calculate by finding a percent of your salary.

If you are offered a 401k with employer matching, try to take advantage of it by contributing at least the amount they are willing to match. If you are able to contribute more, there is a tax benefit to doing so, up to the federally set limit. Adjust how much you are putting in your 401k if your salary changes midyear.

Example: Your employer matches $1 for $1 up to 2% of your salary and $0.50 for every $1 for up to an additional 1%. Up to how much will the employer contribute if your salary is $86,000?

2% of salary: 0.02 × $86,000 = **$1,720**

1% of salary: 0.01 × $86,000 = **$860**

Maximum employer matching:

2% + $\frac{1}{2}$ *of* 1% = **$1,720** + (0.5 × **$860**) = $2,150

Insurance Packages

>>> Calculate the monetary value of insurance benefits the employer offers by seeing how much the *employer* pays for the insurance.

Example: Every biweekly paycheck, Noah's company is paying $200 for health, $50 for dental, and $10 for vision insurance. How much is his company giving him yearly in insurance benefits?

Total per biweekly pay period:	$200 + $50 + $10 = **$260**
Total per year:	26 × **$260** = $6,760

26 is the number of biweekly paychecks in a year. See Lesson 7 if you need reminders on biweekly and bimonthly.

Now, the monetary value is not the only thing to look at.

- What is *your* cost for the insurance for you and, if applicable, your family? Some companies only cover a portion of the insurance cost or do not cover family members.
- Do you like all insurance options?
- Does the job include all the insurance policies you want?

KEEPING PERSPECTIVE

While we are looking at comparing benefit packages between jobs, there is more to picking a job than how much you can make in it. When looking at both careers and specific jobs, evaluate whether or not you can use the way God has wired you to serve Him, bless others, and provide for yourself and your future family.

Comparing Jobs:
Total Compensation & Cost of Living

We saw in the last section that jobs differ in benefits. They also differ in how they pay (such as with a salary, hourly, or by job). **To compare pay, look at something called your total compensation.**

- Your **base pay** is how much you expect to make in a period of time (like a year).
- Your **total compensation** is the sum of your base pay and all the employer is giving in benefits. This pay may be listed on some documents, or you can calculate it yourself.

When calculating the total compensation, we will use only what the *employer* contributes for the 401k and insurances. However, there is an additional tax savings of 401k accounts that we are ignoring here.

Example: Job A offers a $50,000 base pay, 14 days off a year, a 401k match of up to 1% of the base pay, and health insurance that costs the employer $240 a month. Job B offers a $60,000 base pay but no paid days off or 401k. Their insurance contribution is $100 a month. Calculate the total compensation for each job.

Job A

Base pay: **$50,000**

Time off: $\frac{\$50{,}000}{260} \times 14 \approx \$2{,}692.31$

401k: $(\$50{,}000 \times 0.01) = \$500$

Insurance: $\$240 \times 12 = \$2{,}880$

Job B

Base pay: **$60,000**

Insurance: $12 \times \$100 = \$1{,}200$

Total compensation: $\$60{,}000 + \$1{,}200 = \$61{,}200$

Done all as one step: $\$60{,}000 + (12 \times \$100) = \$61{,}200$

Total compensation: $\$50{,}000 + \$2{,}692.31 + \$500 + \$2{,}880 = \$56{,}072.31$

Done all as one step: $\$50{,}000 + (\frac{\$50{,}000}{260} \times 14) + (\$50{,}000 \times 0.01) + (\$240 \times 12) \approx \$56{,}072.31$

≫ Try to solve problems as one step when you can.

Cost of Living

If you are comparing jobs in different areas, be sure to factor in the cost of living in that area. A salary of $80,000 is not going to go as far in New York City as it would in a small town.

Account for any moving costs if they are not covered by the new company.

To compare jobs in different areas, you will want to look at the salary after it is adjusted for cost of living. The company offering you the job might give you that information, or there are "cost of living" calculators online (available with a parent's permission). There are also resources that tell you the % difference in cost of living between two locations, but you need to be careful with these. Some expenses, such as your federal income taxes, do not change when you move. Just because one city has a 20% lower cost of living does not mean that a salary that is exactly 20% less there is comparable. So do not just subtract 20% from your salary to find a comparable one.

Estimates – Cost of living estimates are based on average spending. The actual difference might be slightly more or less for you, but the estimates can give you a general idea.

»» Compare salaries *after* adjusting for cost of living to see how salaries compare.

Virtual Jobs – If you have a virtual job, you can technically live anywhere. Living in a lower cost of living location will free up more of your money for giving and saving.

Example: After adjusting for the cost of living, which job has the greater total compensation?
Job A: $30,000 ($40,000 if adjusted for the cost of living in Job B's location)
Job B: $35,000

$40,000 – $35,000 = $5,000

When adjusting for cost of living, Job A has the greater total compensation by $5,000.

Non-financial Job Considerations

Now, there is a lot more to look at in a job than just the dollars. Here are a few non-financial considerations.

- How much time does the job take to complete?
- Does it have hours that still allow you to be there for your family/other callings on your life?
- How much flexibility does it have?
- How much overtime?
- What is the commute like?
- Where is the job located? Is that where you feel called to live?
- Is it near a good church?
- Is the job one you can do in good conscience, or will you be asked to do things that go against your conscience?
- What is the management like? Is it an environment in which you think you will enjoy working?
- Is it the type of work you enjoy?
- Will the job lead to your future goals?

KEEPING PERSPECTIVE

Keep in mind that whatever job you pick, it is an opportunity to serve the Lord. Whether you work in a secular field or not, your *work itself* is a ministry. And *what you make* can minister too as you give it away.

Self-Employment Income

Now we will take a brief look at some avenues for self-employment income, be that as a supplement to a job or as your main occupation.

Selling Products

There are lots of different types of products that can be sold — and different ways to sell them, such as online, through venues, and via retailers. If you are able to sell items for more than they cost you to make, you are making a **profit** on that item. If you make enough of a profit on all your items put together to cover all your other expenses, then your business itself makes a profit.

>>> On products, *profit = selling price − costs.*

Example: Jill spends $10 in yarn to crochet a baby blanket. She sells the blanket for $45. Calculate her profit.

$45 − $10 = $35

In addition to calculating your profit on a product, it can also be helpful to figure out the **profit margin**, or percent *of* the selling price the profit *is*, for a particular product.

$$\text{profit margin} = \frac{\text{profit}}{\text{selling price}}$$

The profit margin equation is really expressing what we end up with when we calculate the % *of* the selling price the profit *is*.

$$\% \times \text{selling price} = \text{profit}$$

$$\text{Divide both sides by selling price: } \% = \frac{\text{profit}}{\text{selling price}}$$

Replace % with "profit margin" and you have the profit margin equation. **A lot of financial rules are really just customizations on what you already know.**

Example: What profit margin did Jill have in the previous example?

$$\frac{\$35}{\$45} \approx 0.7778 = 77.78\%$$

Jill could have included a fee for her time into the cost of the blanket. But sometimes it can be difficult to track time by product.

A 77.8% profit margin means that 77.8% of each sale is profit. That sounds good, except keep in mind that profit needs to cover Jill's time in making the blanket and general expenses Jill has, such as a website or booth fee to sell the blanket at, credit card fees for taking payment, accountant fees, and more.

Be sure to also evaluate what people will pay. It is not good to make and price a product based on the profit you want alone. You also need to see what similar products cost and make sure yours is priced so that it will sell. If you can not price it to sell *and* make a satisfactory profit, you might want to rethink the product.

Selling Services or Renting

You can also generate self-employment income by selling services or renting out a basement or room.

>>> To calculate how many days it will take you to make money (i.e., turn a profit), determine your total costs and divide that by your anticipated earnings per day. You may have some other math to calculate these values.

Example: Josh figures that in order to get his basement ready to rent, he needs to buy a special lock for $100, another bed for $800, and a few other items for $400. Josh is listing his finished basement online for $60 a day. The website he is using charges 3.5% as their fees. How much will he make per day and after how many days will he turn a profit?

Percent Josh will keep: 100% – 3.5% = 96.5% = **0.965**

Anticipated earnings: **0.965** × $60 = **$57.90**

Total costs: $100 + $800 + $400 = **$1,300**

\# of nights needed to turn a profit: $\frac{\$1,300}{\$57.90} \approx 22.45$

23 days (rounded up, since he will not make more than his expenses until the 23rd day)

Good to know:

- According to many sources, most businesses take at least 2 years to actually be where you can earn a decent amount of money from them. Some find their side business is more of a fun hobby that makes a little side money.
- Self-employment requires self-motivation. You are your own boss — which is great — but it also means you have to motivate yourself.
- Self-employment requires bearing the financial risk. You only make money if you turn a profit. You will likely need to trust God for your needs in very real ways at times.
- Be sure to look into business structure, taxes, and licenses before you begin.

KEEPING PERSPECTIVE

- When setting prices, keep in mind the Bible's call to "Let all that you do be done in love" (1 Corinthians 16:14). Do not charge others more than your services are really worth.
- Make sure your prices are fair and clear in how you present them. The Bible talks about not having a false balance; "a just weight is his delight" (Proverbs 11:1). False balances deceived people into thinking they were getting a different deal than they really were.

And if you make a sale to your neighbor or buy from your neighbor, you shall not wrong one another (Leviticus 25:14).

Review & Case Study: Redirected (My Career Journey)

Hopefully this lesson has given you a few things to think about and equipped you with how to evaluate the financial benefits of different training options and different careers.

Always keep in mind, though, that it is not all about the money. In God's eyes, a janitor has an equally important job as a CEO. Jesus, God in the flesh, did not come "to be served but to serve" (Matthew 20:28). He *chose* to be born in a manger, not a palace — to be born poor, not rich. May your desire be to follow His lead, being content with whatever career (and related finances) God calls you to. If you are blessed with riches, may it be to build up His Kingdom and bless others, not to rely on your own riches and fail to depend on Him for your every need.

> *Remove far from me falsehood and lying; give me neither poverty nor riches; feed me with the food that is needful for me, lest I be full and deny you and say, "Who is the Lord?" or lest I be poor and steal and profane the name of my God* (Proverbs 30:8–9).

As you consider a career, look for one you can be passionate about. Also think about how that career would work with your future family, if having a family is something in your heart. Work hard at what you decide to pursue. But keep your priorities straight and make your aim building God's Kingdom and honoring Him. The case study in this lesson is my own story about how God directed and redirected me in ways I did not imagine; I hope it encourages you to keep your eyes on Him and let Him write your story.

Case Study: Redirected (My Career Journey)

As a teen, I thought I wanted to be a journalist. Then I talked to a journalist, and discovered that journalists have to write fast, which did not fit me. I thought maybe I would give tours of D.C., only we had just moved to the area and thus I did not meet the residency requirements. I was not even sure I wanted a career since I assumed I would get married and be a mom, but Mr. Right was nowhere in sight. I thought of tutoring math and even got students lined up, but they all canceled. I really was not sure what God had for me.

The summer after I graduated, God opened a door for me to learn Spanish. I threw my heart into it, little dreaming that God would use that to guide me to what He had next. Writing a devotional for the sons of my Spanish teacher led to creating a website and selling it to others. My online ministry had been born. I also began teaching her sons piano, and discovered I loved teaching. Soon I had other students too.

At the same time, others encouraged me to start writing about math. I had never pictured myself as a math author, yet seeing how math described God's consistent creation and pointed to the Creator was something I did not want students to miss. I began my first book, saying I would never write a curriculum. But God had other plans. Many curricula and other jobs (such as working at nonprofits and even a bakery) later, I have watched God faithfully close and open doors, redirecting my steps as I put one foot in front of the other. Along the way, I also got a degree for less than $20,000 and no debt, putting into practice some of the saving tips covered in this lesson.

If I could go back and talk to the recently graduated me, I would encourage myself to stop worrying and trust that God really did have a plan. God knows what He has created us to do, and "he will make [our] paths straight" as we trust Him (Proverbs 3:5–6). He guides step by step.

Housing, Part 1

While you probably do not have to worry about paying for housing right now, housing is a big expense in most adults' budgets. Here, we will explore some key principles you will need when it comes time to get your own housing and see how basic math can help you with the process.

*income = giving + saving + **spending***

spending** = taxes + debt + **living expenses

Comparing Rentals

As you get trained and started on a career, you probably will not have finances to buy a house right away. Following are a few general things to look at when deciding on a rental:

Inclusions

Sometimes, rent will include utilities; other times, all utilities (or just certain utilities) will need to be paid for separately. When comparing rental prices, **compare the total costs**. Basic addition can help you total the costs.

Remember, with a parent's permisson, you can use a spreadsheet to help simplify comparisons like this.

Example: You are considering renting an apartment while in college. Calculate the monthly expenses for utilities and rent at two different apartments.

Apartment A

Rent: **$800** per month

Includes: electricity, water, and trash

Internet: **$59.99** a month

Total for Apartment A:
$800 + **$59.99** = $859.99

Apartment B

Rent: **$750** a month

Electricity: **$50** a month

Water: **$100** a month

Trash: **$10** a month

Internet: **$59.99** a month

Total for Apartment B:
$750 + **$59.99** + **$50** + **$100** + **$10** = $969.99

If needed, the skills we looked at with budgeting can help you calculate what the cost would be for a different length of time so as to **compare the cost for the same time periods**.

Example: If a dorm is $3,500 a semester and renting your own small home is $1,000 a month, but you have to sign a year lease, compare the yearly rent if you attend school 2 semesters that year.

Rent of dorm per year: 2 × $3,500 = **$7,000**

Rent of home per year: 12 × $1,000 = **$12,000**

We are figuring out yearly costs to compare them, applying the same skills as when finding yearly amounts for a budget.

Sometimes, you might have a few more steps to the comparison.

> ***Example:*** In the situation from the previous example, if the home rental also requires car costs of $300 a month and parking of $600 a semester, how much more is it than the dorm over a year?
>
> Other costs per year: (12 × 300) + (2 × $600) = **$4,800**
>
> Total cost per year of home rental: **$12,000** + **$4,800** = **$16,800**
>
> Amount home costs more per year: **$16,800** – **$7,000** = $9,800

We first found the yearly cost of these other costs so as to compare the total costs *over the same time period* (a year here).

Contract Length

Rent is for a specified period of time; at the end of that time, either you *or* the property owner can choose to stop renting the property. Some rentals are month to month; others might require a year or more commitment. Some have ways to get out of the contract early but for a fee. Make sure you understand and are comfortable with the commitment you are making.

> ***Example:*** Joan needs to rent a place for a few months while her home is being built. Which apartment will be cheaper and by how much if the home is completed in 5 months?

Apartment A	Apartment B
$1,700 a month	**$1,600** a month
no commitment	6-month commitment, or pay extra month's rent

> 5-month cost at Apartment A: 5 × **$1,700** = **$8,500**
>
> 5-month cost at Apartment B: 5 × **$1,600** + **$1,600** = **$9,600**
>
> Comparing costs: **$9,600** – **$8,500** = $1,100
>
> Apartment A is cheaper by $1,100.

We added an extra month's payment to Apartment B, as that is the fee at this apartment for leaving early. If Joan's home runs late, Apartment B would actually be cheaper. It gives her flexibility in moving too.

KEEPING PERSPECTIVE

As with everything else, there is more than just the financials to consider when picking a place to stay. Look both at the total costs (including utilities) and at what works for you in terms of other factors, such as the living situation, contract length, location, the time it will take you to get to college classes and work, and the proximity to a good church.

Sharing Rent & Saving to Buy

Rent can get expensive. One option to help with rent cost is to share a home with others.

>>> **To see the cost if you share lodging with others,** divide the costs you are sharing by the number of people. Do not divide any costs you will not be sharing.

Be careful about who you decide to share a home with, especially if getting a joint lease (see below).

Example: You could share a $1,000 a month home rental with 3 other students (4 of you altogether). If your car and parking fees of $4,800 a year would *not* be shared, what would your cost be per year, provided you rented the home all 12 months?

Yearly rent: $12 \times \$1{,}000 = \$12{,}000$

Your cost: $\frac{\$12{,}000}{4} + \$4{,}800 = \$7{,}800$

We could have done all this in one step:
$\frac{12 \times \$1{,}000}{4} + \$4{,}800 = \$7{,}800$

This same scenario costs $12,000 + $4,800 = $16,800 a year when not shared — sharing can save a lot. When signing a lease with others, though, think through if you want an individual or joint lease.

- An **individual lease** is an individual contract. If one of your roommates does not make their payments, you are not liable, nor do you have to find a replacement.

- A **joint lease** is a group contract. You are responsible as a group for the payments. If a roommate bails, the rest of you have to cover that amount.

Example: You sign a joint lease for 4 years with 3 other friends. The lease is for $1,000 a month, and each of you pays $250. After the first year, 1 of your friends drops out of college. How much will the remaining 3 of you have to pay until you find a 4th roommate?

Divide total amount of lease by 3: $\frac{\$1{,}000}{3} \approx \$333.33$

Saving to Buy

The key difference between renting and buying is whether or not any of your monthly payment goes toward future ownership of the home (and value in the home if you were to sell). At the end of years of renting, you will still owe rent; at the end of a mortgage, you will own your home.

The tricky part is that homes are expensive to buy. The median (think middle) home price in America in some sample years was in the $400,000s.[39] That is a sizeable amount of money to have saved, which makes home mortgages so popular.

Mortgages are loans with property that can be claimed by the lender if the borrower does not make their payments.

Also, the amount needed to buy a home is a sizeable amount of money for a lender to allow someone to borrow. As a result, to get most types of loans, borrowers have to pay some toward the purchase price themselves rather than taking a loan for the full amount. What the borrower pays up-front is called a **down payment**.

»» The down payment is often expressed as a percent of the home price, and you will typically save money on your mortgage if you can put at least 20% down on a home.

Example: What would a 20% down payment be on a \$430,000 home?

$0.20 \times \$430{,}000 = \$86{,}000$

It takes \$80,000 to make a 20% down payment on the average-priced home in America. It is important to start saving for a future home as soon as you are able.

When buying a house, keep in mind that the more you put down as a down payment, the less the loan amount will be, meaning the less you will pay in interest over time.

»» **The loan amount is found by subtracting the down payment from the total.** You are getting a loan for the *difference* between what you owe and what you pay.

Example: What is the loan amount if you buy a home that costs \$430,000 and you put \$86,000 down?

$\$430{,}000 - \$86{,}000 = \$344{,}000$

You can also use what you know about percents to calculate the percent you would be putting down based on what you have saved.

Example: You are buying a home that costs \$430,000. You have enough to make a down payment of \$86,000. What percent down payment would that be?

% *of* \$430,000 *is* \$86,000: $P \times \$430{,}000 = \$86{,}000$

Divide both sides by \$400,000: $P = \dfrac{\$86{,}000}{\$430{,}000} = 0.20 = 20\%$

KEEPING PERSPECTIVE

Notice that we are continuing to use basic math skills we reviewed early on in the course. Percents, for example, not only help with things like taxes and tips but also with exploring down payments. Remember as you learn about buying homes that it is critical to seek wise counsel and pray through these (and other) decisions.

Affording Housing

Whether buying a house or renting, one big question that needs to be settled is, "How much can I afford?" It is a question that needs answered with lots of prayer and an honest look at the resources God has given you.

Looking at Income

30% of Gross Income – Many suggest not signing up for monthly housing payments (mortgage or rent) that are more than 30% of your gross income.

Example: Use the 30% of gross income guideline to estimate the maximum monthly housing payment someone making $95,000 a year could potentially afford.

yearly gross income

30% of monthly gross income: $0.30 \times \left(\frac{\$95{,}000}{12}\right) \approx \$2{,}375$

monthly gross income (yearly divided by 12)

25% of Net Income – Others suggest not taking on more than 25% of your net income.

Example: Use the 25% of net income guideline to estimate the maximum monthly housing payment of someone whose paycheck deposit is $2,916.67 biweekly.

yearly net income (26 biweekly paychecks)

$$0.25 \times \left(\frac{26 \times \$2{,}916.67}{12}\right) \approx \$1{,}579.86$$

monthly net income (yearly divided by 12)

Multiplying by 26, as 26 biweekly paychecks are in a year.

These percentages are rough suggestions. We will look at how to calculate a more accurate figure later.

Looking at Savings

FORECLOSURE

It is important to not take on more of a housing commitment than you really can afford. If you do not make your payments (meaning you **default** on your loan) for a certain amount of time, the lender can **foreclose** your home — that is, legally make you leave the home. If renting, the landlord can kick you out of their home if you do not pay your rent.

It is very important not to overextend financially when it comes to your housing choice. Not only is it important to keep your commitments as a Christian, but it also is not a good financial situation to get yourself into a monthly payment you can not afford. Failing to pay negatively affects your credit score and your ability to take out another loan or to have another landlord accept you.

The wicked borrows but does not pay back, but the righteous is generous and gives (Psalm 37:21).

Your income is not the only thing that determines how much you can afford in housing. Your savings also plays into it.

- **If you are renting**, consider the need to save up so you can afford a home one day. Make sure your budget includes savings along with your rent.
- **If buying a house**, consider how much you have already saved, as that affects what you can buy.

The amount you have saved affects the down payment you can make, and putting at least 20% down saves significantly. How would you calculate how expensive of a home you could put 20% down on with your savings? You would need to solve for the *base* in the percent equation. The example illustrates this.

> ***Example:*** You want to be able to make a 20% down payment on a home. You have \$80,000 saved for a down payment. What is the maximum home price you could do this for?
>
> % *of* ***base*** *is* ***percentage***
>
> 20% *of* base *is* \$80,000: $0.20 \times \textit{base} = \$80{,}000$
>
> Divide both sides by 0.20: $\textit{base} = \frac{\$80{,}000}{0.20} = \$400{,}000$

You can check yourself by seeing if the percent you were given of the value you found does indeed equal the other value.

> ***Example:*** Check the previous problem.
>
> $0.20 \times \$400{,}000 = \$80{,}000$
>
> Plugging this into a calculator, we see that this is true; we did the math correctly.

Beware of using up up all of your savings to make a down payment. Consider keeping some for emergencies and the unexpected (or expenses you forgot about), as well as for closing costs and moving expenses.

KEEPING PERSPECTIVE

While buying a home may be a long way off for you, understanding how much is required can help you think about starting to save now.

Home Mortgages & Owner's Equity

Now we will examine how mortgages work and how they can help you slowly gain money rather than spending it all on rent.

To review from Lesson 3, each time you make a payment on a loan, a portion of each payment goes to pay interest, and a portion goes to pay off the principal on the loan. Thus, each payment reduces the principal (i.e., what you owe).

This is all shown on an **amortization schedule** (a list of all the payments on a loan and what they are going toward).

Length of Loan:	30 yrs.
Interest Rate (yearly):	4%
Loan Amount:	$100,000
Monthly Rate $\left(\frac{\textit{yearly rate}}{12}\right)$:	$\frac{0.04}{12} \approx 0.0033333$

1st payment — interest is calculated off $100,000
2nd payment — interest is calculated off $99,855.92

Month	Payment	Interest (*monthly rate* × *loan amount* [1st month] or prior month's *loan balance*)	Principal (*payment – interest*)	Loan Balance (*loan amount* [1st month] or *loan balance – principal*)	Cumulative Interest (sum of all *interest* paid so far)
1	$477.42	$333.33	$144.09	$99,855.91	$333.33
2	$477.42	$332.85	$144.57	$99,711.34	$666.19
3	$477.42	$332.37	$145.05	$99,566.29	$998.56

Given that the interest rate on this loan is 4% and the initial loan amount $100,000, you might expect $0.04 \times \$100{,}000 = \$4{,}000$ in cumulative interest at the end of the first year. Instead, it is slightly less than $4,000 (it is $3,967.95) because **as the principal gets smaller after each payment, the interest becomes less, and more of the payment goes to paying off the principal**.

Amortization schedule tools are plentiful online (but be sure to get a parent's permission to search if you want to check them out).

The length of a home mortgage can vary, although 30-year and 15-year home loans are common.

- **The shorter the time period, the less you will end up paying in interest**, as you are borrowing the money for less time.
- **The higher the monthly payments, the shorter the time period**, as you are repaying the loan faster.

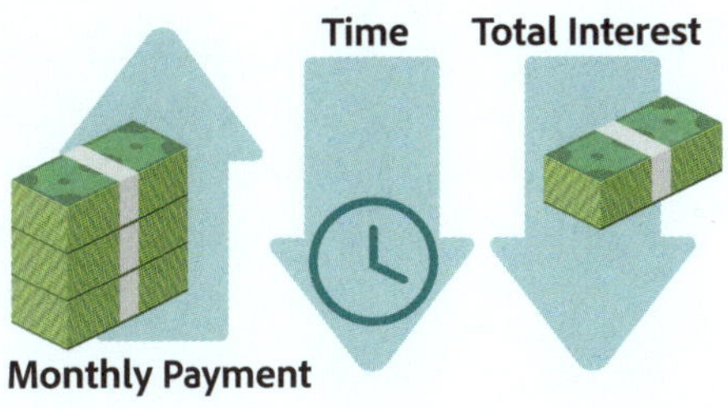

You do not have to keep a home for the entire mortgage length. If you sell the home, you have to pay off any remaining principal. The rest of the money is considered **owner's equity** — the amount that belongs to the seller once the loan is paid off. The owner's equity grows a little bit with each payment, provided the home value stays the same.

>>> The owner's equity is the *difference* between what the home is worth and what you still owe, or
owner's equity = home value – principal on loan.

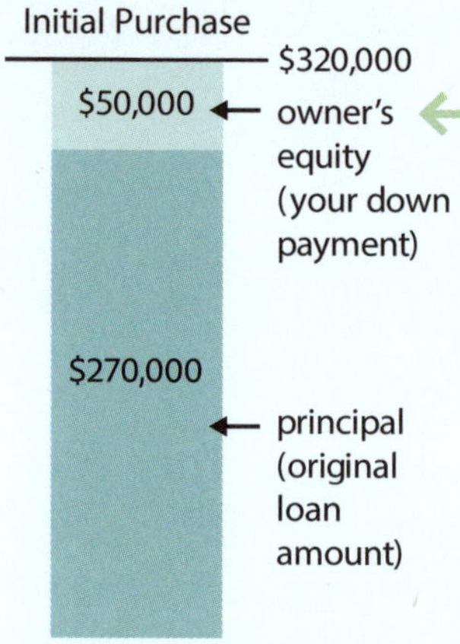

> ***Example:*** You buy a house for $320,000 with a $50,000 down payment and a $270,000 loan. What is your initial owner's equity in the home?
>
> *owner's equity* = $320,000 – $270,000 = $50,000

Notice from the example that, initially, the owner's equity is just the down payment. As time goes on, though, the home may gain or lose value. The loan principal gets paid down as you make payments. Consequently, the owner's equity in the home changes.

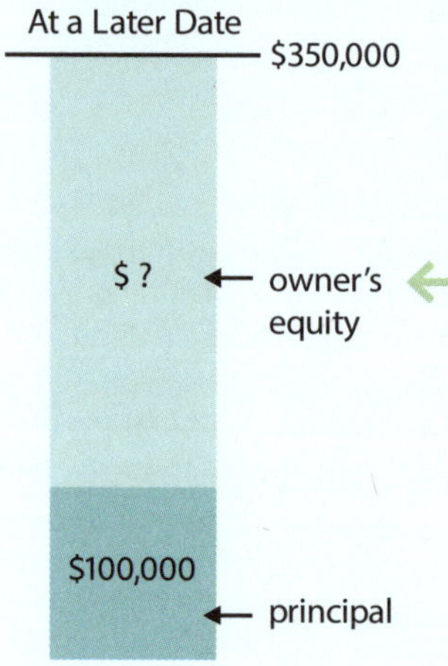

> ***Example:*** Your house is now worth $350,000 and the principal on the loan (i.e., what you still owe) is now $100,000. What is your current owner's equity?
>
> *owner's equity* = $350,000 – $100,000 = $250,000

Good to know:

- The loan amount does not change as the house gains or loses value; all the gain or loss gets reflected in the owner's equity. This means you have the potential to gain money if your home gains value, or to lose money if it does not.
- The loan amount (i.e., principal) *does* get paid off over time, making more of the home value owner's equity. Some of your monthly housing expense each month is *building* value for you.

KEEPING PERSPECTIVE

While car loans are for cars that *lose* value quickly, home mortgages are for houses that you typically expect to maintain or gain value. Even if the house loses value, there is usually enough money you have paid into the house to cover the loss. Thus, if you had to sell unexpectedly, you would expect to be able to pay back the loan, unlike with a car loan. Still, keep in mind that the less you have to borrow, the less in interest you pay and hopefully the sooner you can get out of debt.

Getting Qualified & Approved for a Loan

If you were about to lend hundreds of thousands of dollars to someone, you would want assurance that they would pay you back, right? Mortgage companies want to know that they will get their money back.

- It is common to get a prequalification or preapproval letter from a lender *before* you make an offer to buy a home. This shows the seller you should be able to get a loan.
- After buying a home, you will then have to submit more documentation to get full approval.

Now we will explore some of the key things lenders look at. Many of them are things to keep in mind as you make decisions now so that you are in a good situation to buy a home later.

- **Credit Score** – A good credit score is important — it shows that you have a history of borrowing money and paying it back as promised. We looked at how to check and increase your credit score back in Lesson 3. (See pp. 61–62.)
- **Income** – If you say you are going to make monthly payments, a lender wants to know that you have the income to make good on that. They will look not only at how much you make but also at how steady your employment is.
- **Debt-to-Income (DTI) Ratio** – How much debt do you already have in comparison to your income? This is assessed by something known as your debt-to-income ratio. For this ratio, include the minimum payments required on each credit card as part of your total monthly debt, even though you are hopefully paying off all the cards each month.

>>> *Debt-to-income ratio (DTI)* $= \dfrac{\textit{total monthly debt}}{\textit{total monthly income}}$

Example: Joan currently pays $1,000 in rent, has minimum payments of $50 combined for her credit cards, and owes $350 a month in student loans. Her income is $42,000. Her total debt is $200,000. What is her DTI ratio?

Monthly income: $\dfrac{\$42,000}{12} = \$3,500$

$$DTI = \frac{\textit{total monthly debt}}{\textit{total monthly income}} = \frac{(\$1,000 + \$50 + \$350)}{\$3,500} = 0.40 = 40\%$$

- **Net Worth** – How much money do you have in the bank? How many other assets do you have that you could sell to make a payment? Your **net worth** is the sum of all that you have minus what you owe. A more technical way of saying this is that it is your assets minus your liabilities. **Assets** include anything you have of value, be it money in a bank or 401k, the value of your existing home, cars, expensive jewelry, or a valuable antique. **Liabilities** are anything you owe, be that a loan or your credit card balances.

>> *net worth = assets – liabilities*

A **negative net worth** indicates that you owe more than you have.

Example: If Peter's total assets are $120,000 and his total liabilities are $200,000, what is his net worth?

net worth = $120,000 – $200,000 = –$80,000

Important! **The name "net worth" is misleading.** Your true worth does not depend on what you own at all. You are priceless because your Maker made you in His image and considered you worth dying for. Always remember that.

Good to know:

- **Lenders want a low debt-to-income ratio.** The exact maximum a lender will take varies, but under 36% is generally accepted.
- **If you budget using a a computer or online program or app** (with a parent's permission), it may calculate your net worth for you automatically.
- **Beware of thinking you can really afford what a lender preapproves you for.** Lenders are not factoring in giving — they are looking at the maximum you can afford if you extended your finances to the fullest. A wise loan amount is typically *much* less than what you can get preapproved for.

Getting a Good Rate

While the general rate of interest is influenced by the Federal Reserve's rates, **interest rates offered by lenders still vary some based on how eager a lender is to lend to you**.

And what makes them eager? They want to make a profit off their loan. They want assurance that they will get the interest they are setting out to get, paid promptly and without a hassle. The more your finances assure them you will keep your end of the loan, the more likely they are to give you a loan at a good rate.

KEEPING PERSPECTIVE

Understanding what lenders look at when deciding if they will lend to you (and at what rate) can help you prepare for buying a home. All these things can help you qualify for a home mortgage:

- Build your credit score by paying off every credit card or debt as promised.
- Maintain a steady job.
- Avoid taking on other debt.

Types of Mortgages & Mortgage Insurance

Now we will take a look at some different types of mortgages so you are familiar with them when you encounter them down the road and understand a little better why it is helpful to save up.

Conventional Loans

Privately funded loans are called **conventional loans**. These loans work on a free-enterprise system; companies offer loans for the purpose of making money.

- **If the down payment is less than 20%**, conventional loans require borrowers to pay for **private mortgage insurance (PMI)**, an insurance that covers the *lender* should the borrower not pay as promised.
- **You can stop paying PMI after your owner's equity gets up to about 20% or after you reach halfway through your loan.**

There are numerous online mortgage insurance calculators you can use to find out how much you would be looking at for a specific scenario. If you want to explore them, ask for a parent's permission and help in assessing them.

PMI is charged as a percent of the loan amount. **The higher the buyer's credit score and down payment, the lower the PMI costs.** After all, the better the credit score and down payment, the more confident the insurance company is that the borrower will make the payments.

> ***Example:*** Calculate the PMI cost *per year* on a $415,000 loan if the PMI rate is 1%.
>
> 1% *of* $415,000: 0.01 × $415,000 = $4,150
>
> That is a lot of extra to pay *each year.*

>>> Due to PMI, you save significant money if you can put 20% down on a conventional loan.

Categories of Conventional Loans

Every year, a federal agency (FHFA) set guidelines on mortgages, including a limit on the amount of money thought reasonable to lend for a home loan. Two companies overseen by the government also set certain other requirements, such as a low enough debt-to-income ratio. Lenders can either conform or not conform to these guidelines.

- **Conforming loans** – Follow guidelines.
- **Non-conforming** – Do not follow guidelines. Charge more in interest as a result to make up for the extra risk in not following guidelines.

An example of a non-conforming loan is a **jumbo loan**. In a jumbo loan, consumers can borrow more money than the suggested maximum. In a sample year, the limit for conforming loans was $806,500 (and was even higher in certain areas).[40] That is a lot of money to borrow, and it would allow for a house that costs even more, as it does not include the down payment.

Non-conforming loans are risky. They allow consumers to either borrow a lot of money or borrow more than their finances indicate they should. Many people have gotten in trouble with these loans, overextending themselves to buy a bigger, better home in a high-cost situation. But if housing prices then plummet, they are stuck paying for a home. And if they need to sell it, they may not be able to sell it for enough to pay back their loan.

Government-Backed Loans

What do you do if you can not qualify for a conventional loan? You can wait until you can, or government-backed loans are designed for those in financial need or specific situations. In these loans, the government backs the loan. The table highlights three types of government loans.

FHA Loan	Must pay **mortgage insurance premium** (**MIP** — similar to PMI; it is insurance for the lender) for the first 11 years (longer if <10% down); for those who can not afford conventional loans.
USDA Loan	Helps low-income buyers purchasing in certain areas (especially rural) at a lower rate and with no down payment or insurance.
VA Loan	Helps those in the military and veterans buy a home. No down payment. May have to pay a guarantee fee.

Even though these loans do not require as much of a down payment (if one at all), you should generally still put down what you can. The less you are borrowing, the less you are paying in interest.

Fixed- or Variable-Rate Mortgages

Mortgages are also categorized by whether their interest rates are fixed or variable.

- **Fixed-rate mortgage** – A mortgage where the interest rate stays the same throughout the loan. We are focusing on fixed-rate mortgages in our examples.
- **Variable-rate mortgages** – The interest rate (and payment) changes as interest rates in general go up or down. These mortgages are risky since you could start paying 2% interest and wind up paying 15%.

KEEPING PERSPECTIVE

While there is a lot to mortgages, here are two important takeaways:

- **Save while you are young** so you are able to put more down on a home.
- **Beware of loans that sound too good to be true (like jumbo loans) — you might regret them later.** Make sure you fully understand the risks.

Points & Lender Credits

A **point** is money the borrower pays up-front to lower the interest rate. One point lowers the interest rate by a certain percentage (typically 0.25%) and has a cost (typically 1% of the loan[41]).

>>> Find the benefit of a point by subtracting the % it lowers the rate from the interest rate, and the cost by multiplying to calculate the appropriate % of the loan.

Benefit of 1 point: lower interest rate (typically by 0.25%)

Example: Interest rate of 4% lowers to 3.75%, as 4% − 0.25% = 3.75%.

Cost of 1 point: % of loan (typically 1% *of* the loan)

Example: 1 point on $450,000 loan costs $4,500, as 0.01 × $450,000 = $4,500.

Rather than buying an entire point, you can also buy a portion of a point.

>>> To calculate a partial point's benefit or cost, first calculate the benefit or cost of a point, and then figure out that portion of it.

Example: What does $\frac{1}{2}$ (i.e., 0.5) of a point on a $300,000 loan cost?

Cost of 1 point: 1% *of* $300,000 = 0.01 × $300,000 = **$3,000**

Cost of $\frac{1}{2}$ point is half of the cost *of* 1 point:

$\frac{1}{2}$ × **$3,000** = 0.50 × **$3,000** = $1,500

Double check with the lender if you go to take out a loan. The Consumer Financial Protection Bureau warns that some lenders use points to refer to money that does not affect interest rates and suggest that you ask how any points offered affect the interest rate.[42]

Notice we listed the portion of a point in both fractional and decimal form. That is because 0.5 points is often referred to as *half* (i.e., $\frac{1}{2}$) of a point, but it might be written out as 0.5.

Example: To what rate will buying $\frac{1}{2}$ (i.e., 0.5) of a point lower an interest rate of 7%?

Benefit of 1 point: **0.25%**

Benefit of $\frac{1}{2}$ point: $\frac{1}{2}$ × **0.25%** = 0.5 × **0.0025** = 0.00125 = **0.125%**

Subtract the benefit of $\frac{1}{2}$ point from interest rate:

7% − **0.125%** = 6.875%

- Be careful when rewriting %s as decimals. 0.25% becomes 0.0025 — we divided by 100 by moving the decimal two places to the left.
- Also be sure to subtract the reduction to the interest rate *after* rewriting it in % form.

Deciding if a Point Is Worth Buying

Whether points are worth buying depends on your situation. It takes time for the savings on interest due to a lower rate to add up to the cost of the point.

- If you sell your home quickly, a point probably will not pay for itself.
- If you plan to keep your home the entire length of the mortgage, it will save you money.

>>> **To see how long you would have to keep a mortgage for a point to pay for itself,** compare the cumulative interest columns from two different amortization schedules:

- one at the interest rate with the point
- one at the rate without the point.

Loan: $300,000
Length: 30 yrs.
Cost of Point: $3,000

A	B	C	D	E
Month	Cumulative Interest Without Point (4%)	Cumulative Interest Without Point (3.75%)	Difference (B–C)	
48	$ 46,288.48	$ 43,328.16	$ 2,960.32	Reached a savings in interest of $3,000, the cost of the point. If home is kept 49 months (about 4 yrs.) or longer, the point will save money. If not, you will not recuperate its cost.
49	$ 47,213.61	$ 44,192.66	$ 3,020.95	
50	$ 48,137.06	$ 45,055.52	$ 3,081.54	
360	$ 215,608.52	$ 200,164.84	$ 15,443.68	Total Savings

Look at where the difference in cumulative interest is more than the cost of the point. From there forward, you have saved more than the point cost. By the end of the loan (month 360), you would save $15,443.68 in interest if you bought the point.

Lender Credits

A **negative point** or **lender credit** is money the lender gives you to help pay your closing costs in exchange for *raising* your interest rate. If you are stretched for cash and need help with closing costs, buying a lender point might make sense.

Downside of 1 negative point or lender credit: *higher* interest rate (use 0.25% higher in this course)

Example: Interest rate of 4% *increases* to 4.25%, as 4% + 0.25% = 4.25%.

Benefit of 1 negative point or lender credit: lender gives you % of the loan for closing costs (use 1% in this course)

Example: 1 point on $450,000 loan results in $4,500 of closing costs paid by lender, as 0.01 × $450,000 = $4,500.

KEEPING PERSPECTIVE

You may have noticed that there are a lot of decisions to be made when getting a mortgage. It can all seem overwhelming. Remember that "The Lord is my shepherd ..." (Psalm 23:1). Whatever decision you make, He "will never leave or forsake you" (Hebrews 13:5).

Renting vs. Buying & Interest Rates

We will now look at an overview of two housing options: buying and renting.

Renting	Buying
→ Monthly payment goes entirely to property owner.	→ Monthly payment slowly builds owner's equity in the home; at the end of the mortgage, you own the entire home.
→ Rent could be raised periodically at owner's discretion when lease renews.	→ Fixed monthly payments (provided you have a fixed-interest-rate loan).
→ Owner could decide to stop renting to you at end of contract (usually yearly or less).	→ Home remains yours unless you sell it or foreclose. When you want to leave, you have to sell the home (unless you can afford to keep it).
→ Lower up-front savings required. The first month's rent, a security deposit, and last month's is usually required.	→ Typically requires significant up-front savings for a down payment.
→ You cannot deduct rent on taxes.	→ You may be able to deduct interest paid on taxes.
→ Most repair work and sometimes maintenance is covered by owner (although you will have to wait for them to do it); you may be responsible for things like mowing the grass.	→ Home maintenance repair falls entirely on you.
→ All changes must be approved by owner.	→ You can make changes and improvements to your home.
→ Utilities may or may not be part of rental package.	→ You can sometimes select your utility company, but you have to pay for them all yourself.

There is a time to rent and a time to buy. Being familiar with the strengths and weaknesses of each can help you prayerfully make a good decision.

Interest Rates

Interest rates change daily. The rate you are quoted when you first start looking for homes may not be the same rate you will get when you actually close on one.

After you have selected a home and have a contract, you can lock in your interest rate if you want. This means that even if the interest rates rise before you close, your rate will stay at the existing rate. For many loans, it also means that if the interest rates lower before you close, you will still have to pay the higher rate. There may or may not be a fee for locking in the rate.

- If interest rates are trending up, you might want to lock in right away.
- If the trend is going down, you might decide to wait.

Graph of Interest Rates from 1971 through early 2025[43]

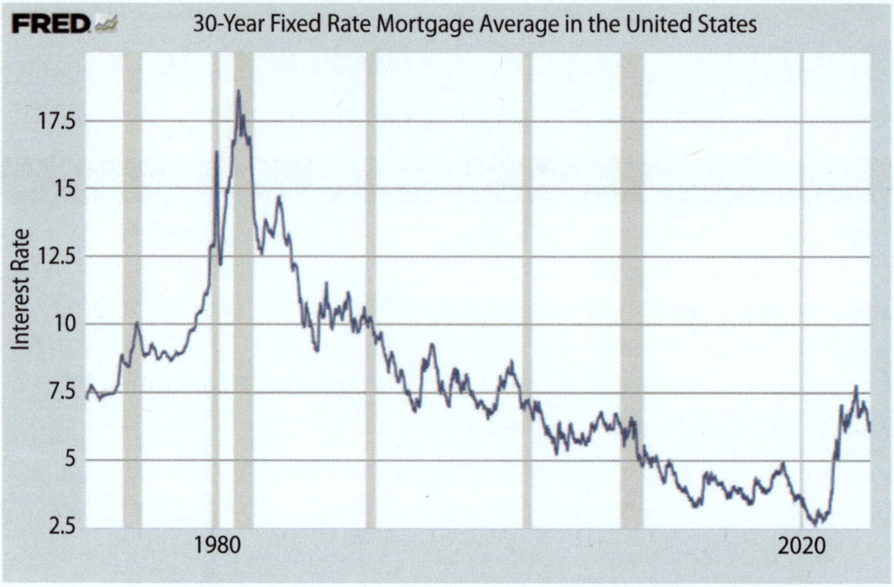

As you can see from the graph, interest rates have varied greatly.

- **The interest rate depends on the type of loan.** The graph shows the rate for 30-year fixed mortgages.
- While the graph shows the averages, **individual lenders compete with one another to get your business**, so at any point on this graph you might be able to get a 30-yr. loan at a slightly higher or lower rate. Think of interest rates like gas prices. Gas prices at every station tend to go up or down based on the price the stations have to pay for gas, but there are still some stations who will charge less than others. In the case of interest rates, your credit score and other qualifications also come into play in the rate you will pay.

Remember: Debt Adds Up

Whatever the interest rates, remember that debt adds up quickly. The less of a loan you can take out, the less you will pay in interest. When picking a home, be cautious about how much debt you take on.

KEEPING PERSPECTIVE

There are seasons in your life when you might need to rent — and seasons where you might need to buy. Remember that in all seasons, God never changes. Rely on Him for wisdom to guide and sustain you. Start today with the smaller decisions you face. That way, you will be better prepared for these bigger decisions.

Review & Case Study: Housing Bubble Collapse

We have looked at a lot of details about housing, and we will be looking at more in the next lesson. Here are two key things to remember:

- While buying a home may be years away for you, **saving for one can begin as soon as possible**. If you believe God has called you to have a family one day, it is not too early to begin taking steps toward preparing for that. While you do not have housing expenses is the perfect time to set money aside to help you when you do.
- When choosing housing, **there is more to consider than the finances alone**. Selecting a place to live can be a journey into seeking God, of watching Him change your heart and mind, and of discovering His provision.

Case Study: Housing Bubble Collapse

By 2005, many cities had seen a huge spike in housing prices. The chart shows the *ratio* of house prices compared to the **per capita income** (that is, the total income of an area divided by the total number of people). In other words, it is a graph of $\frac{\textit{housing price}}{\textit{income}}$.

Focus on the greenish teal curve representing housing prices in Los Angeles. In 1974, this line was hovering down by the U.S. average (shaded section) at around 6x (i.e., 6 times). So, if someone made $100,000 on average, they could buy a home for about 6 × $100,000, or $600,000. By 2007, the ratio was more than 15x — so if someone made $100,000 on average, they would have to spend 15 × $100,000 = $1,500,000 for a house.

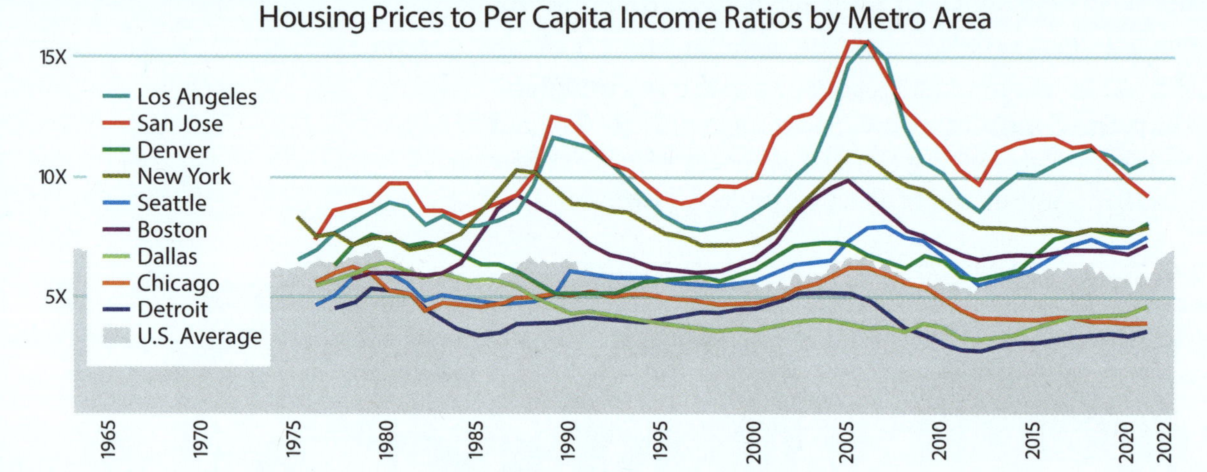

What happened? Over the next few years, prices fell steeply down, closer to the 10x line. But think about what happened to people who bought homes during that peak.

Many got jumbo loans, as they could not qualify for a conventional loan at those prices. Not surprisingly, after a few years, many had to foreclose, as they could not sustain the payments. They could not sell the home for enough money to pay back their loan, as the homes were no longer worth the inflated prices. The economy had taken a turn for the worse, too. California was not alone in this. Other areas in the United States faced foreclosures as well.

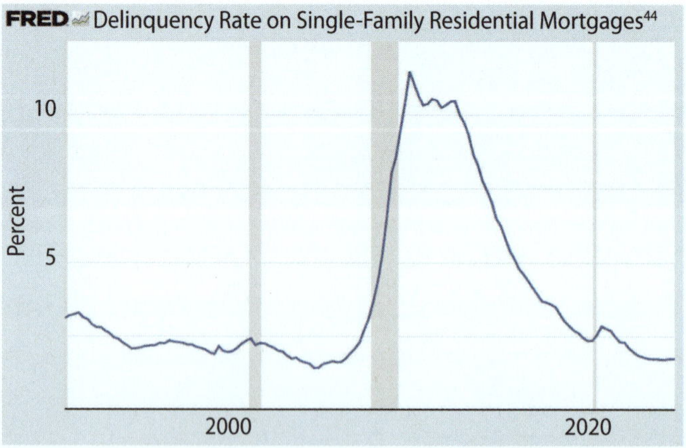

Lessons learned:

- Nonconventional loans can be dangerous, as can buying what you really can not afford.
- Beware of the pull to buy what you do not have money for, whether it is a house whose monthly payment you can not afford or something else altogether.

Housing, Part 2

income = giving + saving + ***spending***

spending *= taxes + debt +* ***living expenses***

There is a lot involved in buying a home. Understanding the total cost, not just the cost of the mortgage, is important. In this lesson, we are going to dig into additional home costs, as well as look at some terms and hidden costs of home ownership that are good to be aware of ahead of time.

Realtors & Housing Terms

As you may have noticed, buying a home is complicated. A realtor specializes in helping you through that process. When looking for your first home, your first step may be to find a realtor. How much will a realtor cost you? Hopefully nothing when you are buying.

Currently, the realtor fee for both buyer and seller is usually paid by the seller.

A realtor helps you understand the current market. Market knowledge helps you avoid paying too much for a home or selling one for too little.

In housing, the **market** refers to all the homes for sale.

- In a **buyer's market**, there are more sellers than buyers, giving buyers an advantage and causing house prices to drop.
- In a **seller's market**, there are more buyers than sellers, giving sellers an advantage and causing house prices to rise.

Realtor fees when you go to sell are one big reason why if you do not plan to stay somewhere very long, renting might be your better option. Realtor fees are calculated as a percent of the selling price, and they can be quite costly.

Example: If you owe 5% in realtor fees, how much should you expect to pay in realtor fees if selling a $350,000 home?

5% *of* $350,000: 0.05 × $350,000 = $17,500

Selling a home yourself or with a selling service can reduce realtor costs (you will probably still need to cover a buyer's realtor) when selling a home — or it can be a surprisingly costly approach if you are not familiar with the housing market, as you could significantly undervalue your home or make some other costly mistake a realtor would have caught.

Homes for sale are referred to as **listed** on the market.

Pay close attention to the way a realtor's contract is worded to see if there are situations where you will have to pay the realtor fee as the buyer.

- When selling a home, it is wise to talk with at least 3 different realtors before making a decision.
- Make sure that your selling realtor is contracted to pay the buying realtor the standard rate in your area.

Housing Terms

The following are some terms you will encounter as you walk through buying a home. We will refer to them as we go forward.

- **Appraisals** – Since home mortgages are typically for substantial sums of money, lenders want to make sure that what you are paying for the home is reasonable. Buyers often pay a small fee for an **appraisal**, which is an independent evaluation of what the home is worth.
- **Inspections** – As a buyer, you have the option to get a home **inspection**, where a trained inspector inspects the home for issues. The small inspection fee could save you significantly in identifying issues, many of which the seller might then fix for you.
- **Earnest money** – When you make an offer on a home, a seller wants to know that you are serious about the offer and will not back out. Thus, you might put down a percent of the house's price as **earnest money** — money the seller gets to keep should you back out of the deal for any reason except those which are allowable (such as a failed inspection). The earnest money gets returned to you at closing, and you can use it toward your down payment or other closing costs.

Example: Calculate the earnest money a buyer owes if they need to put down 1.5% of a $450,000 home.

1.5% *of* $450,000: $0.015 \times \$450{,}000 = \$6{,}750$

KEEPING PERSPECTIVE

Where there is no guidance, a people falls, but in an abundance of counselors there is safety (Proverbs 11:14).

Buying a home is a big decision. Getting input from others provides a safety net for you. A realtor, appraisal, and inspection are all "counselors" that can help you see potential issues. Be sure to also seek counsel from godly people in your life — and from God Himself. Jesus really is a "Wonderful Counselor" (Isaiah 9:6). Remember that Jesus is with you through each step of life, be it a big step or a small one.

Closing Costs

You are ready to buy your first home. As you work through the loan details, you are going to find out about a lot of additional costs called closing costs. **Closing costs** are the costs due when you complete the purchase or sale on the home, which is referred to as *closing* on the home.

The closing costs you will pay vary based on the home's location as well as other factors, but it is often a significant amount. Since the seller typically pays the realtor fees, closing costs are typically higher for the seller.

Some closing costs buyers might expect are:

- Underwriting fees – processing and administrative closing fees
- Property taxes
- Homeowners Association (HOA) fees
- Switching fees (to update you as the owner with the HOA, county, or other places)
- Escrow (see next page)
- Points
- Payment on your mortgage for portion of a month you will own the home
- Insurances – mortgage insurance, homeowner's insurance, title insurance

Closing costs are in addition to a down payment, which is also due at closing. Take closing costs into consideration when figuring out how much money you can put down on a house. If you only have $50,000 saved for a home and expect $4,000 in closing costs, you will only be able to put down $50,000 – $4,000 = $46,000 as a down payment.

When negotiating a home contract, you can sometimes negotiate extra money for closing costs.

Closing costs will all be calculated for you. However, it can be helpful to understand a little about *how* they are calculated.

Many of the costs, such as property taxes, HOA fees, and interest, are prorated. A **prorated amount** is the total due for a smaller period of time than the bill normally covers.

For example, property tax is usually paid by the year. If the seller has already paid the property tax for the year, the buyer needs to reimburse them for the portion of the year they will own the house.

>>> To calculate the amount due, first find the days or months owed using subtraction. Then calculate the cost per day or month by dividing the yearly cost by 365 or by 12, and multiply that amount by the number of days or months owed.

Example: The seller paid $1,280 in property tax on 123 Pleasant Lane for 1 year (365 days). 80 days into the year, they sell the home. Calculate the prorated amount the buyer owes the seller.

Days owed: $365 - 80 = 285$

Cost owed: $\frac{\$1,280}{365} \times 285 \approx \$999.45$

cost per day calculated by dividing yearly cost by days in a year

Understanding how to prorate amounts can also help you in other ways. If you cancel a service midyear and they prorate the yearly fee, you will get a portion back.

Escrow

Escrow is like a bank account the lender holds on your behalf. Your earnest money goes into an escrow account for you until closing. Just like that money was held on your behalf, so your lender may collect money toward property taxes, homeowner's insurance, and/or HOA fees at closing and then again each month as part of your monthly payment, holding this money in an escrow account for you and paying those fees on your behalf when due.

You can waive escrow (i.e., have the lender *not collect* that money) if you have put 20% down on the home and have excellent credit. If you do, you will get to keep the money yourself (and hopefully earn interest on it) until those bills are due and then make the payments yourself. This can free up some money at closing that you can put toward your down payment instead. However, make sure you budget for property taxes, homeowner's insurance, and HOA fees, as they can be a sizeable amount.

If you are using escrow, take that into account when looking at a loan amortization. You will need to use a tool that accounts for that or else evaluate your loan without that part included, knowing it will be added.

Payment
$3,560

For example, this payment amount could or could not include escrow, depending on how it was generated.

Example: Your yearly property tax is expected to be $3,421 and the HOA fee $300. If you need to deposit 2 months' worth of these fees into escrow at closing, how much will that be?

Total yearly cost: $\$3,421 + \$300 = \$3,721$

Cost of 2 months' worth: $\left(\frac{\$3,721}{12}\right) \times 2 = \$620.17$

Divided to calculate monthly cost

Calculating the monthly cost $\left(\frac{\$3,721}{12}\right)$ would also tell you how much to set aside each month if you were not using escrow.

KEEPING PERSPECTIVE

Your closing company will calculate all the closing costs for you and give you a detailed list. A good realtor will have these estimated for you long before closing. Just make sure you understand the basic concept of prorating and know that there are a lot of different fees you should expect to see on your closing papers. (I hope seeing the different expenses are encouraging you to start saving early for a house.)

Costs of Location & Fixer Uppers

Pretty much everything about a house can change *except* its location. When it comes time to buy a home, it is important to think through where you really want to live and if it will be easy to resell if you need to move.

Math can also help you evaluate the costs of different locations.

Commutes

Homes further away from your work will result in more gas, wear and tear on your car, and higher car insurance due to longer commutes. You can approximate these expenses using the government's standard mileage reimbursement rate, which was $0.21 in a sample year.

House without power lines is likely to sell better.

>>> *additional commuting cost = additional miles × mileage rate*

Example: You are debating between a home that is 10 miles from your work and one that is 40 miles away. You plan to work 5 days a week, which would be 260 days a year. About how much additional commuting cost will you have each month to make the longer commute?

Difference each way: 40 mi. – 10 mi. = **30 mi.**

Yearly extra miles, making trip there and back for 260 days: (2 × **30 mi.**) × 260 = **15,600 mi.**

Additional commuting cost: **15,600 mi.** × $0.21 = **$3,276**

Additional monthly cost: $\frac{\$3,276}{12}$ = **$273**

Use $0.21 as the mileage rate in this course.

The 260 days comes from 5 (number of workdays) times 52 (weeks in a year).

>>> Add the monthly additional commuting cost to the monthly payment of the home from which you would be commuting when comparing monthly payments.

Example: The home 10 miles from your work would have a monthly payment of $1,918.56, while the one farther away would be $1,798.65. Which house costs more once you account for the additional commuting cost?

Close house: **$1,918.56**

Far house: $1,798.65 + **$273** = **$2,071.65**

When we account for the extra commute, the close house is actually less expensive by $153.09, as **$2,071.65** – **$1,918.56** = $153.09.

This same principle applies to rentals (including college housing) — if you live farther out, you might save on rent, but be sure to think through the cost (and extra time) of a longer commute.

Other Location Costs

Different locations can have different fees as well. For example, some communities have Homeowners Associations, or HOAs, which come with fees. Different counties can also have different property tax rates. When comparing homes, look at the total cost after accounting for these differences.

HOA
HOME OWNER'S ASSOCIATION

Example: House A has a monthly mortgage payment of \$1,000 and property taxes of \$2,000 each year. House B has a monthly mortgage payment of \$1,200 and property taxes of \$1,100 per year. Compare the monthly commitment for each house.

House A:

$\$1{,}000 + \frac{\$2{,}000}{12} \approx \mathbf{\$1{,}166.67}$

House B:

$\$1{,}200 + \frac{\$1{,}100}{12} \approx \mathbf{\$1{,}291.67}$

$\mathbf{\$1{,}291.67} - \mathbf{\$1{,}166.67} = \$125$ or $\mathbf{\$1{,}166.67} - \mathbf{\$1{,}291.67} = -\$125$

House B's commitment is \$125 more each month, or House A's is \$125 less.

Notice that we divided the yearly fees by 12 so we were determining monthly amounts.

Fixer Uppers

Homes that need repairs, affectionately called "fixer uppers," tend to cost less. But not all homes that need work are priced right. A little math can help you evaluate the deal.

>>> *total home cost = purchase price + work needed*

Ideally, you want your *total home cost* to be *less than* what the house will be worth when you finish the work. After all, doing that work takes time, money, and energy.

Example: You are considering a home that is listed for \$315,000. It needs \$20,000 worth of work. Other comparable homes in the area not needing work are selling for \$325,000. How does the total home cost compare with other homes not needing work?

total home cost = \$315,000 + \$20,000 = \$335,000

This is more than the other home's price of \$325,000.

(If we needed to see how much more, we would use subtraction to calculate the difference: \$335,000 – \$325,000 = \$10,000.)

This home really is not a good deal. You would not save in the end.

KEEPING PERSPECTIVE

When buying a home, think about non-financial aspects, such as nearness to college, family, church, work, and/or ministry, and how you like the area. The most important things *are not* financial. (This holds true for other decisions you might be making now too.)

Costs of New Homes vs. Resale

When buying a home, you could either buy a brand-new one or one that had a previous owner (called a **resale** home). Following are a few considerations with each:

	Advantages	Disadvantages
New Home (one you build or that is built but not lived in yet)	▪ Everything is new and under warranty, so no anticipated major replacement costs initially ▪ More potential for customization (if not built yet)	▪ Potential costs in getting house and yard settled (blinds, sod, towel holders, and more) ▪ May have to wait for home to be built ▪ Potential for extra costs (if custom built) ▪ May have to deal with builder on anything not done right
Resale (one that has had a previous owner)	▪ Ready to move into sooner ▪ Already settled and issues addressed ▪ Neighborhood is established ▪ Yard and interior fixtures already established/installed	▪ Not easy to customize ▪ Potential for major appliances and home features to need replaced as home ages

When comparing home prices, it can be helpful to look at the **price per square foot**. Notice the *per*. Just like in a unit price we divide the cost by the units, we will divide the price by the square feet.

$$\text{price per square foot} = \frac{\text{home's total price}}{\text{home's square feet}}$$

Example: Calculate the cost per square foot of a $260,000 home that is 980 square feet.

$$\frac{\$260{,}000}{980 \text{ sq. ft.}} \approx \$265.31 \textit{ per} \text{ sq. ft.}$$

More About New Homes

New homes fall into two general categories:

Method	Advantages	Disadvantages
Subdivision/ Track Homes	▪ Less expensive ▪ Building process streamlined, resulting in $ savings	▪ Limited in customizations
Custom Homes	▪ Full customization	▪ More expensive ▪ Building process more involved

Keep in mind that builders will sometimes offer incentives *if you use their lender*. For example, they may offer to pay a point on your mortgage or even all your closing costs. Take these into account when comparing their price with other homes *if you plan to use their lender*. Math can help you evaluate these incentives.

If the builder offers a point, that value is the value of the cost of a point.

Example: Builder B is offering new homebuyers 1 point (worth 1% of the loan) if they use their lender. How much is this worth on a $425,000 loan?

1% *of* $425,000: 0.01 × $425,000 = $4,250

For any kind of new home, keep in mind there are certain new-home costs that you might not have in a resale home. For example:

Examples of Hidden New-Home Costs

Blinds and window coverings

Fence

Ceiling fans and more lighting

Landscaping

Towel and toilet paper racks

Short on cash but love the house? If the price of the home supports it, you can sometimes pay extra for the home, getting the extra back at closing that you can use to fix the house. For example, you might pay $410,000 on a $400,000 home, with a stipulation that you get $10,000 to fix up the home. Your mortgage will be higher, but you will have the cash you need for repairs.

Check the Details

Builders often have packages and options, each of which costs different amounts. Figure out the total price of the home based on the options you select.

More on Resale Homes

When you buy a resale home, it typically comes with things like window coverings. However, it also comes with older everything. Since we live in a fallen world where things fall apart, it can be helpful to look at how soon the used home you are considering may need certain major items replaced. Budget extra for home repair when buying an older home, especially if you know of things that will need to be replaced soon.

KEEPING PERSPECTIVE

Whether you are led to buy a new home or a resale home in the future, think through *all* the costs of that home, not just the purchase price. Make sure that you have saved enough money for the things that need done right away and a buffer for any unexpected expenses.

The principle of thinking through *all* the costs applies to any big decision you make. Thinking about a trip? Buying a car? Considering college? Look at all the costs.

Homeowner's Insurance

We live in a fallen world where a lot of things can go wrong.

- Hail can damage a roof.
- Fires and hurricanes can destroy the home and everything in it.
- Floods can result in tens of thousands of dollars of damage.
- People can steal valuables.

Because of this, lenders require homeowner's insurance. Homeowner's insurance works much the same way as other insurances. You pay a set fee, or premium, periodically for the insurance. In return, they will cover replacing or repairing your home and belongings up to certain maximums (called the **liability limit**) after you pay a deductible. A **claim** is when you have to ask the insurance to pay for something.

When you get homeowner's insurance, the insurance company will calculate how much they think it would take to rebuild your home. This might be more than you paid for the home, as it may cost more to rebuild it.

You can ask for more coverage than the company calculates. But there is a cost if you do. You will have to decide if extra is worth the extra **premium** (i.e., your periodic payment for the insurance) cost.

Your deductibles also affect your premiums. Remember, your **deductible** is what you pay *before* the insurance pays. If you have a $1,000 hail deductible and your roof needs replaced after a hailstorm, you will pay the first $1,000 and the insurance will pay the rest, up to the liability limit for that situation.

It makes sense that higher liability limits and lower deductibles result in higher premiums. Both cause the insurance company to pay more should there be a claim.

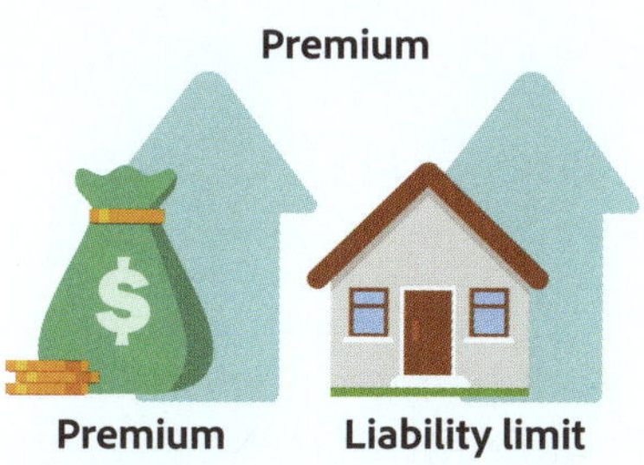

»» The *higher* the liability limits (the more coverage you would get if there is a claim), the *higher* the premium. We call this a **direct relationship**.

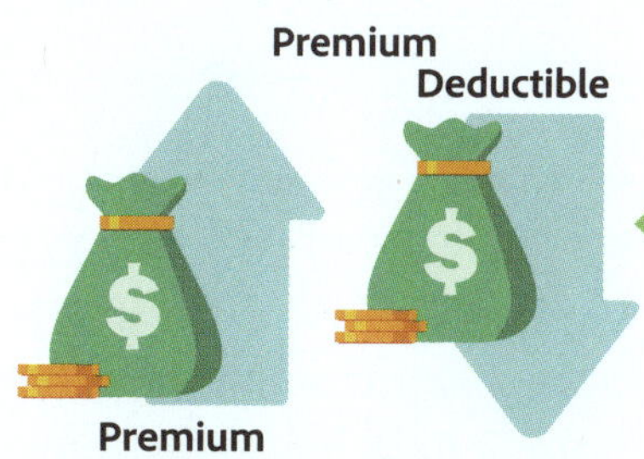

»» The *lower* your deductible (the less you have to pay if there is a claim), the *higher* the premium. We call this an **inverse relationship**.

ALGEBRA Connection

In algebra, we explore how to describe direct and inverse relationships mathematically and what they look like graphed on a coordinate system.

Direct Relationship

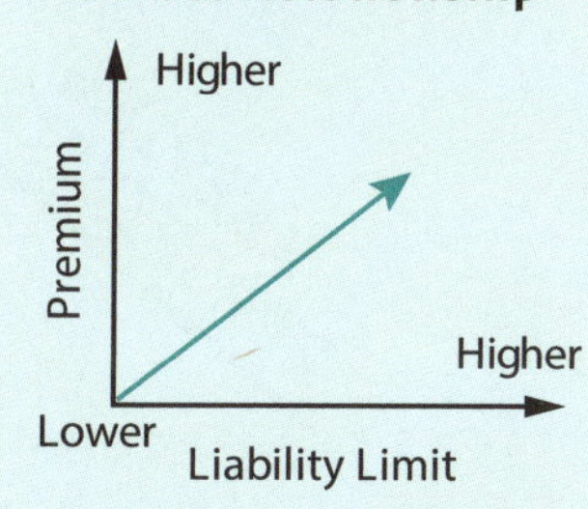

Inverse Relationship

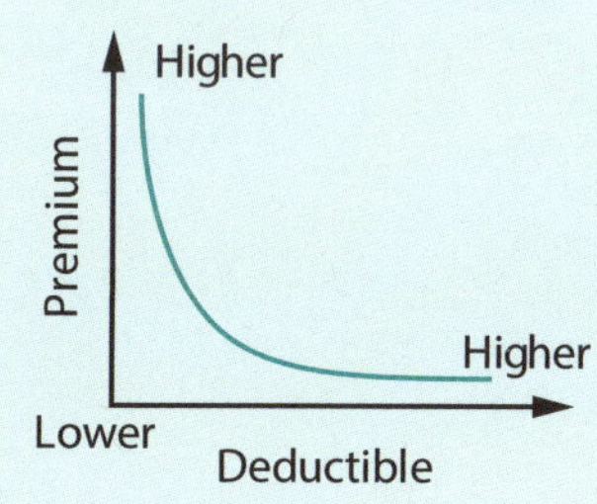

Rental Insurance — Rental insurance is similar to home insurance, except it only applies to your belongings, not the building. As a result, it is much less expensive.

Homeowner's Insurance	**Rental Insurance**
Covers home and belongings.	Covers belongings only.

Check the Details

Insurances may have different coverage for different issues. They may also have exclusions. Make sure you understand what is covered and what is not.

It is common to have exclusions for floods and hurricanes, and a separate policy for those if desired.

- **Named peril** policies will only cover damages resulting from specific things that are mentioned in the policy. If hail was not specifically mentioned and your home gets hail damage, the policy will not cover anything.
- **All risk** policies will cover all damages resulting from anything that is not specifically excluded. These are generally safer since you know up-front what is excluded.

Good to know:

- Some companies offer a discount if you pay your premiums yearly.
- Some offer discounts if you use the same company for both your homeowner's and car insurance.
- You will be asked to estimate your belongings when getting homeowner's or rental insurance, or you can go with the standard coverage. If you have a lot of expensive antiques or other valuable items, you might need more than the standard.
- Ask what discounts are offered.
- Shop around every few years to see if you could save money by switching insurance companies.

KEEPING PERSPECTIVE

You have already learned all the math skills needed to understand homeowner's insurance. You will get a chance to review some of them as you explore homeowner's insurance on your exercise.

As you do, remember that while insurance can be a prudent way to provide for yourself and your future family in a catastrophe, ultimately the Lord is your true provision. You will never be able to insure against everything, but you can be confident that God will get you through whatever trials He allows. You can take reasonable precautions, but do not let fear into your heart.

[F]or God gave us a spirit not of fear but of power and love and self-control (2 Timothy 1:7).

Home Maintenance

Homes need to be maintained. As life changes, they need to be modified. Once you get a home, home maintenance and possibly improvement should be part of your monthly budget.

Home Maintenance

In this fallen world, homes break down and need care. Here are some examples of some ongoing maintenance a home might need:

- outside maintenance (such as mowing, fertilization, aerating, raking leaves, trimming trees and bushes, pest control, and cleaning gutters)
- indoor maintenance (such as changing air conditioning and heating filters, cleaning out ducts, and replacing batteries in fire alarms).

More expensive maintenance also needs done periodically (some items every 5–15 years, with others, like a roof, less frequently). You can visually inspect some of these items when buying a home, and a good inspector can let you know their condition:

- air conditioner
- carpet
- dishwasher
- heater
- paint (interior and exterior)
- refrigerator (if included with the house you are buying)
- roof
- water heater

So how much does home maintenance cost? A lot depends on how much you do yourself versus hiring a service.

>>> Some sources suggest **budgeting at least 1% of the home price for maintenance** but possibly much more.

Example: If budgeting 1% of the home price yearly for maintenance on a $300,000 home, what should you budget monthly?

Yearly amount: $0.01 \times \$300{,}000 = \$3{,}000$

Monthly amount: $\frac{\$3{,}000}{12} = \$250$

Shown as one step: $\frac{(0.01 \times \$300{,}000)}{12} = \$250$

Mental Math Reminder
Remember, 1% equals 0.01 or $\frac{1}{100}$. Multiplying by 0.01 is the same thing as dividing by 100, which can be done by moving the decimal 2 places to the left.

$0.01 \times \$300{,}000 = \$3{,}000$

For the 1% rule to hold true, **more expensive homes tend to cost more to maintain**. This makes sense when you think about how more expensive homes might be bigger, have a bigger yard, have extra amenities to maintain, be in more expensive areas where repairs cost more, and require fancier materials to maintain an upscale look.

The 1% guideline does not always work though. You may be able to do the maintenance for less than that estimate — or you might end up paying more.

If you plan to hire services for ongoing maintenance, price out the costs in your area and use those estimates, plus some extra padding for the unexpected.

Ever wonder why your parents have certain rules around the house? Some might have to do with trying to make sure maintenance costs stay down! They do not need extra costs of replacing something due to careless damage.

Home Improvement

Life changes, and there is always more that could be improved or adjusted with a home. Thankfully, most home improvement items are things you can budget and save for, waiting until you have the resources to pay for them. Just know that home improvement projects may enter your budget as they come up. And when considering a home, evaluate its current condition and plan for changes needed immediately before you buy.

KEEPING PERSPECTIVE

Your mortgage is not the only expense of home ownership.

Understanding this can help you evaluate homes to see what you can afford. If considering a home with a huge yard, for example, plan on extra expenses to maintain it. (You might need to budget for a riding lawn mower or plan to pay much more for lawn services than with a smaller lot.) Remember to budget for any immediate home improvement projects you want to make, knowing more will come up over time.

Evaluating Total Costs & Affordability

We have looked at a lot of different costs related to homeownership to help you better:

- realize that it is important to start saving for a home
- evaluate whether you can afford to buy or rent a home, thus avoiding the pitfall of overextending yourself
- budget correctly once you do.

There is a lot to consider, but there are tools to help.

- **Spreadsheets** are great for listing all the costs you can think of and calculating yearly or monthly totals.
- **With a parent's permission, online tools can help you evaluate what you can afford.** Some home searching platforms will even list approximate mortgage, fees, taxes, and more right there for you.
- Simply being familiar with the different types of costs we have covered can help you set a realistic estimate for what you can afford. You can then double check more accurate estimates for a specific house you are thinking of buying.

Getting a Rough Estimate Based on Down Payment

The spreadsheet shown could help evaluate how different home prices compare with the down payment you want to make.

- Based on what is input in the highlighted cells, other fields update.
- The formulas in the fields are written in the left column, next to the explanation.

Using a spreadsheet like this, you could type in different home prices and see approximately how your down payment would measure up to being 20% down. If you are unable to meet 20% (as in the $412,000 home price sheet shown, where row 8 is negative), you could decide to save more or to look in a lower price range. In the $250,000 sheet shown, you could put 20% down and still have $2,500 left over.

Other moving costs in B5 could include things like a moving truck and boxes — and/or changes you want to make to the home you buy before moving in.

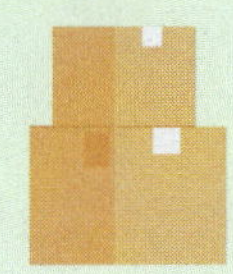

For $412,000 Home

	A	B
1	Home Price	$ 412,000.00
2	Current Savings	$ 80,000.00
3	Desire to Keep in Savings	$ 20,000.00
4	Closing Cost (=0.01*B1)	$ 4,120.00
5	Other Moving Costs	$ 5,000.00
6	Savings Available for Down Payment (=B2-(B3+B4+B5))	$ 50,880.00
7	20% Down (=0.20*B1)	$ 82,400.00
8	Amount Available Savings Is Under or Over 20% Down (=B6-B7)	**$-31,520.00**

For $250,000 Home

	A	B
1	Home Price	$ 250,000.00
2	Current Savings	$ 80,000.00
3	Desire to Keep in Savings	$ 20,000.00
4	Closing Cost (=0.01*B1)	$ 2,500.00
5	Other Moving Costs	$ 5,000.00
6	Savings Available for Down Payment (=B2-(B3+B4+B5))	$ 52,500.00
7	20% Down (=0.20*B1)	$ 50,000.00
8	Amount Available Savings Is Under or Over 20% Down (=B6-B7)	**$ 2,500.00**

Figuring Out the Monthly Payment

Before deciding how much you can or can not afford, you also need to sit down with your budget and see what kind of a payment you are comfortable with.

income = giving + saving + spending

1. **Calculate your monthly income.**
2. **Determine what percent you want to set aside for *giving*.**
3. **Figure out how much to put in *saving*** (for home and car repairs, as well as life's unexpected, fallen-world turns).
4. **Estimate all your other *spending*.** Do not include expenses like rent that would stop once you bought a home.
5. **See what is left over that you could use for a home expense.** Your home expense includes *all* home ownership expenses, not just your mortgage. You need to also include property taxes, HOA fees, homeowner's insurance, and maintenance.

Look at your current spending. Is your rent amount working out well in your budget? Too much? Do you have extra to go higher? If you do not currently pay anything for housing, what are you saving that could go for a house payment instead?

Example: Your net income is $6,000 a month, and you spend $2,000 on other expenses (besides your current rent payment). If you want to give $800 (which is 10% of your gross income) and put 10% of your net income in savings, how much could you afford for home expenses per month?

Calculate 10% of net income to figure out the savings amount:
0.10 × **$6,000** = **$600**

Calculate amount left after *giving*, *saving*, and *spending*:
$6,000 – ($800 + **$600** + $2,000) = $2,600

KEEPING PERSPECTIVE

While we are focusing on the mathematical side of home buying, remember that there are a lot of non-financial considerations too. The most important thing is figuring out where God wants you. It can be a process to figure this out, but God will be faithful to lead if you wait on Him and ask Him to show you. He wants to build your faith in Him through this process.

Extra Payments & Refinancing

We will end our look at housing with two things about mortgages that do not apply until after you have bought a home but are helpful to know about ahead of time.

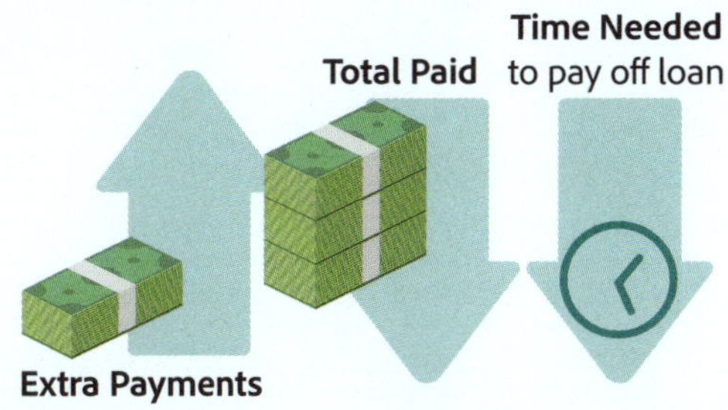

Extra Payments

Extra payments lower the total you pay *and* the time needed to pay off your loan. Every year you have a mortgage, you are paying the interest rate on the principal you have left to pay. If you can pay off more of that principal early, it will reduce how much interest you pay overall and how long it takes you to pay off the loan. Extra payments lower the principal off which interest is calculated for *every* future payment, so savings can add up.

>> To see how much money and time on the loan extra payments will save, use an amortization schedule where you can enter extra payments and see how the cumulative interest at the end of the loan and the length of the loan changes.

Example: The cumulative interest for a 30-year loan of $250,000 at 4.5% is shown two ways: one if no extra payments were made, and one if an extra payment of $1,266.71 was made in the second month of the loan. How much did that one-time payment save in interest?

$206,018.21 – **$202,446.97** = $3,571.24

Cumulative Interest (no extra payment)	Cumulative Interest (extra payment)
$206,018.21	**$202,446.97**

Example: By how much will paying an extra $283 a month reduce the length of a 30-year, $300,000 loan at 5.5% interest?

360 mos. – **258** mos. = 102 mos. = 8.5 yrs.

Final month (no extra payment)	**360**
Final month (extra $283 a month)	**258**

Note: The extra $283 a month would also save around $100,000 in interest.

Refinancing

When you get a mortgage, your interest rate is affected by your personal financial situation and the type of loan you get, but it is also controlled in general by the going interest rates in the economy. Look back at the interest rate chart on page 228. Notice that there was a huge spike in interest rates in the 1980s that slowly tapered off to a low in 2020.

Since some mortgages last 30 years, for most loans, **you continue paying the interest rate you signed up for**, even if the interest rates go down. Thus, someone who got a 16% interest mortgage back in the 1980s would still be paying 16% of their principal in 2004 — unless they refinanced their loan.

A	B	C	D
Month	Cumulative Interest @ 8.75%	Cumulative Interest @ 3.5%	Difference (B – C)
1	$3,004.17	$1,201.67	$1,802.50
2	$6,006.60	$2,401.44	$3,605.16
3	$9,007.30	$3,599.32	$5,407.98
4	$12,006.25	$4,795.30	$7,210.95
5	$15,006.25	$5,989.37	$9,014.06
6	$17,998.82	$7,181.52	$10,817.30
7	$20,992.43	$8,371.76	$12,620.67
8	$23,984.23	$9,560.07	$14,424.15
9	$26,974.21	$10,746.45	$16,227.75
10	$29,962.36	$11,930.90	$18,031.46

Refinancing is essentially closing on your home again. It typically involves several thousand dollars in closing fees. Math can help you see if it makes sense to refinance.

- Treat your current loan principal as a new loan (as it will be if you refinance) and plug that into an amortization calculator, once at your current interest rate and once at the new interest rate.
- Compare the cumulative interest columns to see after what payment the difference would be greater than the refinancing costs.

For example, there is $412,000 left on your loan at 8.75% interest. The rates have dropped to 3.5%. Following the steps above gives us the comparison shown. In this case, by payment 10, refinancing will have saved $18,031.46. If the refinancing itself cost $3,000, refinancing would have saved money in month 2, as that is where column D is more than $3,000.

Good to know:

- When you choose a lender, ask if there is any penalty for paying the loan off early. Even if you do not think you will be able to, it is nice to have the option should your finances change.
- Refinancing might also make sense in other situations besides interest rate changes, such as if you are struggling to make your payments. Switching to a new 30-year loan could lower your monthly payments.
- It is possible to refinance at the same interest rate but get some of your owner's equity out of your home. This *increases*, not reduces, your debt. Banks also offer home equity lines of credit, which are additional loans against the equity in your home. While there are situations when it might be necessary, be cautious about refinancing that increases debt.

KEEPING PERSPECTIVE

Math can help you figure out real-life decisions you will face. Do not try to memorize every detail we are covering.

Here are the key ideas for today:

- Paying off your loan more quickly saves you money in the end.
- Make sure you understand and think through major steps like refinancing.

Review & Case Study: God Leads (My Family's Home Journey)

While there is a lot to buying a home, you now know some of the aspects to consider. While you are not expected to remember all the details, do keep in mind the following:

- There are a lot of costs and factors to consider when buying a home. Just as with rentals, be sure to look at the **total costs** when comparing homes or deciding what you can afford.
- It is important to think through various aspects before jumping into a home purchase.

As you get older, you are going to have to make more and more decisions whether they are big, like buying a home, or small, like deciding how much to spend at a restaurant.

- Understanding the big picture of *income = giving + saving + spending* can help you make wise choices. If you limit your *spending*, you will have more to save for things like a home.
- Knowing how to use basic math skills can help you evaluate the financial side of many decisions you face.
- Remembering that God has a plan and is your Shepherd is key. Trust God today in the small decisions, as they are what prepare you for the bigger ones.

This lesson's case study shares how I watched God guide my parents through some housing decisions. It illustrates that God knows what He has for each of us each step, and that what He has may not be what we think.

Case Study: God Leads (My Family's Home Journey)

As a teenager, my dad's job moved us from Colorado to Texas, where we struggled to find a home. To complicate matters, our home in Colorado had not sold, which limited the funds available for a down payment on a new home (and thus the price home we could look at).

After a long search, we found the perfect home. My parents prayerfully offered to buy the home for a price that stretched them, but that they thought they could do. Only they could not afford the 3 acres next door with the pond my brother and I really wanted. But God provided. The appraisal valued the house lower than their offer, and the sellers had to throw in the 3-acre lot to make the deal.

A year later, my dad's company unexpectedly decided to move him again, this time to Massachusetts. My parents had no idea how the house would sell; it was a large home in the middle of nowhere, and the house across the street had sat on the market for a long time already. Yet a family from Holland bought it before it even got listed. God had provided again.

We struggled once again to find a home in Massachusetts. Every home my mom even remotely liked sold before she could place an offer. The closed doors led my parents to rent, and that rental not only led us to get to know a dear friend, but also allowed us to move easily when my dad's job unexpectedly moved us again less than a year later. While the rental had its challenges, God worked them all to help sanctify and draw us to Himself.

At the time, those moves included lots of moments of frustration and wondering what God had for us. Yet He had a plan and led step by step. Even more than the end result, though, He used the process to grow us in trusting Him. The decisions we face — whether they be housing or smaller decisions — are all opportunities to grow in trusting Him and letting Him sanctify us.

For this is the will of God, your sanctification (1 Thessalonians 4:3).

Decisions & Review

In this quarter, we have looked at a lot of big decisions, from buying a car to choosing a career to deciding where to live. As we review the quarter, we are going to take a closer look at some examples in the Bible of people making decisions. We will also touch on a few other concepts.

Decision Introduction & Case Study: Seeking God (Jehoshaphat)

We will begin with three key principles that can help you make the decisions in your life now and those ahead of you. We will look at others as we continue.

Have your mind renewed. Do you ever wish God would just write a neon sign up in the sky with exactly what He wants you to do? Well, in a way, He has. The Scriptures can make us "complete, equipped for every good work" (2 Timothy 3:17). If you want to know God's will, you need to focus on the "renewal of your mind" in the Word of God (Romans 12:2). As you make a habit of replacing your own way of thinking with what God's Word says instead (thereby renewing your minds), you will find yourself guided in decisions — be they big or small — by God's Word instead of just your own wishes or the world's ideas.

Do a heart check. It is easy to focus on the decision itself, but God cares about your heart in the process. Are you acting out of fear? Greed? Love? What is your motivation? Take your heart to the Lord and surrender your own desires and fears so you will be able to see clearly to make a wise decision.

Make the even bigger decisions. Financial decisions get easier once you have made overall budget decisions. If you have already set a lifestyle cap and determined how much you have to spend versus give and save, that will help drive the big financial decisions you make.

Case Study: Seeking God (Jehoshaphat)

Second Chronicles 20 tells about a king who did not know what to do. An army was attacking King Jehoshaphat's people, and he was afraid (v. 3). So, what did he do?

He had the people fast and pray. He publicly cried out to God, acknowledging God's might and rule and that this situation was not too hard for God. Jehoshaphat explained that these men wanted to drive him out of the inheritance God had given.

His prayer concludes with this: "O our God, will you not execute judgment on them? For we are powerless against this great horde that is coming against us. We do not know what to do, but our eyes are on you" (v. 12).

When afraid of a situation and faced with a big decision (like leading a nation against an invasion), Jehoshaphat acknowledged that he did not know what to do. He put his eyes on God and the truth God had revealed (that the land belonged to Israel). And God answered and defeated their enemies (vv. 22–23).

Next time you are faced with a situation where you do not know the right decision to make, be it leading a nation or something more personal, like picking a home or deciding how to handle a situation with a friend, remember that God knows what to do. Put your eyes on Him and think about His power and care.

Pray. It goes without saying that we need to pray over decisions. God wants you to be "casting all your anxieties on him ... " (1 Peter 5:7). Nothing is too big or small for Him.

KEEPING PERSPECTIVE

In preparation for Test 3, we will briefly review budgeting for and purchasing a car. When you apply these skills to evaluate different cars, do so while keeping your eyes on God and remembering that He is your Shepherd and big enough to direct your steps.

Umbrella Policy & Case Study: Rehoboam & Counsel

Since we covered car and home purchases this quarter, we will briefly look at something called an umbrella policy. An **umbrella policy** is like an umbrella on top of a raincoat — it goes over your regular car and homeowner's policy, providing further protection.

Homeowner's and car insurance policies only cover up to certain limits, and at some point, you may want more coverage. If people know you have more money (or assets — such as a lot of equity in a home or a high salary), they can sue for extra if in a car accident — or if they got offended by what you said. **Umbrella policies are designed to provide extra protection.** They apply to everyone but are especially important for those

- with a higher net worth
- with a high salary
- in fields or engaged in activities where lawsuits or issues outside of your other insurance coverage are more likely.

>>> If you lose a lawsuit with something covered on an umbrella policy, all you owe is your deductible.

We will not go into a lot of detail about an umbrella policy in this course, but know that there are various types of insurances available, and that math can help you evaluate them.

Example: Is it cheaper to pay $2,400 a year for a policy, or $250 a month?

Monthly cost if pay yearly: $\frac{\$2,400}{12} = \$200$

$200 is less than $250, so paying yearly is cheaper.

Case Study: Rehoboam & Wise Counsel

When making big decisions like the ones we have looked at this quarter, it is prudent to seek wise counsel. Our hearts can easily deceive us, but God can use others to help us see things we might not otherwise.

Rehoboam was the son of King Solomon, the wisest man who ever lived (with the exception of Jesus, of course). Solomon's proverbs had warned him to pay attention to his words.

> *Hear, O sons, a father's instruction, and be attentive, that you may gain insight* (Proverbs 4:1).

King Solomon had also warned of the blessing of listening to wise counsel and the dangers of becoming a companion of fools.

> *Whoever walks with the wise becomes wise, but the companion of fools will suffer harm* (Proverbs 13:20).

When he inherited the kingdom from his father, Rehoboam faced a big decision. The people asked him to lighten their load. He told them to give him three days. During that time, he sought the counsel of the old men who had advised his father. They advised him to speak kindly to the people. Instead, he listened to the young men around him who told him to tell the people he would be harsher, not kinder, than his father. As a result, Rehoboam lost most of his kingdom to Jeroboam (1 Kings 12:1–20).

Seek wise counsel. Asking for counsel is humbling. But God put older people around us for a reason. Getting good counsel shows wisdom, not weakness.

> *The way of a fool is right in his own eyes, but a wise man listens to advice* (Proverbs 12:15).

KEEPING PERSPECTIVE

You will be reviewing Lesson 12 (Career Preparation) today; remember to seek wise counsel as you consider your own future career and make decisions.

There are a lot of different types of insurance out there. What you need may vary at different times in life.

Struggles & Case Study: Redirection/Difficulties (Paul)

It is tempting when we make a decision to assume that if things turn out poorly, we must have made the wrong decision. The Bible, though, does not ever promise that the right decision leads to the most successful path from a human perspective.

While there certainly are times where we face consequences for bad decisions, there are other times where hardships are part of the path God has chosen. When you face challenges and wonder if you made the right choice, certainly examine your heart and repent of any sin. But take comfort in knowing God wants to work good right where you are — and that tough does not necessarily mean you made a wrong turn.

> *And we know that for those who love God all things work together for good, for those who are called according to his purpose* (Romans 8:28).

Remember that God is over your decisions and trust Him to redirect as needed.

Today's case study illustrates this, along with another key principle.

Key Decision-Making Principles

- Have your mind continually renewed so you can use wisdom and make decisions in accordance with God's principles and revealed will.
- Do a heart check to make sure your heart is loving God.
- Make the even bigger decisions of a budget and lifestyle cap.
- Pray.
- Seek wise counsel.
- Remember you are just a servant.
- Trust God to redirect your decisions as needed.
- Keep focused on eternity.

Case Study: Redirection/Difficulties (Paul)

The Apostle Paul, like the rest of us, faced many decisions in his lifetime. He trusted God to direct his path — and we see evidence that God did.

One time, he was planning to go to an area called Bithynia, but the Spirit did not let him. Instead, God used a vision to redirect him and his party to Macedonia (Acts 16:6–10).

Everything went great in Macedonia, right? After all, they had listened to the Lord's leading.

Just as God redirected Paul's path, God is able to get us where He wants us.

Wrong. Paul ended up in prison in Macedonia — but first the church of Philippi was born. Even Paul's imprisonment resulted in the jailer's and his family's salvation. (See Acts 16:11–34.)

When in a Roman prison, Paul wrote the church of Philippi this:

> *[F]or I know that through your prayers and the help of the Spirit of Jesus Christ this will turn out for my deliverance* (Philippians 1:19).

Paul recognized that God worked *in* difficulties, not without them. Indeed, God used Paul's Roman imprisonment to have him write much of the New Testament. God had a bigger plan.

Paul viewed himself as God's servant. He did not see life as about his temporal comfort but as about God's purposes. He lived surrendered to God's bigger plan.

Remember you are just a servant. All things belong to God; He made us, and if we are in Christ Jesus, we are His servants, or slaves (Romans 6:22). Decisions then should not be about what *we* want or what will make *us* happy as much as about *God's glory* and building *His Kingdom*.

KEEPING PERSPECTIVE

God promises that if you trust in Him and do not lean on your own understanding, He *will* "make straight your paths." That does not mean everything will turn out the way you want (it may be God's will for you to go through some kind of hardship), but it does mean God will direct you according to His purposes.

> *Trust in the Lord with all your heart, and do not lean on your own understanding. In all your ways acknowledge him, and he will make straight your paths* (Proverbs 3:5–6).

Skill Sharpening: More with Negative Numbers & Graphs & Case Study: Temporal or Eternal Wealth (Lot)

We have seen several times in this course that negative numbers help in working with money, such as on a budget showing categories where the actual amounts were more than the budgeted amounts (i.e., the *opposite* of having money left over). In the next quarter, we are going to look at investments and see that sometimes they do the *opposite* of gaining money — that is, they *lose* money. As we do, we will use negative numbers.

	Budgeted	Actual	Difference
Groceries	$ 100.00	$ 120.00	$ −20.00

>>> When multiplying or dividing where only one of the values is a negative number, your answer will also be negative. Your calculator will do this for you, but **make sure you input the negative signs.**

Example: Calculate $\frac{-\$60}{\$100}$.

Dividing a loss by a gain.

$\frac{-\$60}{\$100} = -\$60 \div \$100 = -0.60$

Example: Calculate $0.01 \times -\$6$.

Calculating 1% of a loss.

$0.01 \times -\$6 = -\$0.06$

We will also see that graphs apply a lot in exploring investments. While we will see a lot of different-looking graphs, keep in mind that:

- The horizontal axis (or line) tracks one thing and the vertical axis tracks another.
- They both start at a 0 point, and the farther away from that point you get, the greater that value.

The graph in the sidebar shows the price of an investment losing and gaining value over time. At any point along the curve, the horizontal value is the months that have passed, and the vertical value is the price at that point in time.

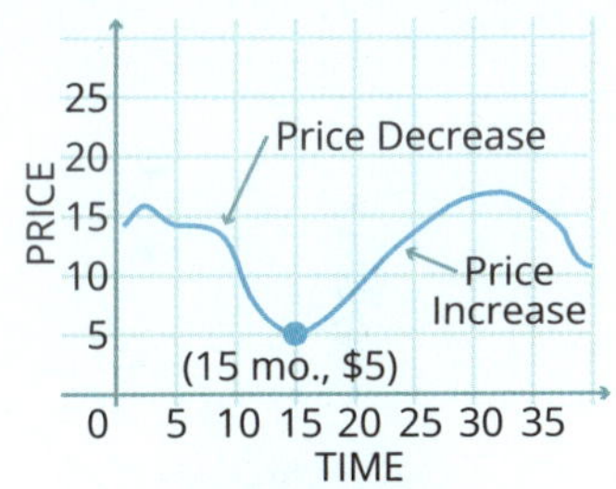

ALGEBRA Connection

In algebra, lots of rules for working with negative numbers and graphing equations are presented. These skills apply — including with your finances.

Case Study: Temporal or Eternal Wealth (Lot)

Lot started out making some great decisions. He went with Abraham into a strange land. The Bible calls him a righteous man (2 Peter 2:7).

Yet Lot made one decision that cost him dearly. When Abraham (called Abram at the time) suggested they separate, Lot did not beg to remain with Abraham to whom God had promised to bless all nations. He did not ask wise counsel (such as Abraham's). We do not see any record of him praying. Instead, he "lifted up his eyes and saw that the Jordan Valley was well watered. ... So Lot chose for himself all the Jordan Valley" (Genesis 13:10–11).

He appears to have made his decision based solely on what looked good financially — at what would be best for the here and now. Good land meant good grazing for his flocks and herds.

But there was a problem: "the men of Sodom were wicked, great sinners against the LORD" (Genesis 13:13).

Lot ended up surrounded by wicked men. Angels had to literally drag him out of the city.

> *But he lingered. So the men seized him and his wife and his two daughters by the hand, the LORD being merciful to him, and they brought him out and set him outside the city* (Genesis 19:16).

In the end, his wife became a pillar of salt and his daughters went through a time of faithlessness. God still preserved him, but his decision proved costly.

Keep focused on eternity. Do not forget that temporal things like wealth are just that: temporal. None of it lasts. Lot lost pretty much everything he had in one day. What good had the rich ground really done for him?

KEEPING PERSPECTIVE

We know from the Bible that one day everything here will burn up (2 Peter 3:7). We know we will one day die. Things here and now are temporal — like a vapor.

> *[Y]et you do not know what tomorrow will bring. What is your life? For you are a mist that appears for a little time and then vanishes* (James 4:14).

As we make decisions, we would be wise to look beyond the temporal implications to the eternal ones — to live for God's coming Kingdom and to keep our eyes on eternal rewards.

> *But as it is, they desire a better country, that is, a heavenly one. Therefore God is not ashamed to be called their God, for he has prepared for them a city* (Hebrews 11:16).

Quarter 4: More Applications, Investments, & Final Review

In our last quarter, we will be doing the following:

- applying math skills to daily life, such as in the kitchen, when traveling, and with health
- looking at a broad overview of some investment options for money you have saved
- touching on some more job-related ways math applies
- reviewing a lot of what we have learned

Skill Review: Working with Fractions

As we use math in the kitchen and other daily-life scenarios, we are going to encounter fractions. Because of this, here is a quick review of how to work with them. We will be working with easy fraction problems so are only reviewing some basics.

>>> The **numerator** refers to the top number; the **denominator** refers to the bottom number. $\frac{\textbf{numerator}}{\textbf{denominator}}$
In $\frac{2}{3}$, 2 is the numerator and 3 is the denominator.

>>> To rewrite fractions with a different denominator, multiply or divide both the numerator and the denominator by the same amount.

Example: Rewrite $\frac{1}{2}$ by multiplying the numerator and the denominator by 2.

$$\frac{1}{2} = \frac{1 \times 2}{2 \times 2} = \frac{2}{4}$$

Example: Rewrite $\frac{2}{4}$ by dividing the numerator and the denominator by 2.

$$\frac{2}{4} = \frac{2 \div 2}{4 \div 2} = \frac{1}{2}$$

>>> Some fractions can simplify to whole numbers. Remember, fractions mean division, so you can always complete the division.

Example: Simplify $\frac{8}{4}$ to a whole number.

$\frac{8}{4} = 8 \div 4 = 2$

Example: Simplify $\frac{4}{4}$ to a whole number.

$\frac{4}{4} = 4 \div 4 = 1$

>>> When you add or subtract fractions, denominators must be the same; then add or subtract the numerators. If the denominators are not the same, first rewrite the fraction(s) so they are.

Example: $\frac{3}{4} + \frac{3}{4}$

$\frac{3}{4} + \frac{3}{4} = \frac{6}{4}$

Example: $\frac{1}{2} + \frac{1}{4}$

We have to first rewrite so that the denominators are the same. We can do this by thinking through what one denominator would have to be multiplied or divided by to equal the other. In this case, 2 multiplied by 2 equals 4, so we will multiply the first fraction's numerator and denominator by 2.

$\frac{1}{2} + \frac{1}{4} = \frac{1 \times 2}{2 \times 2} + \frac{1}{4} = \frac{2}{4} + \frac{1}{4} = \frac{3}{4}$

>>> When you multiply fractions, you multiply the numerators and the denominators.

Example: $\frac{1}{2} \times \frac{3}{4}$

$\frac{1}{2} \times \frac{3}{4} = \frac{3}{8}$

>>> When you multiply a fraction by a whole number, treat the whole number like a numerator.

Example: $2 \times \frac{3}{4}$

$2 \times \frac{3}{4} = \frac{6}{4}$

We can work with fractions because of the consistent way God holds creation together. Fractions represent division problems, and division operates consistently. In Jeremiah, God reminds His people that He keeps His "covenant with day and night and established the laws of heaven and earth" (Jeremiah 33:25; NIV). He is keeping the world together the way He established, and He will keep His Word to us.

Applying Math Skills to Daily Life

income = giving + saving + ***spending***

How do you double a recipe? How much does something cost in a different country's money system? How do you convert between different measurements and temperature systems? We will explore all these and more in this lesson while reviewing some of the main math skills we have used throughout the course and preparing you to use them in other situations you might encounter.

Doubling or Halving a Recipe Plus Skill Review: Mixed Numbers

You want to double your mom's famous peach cobbler recipe in order to bring it to a church potluck. Or you need to halve a potluck recipe for use at home.

How do you figure out how much of each ingredient to use (and thus how much you need to buy at the store)?

>>> To adjust a recipe, multiply each ingredient by the same amount.

- **To double a recipe**, multiply each ingredient by 2.
- **To halve a recipe**, multiply each ingredient by $\frac{1}{2}$ (or divide it by 2 — both mean the same thing).

Example: If a recipe you are doubling calls for $\frac{3}{4}$ c. of flour, how much should you put in?
Doubling by multiplying by 2: $2 \times \frac{3}{4}$ c. $= \frac{6}{4}$ c.

Skill Review: Mixed Numbers

Now, knowing you need $\frac{6}{4}$ c. of flour is not all that helpful, as you do not have a $\frac{6}{4}$ c. measurer. To see what to measure, you will need to rewrite as a mixed number.

>>> A mixed number has a whole and fractional portion. To rewrite a fraction as a mixed number, divide the numerator by the denominator, leaving any remainder as a fraction. On the everyday fractions we will be looking at, we will do this mentally.

Common U.S. Kitchen Measurements

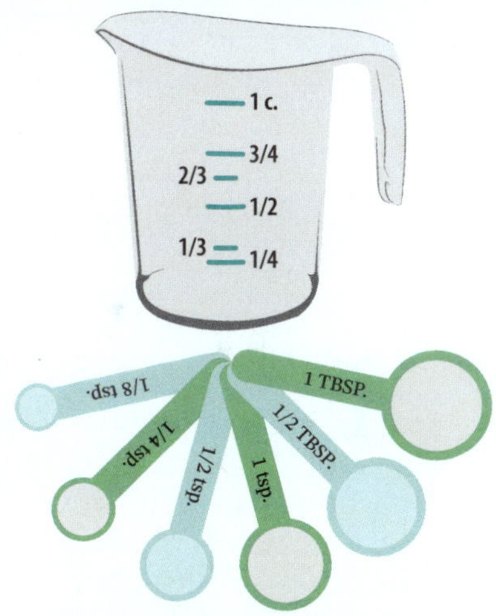

>>> When asked to work with kitchen measurements in this course, **give your answer in measurements you would find in a kitchen.** (See above and in the **Reference Section**.) Mentally dividing the fraction by $\frac{2}{2}$ is often a good way to rewrite a measurement.

Example: Rewrite $\frac{6}{4}$ c. as a mixed number.

Just think about separating 6 into 4 piles. You would be able to put 1 in each pile, and then you would have 2 left over that you still need to divide into 4 piles. The answer is $1\frac{2}{4}$.

Now, you do not have a $\frac{2}{4}$ measurer, so you still need to rewrite the fractional portion to have a different denominator.

Example: Rewrite $1\frac{2}{4}$ so the fractional portion is a common measuring cup.

Divide the fractional portion by $\frac{2}{2}$. Do it mentally.

$1\frac{2}{4} = 1\frac{2 \div 2}{4 \div 2} = 1\frac{1}{2}$

Notice that if we had thought through dividing the remaining 2 circles into 4 piles, we would have put $\frac{1}{2}$ a circle in each pile.

Example: If a recipe you are halving calls for $\frac{2}{3}$ c. of milk, how much should you put in?

Calculate half by multiplying by $\frac{1}{2}$: $\frac{1}{2} \times \frac{2}{3}$ c. $= \frac{2}{6}$ c.

Mentally rewrite as a common measurer: $\frac{2}{6}$ c. $= \frac{2 \div 2}{6 \div 2}$ c. $= \frac{1}{3}$ c.

⋙ When working with mixed numbers, mentally multiply the whole and fractional portions separately.

Some ingredients, like eggs, can be problematic to halve. How do you get half an egg? One way is to whisk it and measure out about half of the egg (about 2 tablespoons).

Example: Double a recipe that calls for $2\frac{1}{4}$ cups of flour. Mentally calculate the flour needed.

Double whole amount (2 c.): $2 \times 2 = 4$

Double fractional amount ($\frac{1}{4}$ c.); rewrite as a common measurer:

$2 \times \frac{1}{4} = \frac{2}{4} = \frac{2 \div 2}{4 \div 2} = \frac{1}{2}$

Add together: $4 + \frac{1}{2} = 4\frac{1}{2}$ cups of flour.

KEEPING PERSPECTIVE

The more you practice halving or doubling recipes, the easier it will become to do them mentally. Keep in mind that we can mentally work with fractions because fractions operate consistently, reminding us that God created and sustains a consistent world.

Changing Pan Sizes

A peach cobbler recipe says it should go in an 8-in. square pan, but you do not own that size. Will your 9 in. x 13 in. pan size work? Or a 10-in. circular pan? Adjusting recipes can be complicated, as what matters in one recipe may not be what matters in another. But basic math can help you decide if and how to adjust.

If you want your recipe to fit in a new pan to the same depth as the original, you will want to compare the area of the pans and use that to adjust the recipe if needed. Still, keep in mind that the recipe might cook differently in a small pan versus a large pan.

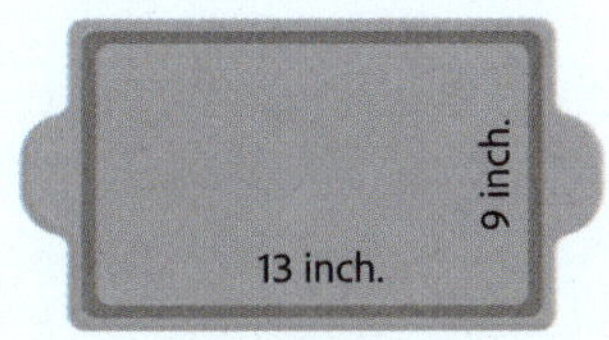

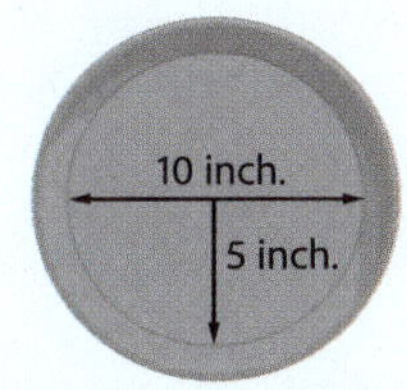

Square Pan	Rectangular Pan	Circular Pan
Area = *side*^{2}	*Area* = *length* × *width*	*Area* = 3.14 × *radius*^{2}
= $(8 \text{ in.})^2 = 8 \text{ in.} \times 8 \text{ in.}$ = **64 in.^2**	= $13 \text{ in.} \times 9 \text{ in.}$ = **117 in.^2**	= $3.14\,(5 \text{ in.})^2 = 3.14 \times 5 \text{ in.} \times 5 \text{ in.}$ = $3.14 \times 25 \text{ in.}^2$ = **78.5 in.^2**

Notice that we put parentheses around the (5 in.). This is to show that we are multiplying 5 in. by itself 2 times. This results in multiplying both 5 × 5 (getting 25) and in. × in. (getting in.^{2}).

»» See how much bigger or smaller one pan is than another by comparing the areas with division: $\frac{\textit{area of pan want to use}}{\textit{area of pan recipe is written for}}$.

»» Multiply each ingredient in the recipe by the result to get the recipe to fit inside to the same depth. Depending on the recipe, you could round to the nearest 0.5 (so 0.5, 1, 1.5, 2, 2.5, and so forth).

Example: You want to use the rectangular pan, but your recipe was written for the square pan. By what should you multiply each ingredient?

$$\frac{\textit{area of rectangular pan}}{\textit{area of square pan}} = \frac{117 \text{ in.}^2}{64 \text{ in.}^2} \approx 1.83 \approx 2$$

In the above example, you should multiply each ingredient by 1.83 (which you might round to 2 in real life, doubling the recipe).

Example: You want to use the circular pan, but your recipe was written for the rectangular pan. By what should you multiply each ingredient?

$$\frac{\textit{area of circular pan}}{\textit{area of rectangular pan}} = \frac{78.5 \text{ in.}^2}{117 \text{ in.}^2} \approx 0.67 \approx 0.5$$

We rounded to the nearest 0.5, which is the same as rounding to the nearest half dollar (see Day 61).

As you modify recipes, keep in mind the recipe you are making. For example, putting zucchini bread in muffin tins instead of a loaf pan will cause it to cook faster, so you will need to adjust the cooking time.

In the previous example, we put the area of the rectangular pan in the denominator, as it is the pan the recipe was written for this time. Notice we rounded down to 0.5 — 0.67 is closer to 0.5 than to 1. If making this recipe, we might want to halve the recipe (0.5 is $\frac{1}{2}$).

Example: You want to use the circular pan, but your recipe was written for the square pan. By what should you multiply each ingredient?

$$\frac{\textit{area of circular pan}}{\textit{area of square pan}} = \frac{78.5 \text{ in.}^2}{64 \text{ in.}^2} \approx 1.23 \approx 1$$

In the above example, since the circular pan is only slightly more area than the square one, you might be able to round down to 1, which would mean using the circular pan without changing the recipe.

>>> To actually adjust the recipe, multiply each ingredient by the value you found, rounded to the nearest 0.5.

Example: What would you adjust $\frac{1}{2}$ c. of milk to so the recipe would fit in the rectangular pan to the same depth as the square pan (see 1st example)?

Multiplied by 2, as that is rounded value found in first example: $2 \times \frac{1}{2}$ c. $= \frac{2}{2}$ c. $= 1$ c.

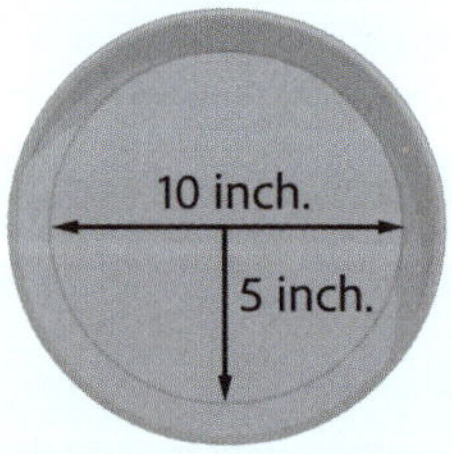

Labeling Pans

A square pan is labeled by the length of its sides. An 8-in. square pan has 8-in. sides.

A circular pan is named by its diameter. If you have a 10-in. pan, that means the diameter is 10 inches, which would make the radius half that, or 5 in.

KEEPING PERSPECTIVE

Math helps us with real-life tasks — including cooking in the kitchen. Why? Because math is really a way of describing the orderly way God governs creation. As you apply math to kitchen settings, remember that math works consistently outside of a textbook because "*all things* are [God's] servants" (Psalm 119:91).

Substituting Measurers & Familiarity with Common Measurements

Sometimes the measuring tool you need is in the dishwasher or missing. To figure out what measurer you could use instead, you often need to add fractions together.

Example: You need to measure $\frac{1}{2}$ cup of flour, but your $\frac{1}{2}$ cup measurer is missing. You have a $\frac{1}{4}$ cup measurer. How many times would you need to fill it up to get $\frac{1}{2}$ cup?

Every time we fill the $\frac{1}{4}$ cup measurer, we are adding $\frac{1}{4}$ cup. Since the denominator will stay the same (4), we can easily do the addition in our heads, adding the numerators.

$\frac{1}{4} + \frac{1}{4} = \frac{2}{4}$

Now we just need to realize that $\frac{2}{4}$ simplifies to $\frac{1}{2}$ if we mentally divide both the numerator and denominator by 2.

$\frac{2}{4} = \frac{2 \div 2}{4 \div 2} = \frac{1}{2}$

If we fill the $\frac{1}{4}$ cup twice, we will have $\frac{1}{2}$ cup.

Familiarity with Common Measurements

Some Good Relationships to Know

3 teaspoons (tsp.) = 1 tablespoon (TBSP.)

4 tablespoons (TBSP.) = $\frac{1}{4}$ cup (c.)

Example: You have a recipe needing $\frac{1}{4}$ cup of oil but can only find a tablespoon. How many times would you fill it up?

4 TBSP. = $\frac{1}{4}$ c., so we would have to fill the tablespoon 4 times.

≫ Sometimes, especially if halving or quartering recipes, you may not have an exact measurement and will have to substitute the closest you can, filling the measurer slightly extra or less. In U.S. Measurers, all the numerators are 1. The greater the denominator, the smaller the size.

This makes sense, as fractions mean division. $\frac{1}{4}$ c. is the amount you would get in each pile if you took 1 cup and divided it by 4, while $\frac{1}{3}$ c. is the amount you would get if you took 1 c. and divided it by 3. Dividing by 3 instead of 4 would make each pile bigger.

Example: You halve a recipe and now need $1\frac{1}{3}$ tsp. of something. You do not have a $\frac{1}{3}$ tsp. measurer. What common measurer could you use, and how would you fill it?

3 is in between 2 and 4, so $\frac{1}{3}$ tsp. is between $\frac{1}{2}$ tsp. and $\frac{1}{4}$ tsp. We could use the $\frac{1}{4}$ tsp., filling it up extra, or the $\frac{1}{2}$ tsp., filling it up less.

When adjusting recipes or working in the kitchen, you might need to switch back and forth from decimal to fraction form. While you can always divide the numerator by the denominator to rewrite a fraction as a decimal, memorizing the decimal equivalent of the common fractions listed will save you time. Thinking of the decimal equivalent can also help you figure out whether to fill up a measurer more or less.

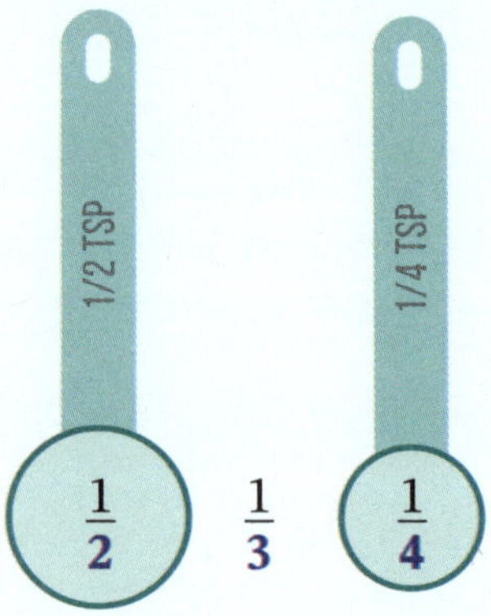

Decimal Equivalent of Common Fractions

$\frac{1}{8} = 0.125$	$\frac{1}{4} = 0.25$	$\frac{1}{2} = 0.5$	$\frac{3}{4} = 0.75$
$\frac{1}{5} = 0.20$	$\frac{1}{3} \approx 0.33$	$\frac{2}{3} \approx 0.66$	

Example: You see that you need $\frac{2}{5}$ tsp. of something. You do not have a $\frac{2}{5}$ tsp. measurer. What common measurer could you use, and how would you fill it?

Decimal equivalent of $\frac{2}{5}$: $\frac{2}{5} = 2 \div 5 = 0.40$

If you know that $\frac{1}{4}$ is 0.25 and $\frac{1}{2}$ is 0.5, you can tell 0.40 is closest to $\frac{1}{2}$. You would use your $\frac{1}{2}$ tsp., filling it less.

KEEPING PERSPECTIVE

I hope throughout this course you have grown in appreciation for how much math applies in everyday life. Do not take that fact for granted. Not only does it rely on God's consistency holding creation together, but it also relies on Him giving us the ability to explore and have "dominion" over His world (Genesis 1:26–28). The Bible gives us a framework for *why* we are able to use math.

Unit Conversion: Review & Shopping

In Lesson 2, we looked at finding the unit price of an item. We have seen how finding the unit price can help you determine which is really the best deal.

\$4.99 for 120 fl. oz.

Unit price: $\frac{\$4.99}{120 \text{ fl. oz.}}$

$\approx$ \$0.04 per fl. oz.

\$3.99 for 70 fl. oz.

Unit price: $\frac{\$3.99}{70 \text{ fl. oz.}}$

$\approx$ \$0.06 per fl. oz.

Cheaper per fl. oz.

When comparing prices, make sure you are comparing like units of measure. Be especially careful of this when shopping online (with a parent's permission), where you can not see the size as easily. Online, a gallon and a pint drink might appear the same size.

\$3.75 for 1 gal.

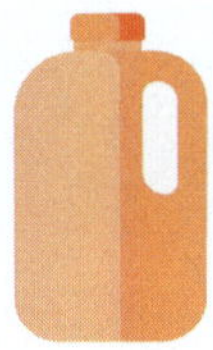

\$1.75 for 1 pt.

Before comparing these prices, we first need to **rewrite one of them so we will have the same units**. We could pick either unit. We will rewrite the gallon as pints.

$1 \cancel{\text{gal.}} \times \frac{8 \text{ pt.}}{1 \cancel{\text{gal.}}} = 8 \text{ pt.}$

Now we can check the unit prices.

$\frac{\$3.75}{8 \text{ pt.}} \approx$ **0.47 per pt.** **\$1.75 per pt.**

It is now obvious that the gallon is the much better deal.

Liquid Measurements

1 gallon (gal.)
= 4 quarts (qt.)
= 8 pints (pt.)
= 128 fluid ounces (fl. oz.)

Dry Measurements

16 ounces (oz.)
= 1 pound (lb.)

If you need a refresher on unit conversion, see Lesson 9.

⋙ When needing to decide which is less expensive per unit, sometimes you do not need to determine the unit prices but can round and look at how the sizes compare, and then use that to see how the prices compare.

Example: Which costs less per unit: a \$3.99 10-oz. bag or an \$8.99 1-lb. bag?

We know 16 oz. = 1 lb.

If we round 1 lb. to 15 oz., we can see that 10 oz. is a little less than $\frac{2}{3}$ of a lb. The 1-lb., or 16 oz., bag is about \$9.

$\frac{2}{3}$ of that would be \$6, since \$3 + \$3 + \$3 = \$9.

Since the 10-oz. bag is well under \$6, it is the better deal.

10 is $\frac{2}{3}$ of 15.

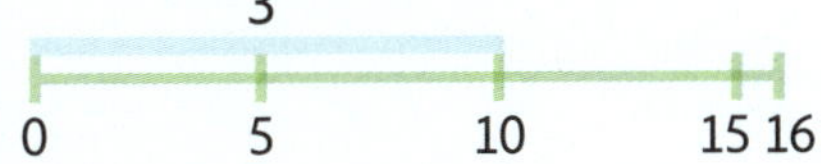

Remember, while the larger size is sometimes the best deal, it is not always. Doing the math will help you identify what is really less expensive.

KEEPING PERSPECTIVE

Often, it makes sense to purchase whatever has the lowest unit price. But there may be times where you decide it is worth buying the more expensive one. For example:

- You might decide that an organic vegetable is worth the extra price per ounce.
- You might decide you should only buy a smaller package even though it is more per unit because you know you will never eat the larger one before it goes bad.

Unit Conversion: Metric Measurements

In other parts of the world, the metric system is used to measure. Speed limit signs in Canada, for example, are given in kilometers instead of miles. Milk is sold in liters.

Metric measurements are also used in scientific fields in the United States. They are important to know.

One helpful thing about the metric system is that it is all based on 10, making it easy to convert between units within the metric system.

The main unit for distance is the **meter** (m.), for liquid capacity the **liter** (l.), and for mass (similar to weight) the **gram** (g.). Prefixes in front of meter, liter, and gram are used to specify smaller and greater units.

Stop and look at the metric chart.
If you start at the bottom of the chart, each value is 10 times the previous one (for example, $\frac{1}{100}$ is $10 \times \frac{1}{1{,}000}$).

Because the metric system is based on 10, it is easy to mentally convert between metric units.

>>> When converting within the metric system, you multiply or divide by 10 or a power of 10 (like 100 or 1,000).

Example: Convert 40 kilometers to meters.

We can see in the chart that 1 km. = 1,000 m.

$40 \cancel{\text{km.}} \times \frac{1{,}000 \text{ m.}}{1 \cancel{\text{km.}}} = 40{,}000 \text{ m.}$

This multiplication could have been done mentally by moving the decimal place to the *right* 3 places.

Three 0s

40 × 1,000 = 40,000 Decimal place moves to the right 3 places.

In the example, we ended up multiplying 40 km. to m. We were going from a larger unit to a smaller, and if you are measuring something in a large unit, it is going to take *more* small units to measure the same thing (so *multiply* a larger unit to convert to a smaller one). Conversely, if you are measuring something in a small unit, it is going to take much *less* of a large unit to measure the same thing (so *divide* a smaller unit to convert to a larger one).

- larger unit to smaller >>> end up multiplying
- smaller unit to larger >>> end up dividing

Metric Chart

	Distance
Each unit is 10x previous row	1 *kilometer* (km.) = 1,000 meters (m.)
	1 *hectometer* (hm.) = 100 meters (m.)
	1 *dekameter* (dam.) = 10 meters (m.)
	meter (m.)
Each unit is previous row divided by 10 (i.e., $\frac{1}{10}$ of previous row)	1 *decimeter* (dm.) = $\frac{1}{10}$ of a meter (m.); 10 dm. = 1 m.
	1 *centimeter* (cm.) = $\frac{1}{100}$ of a meter (m.); 100 cm. = 1 m.
	1 *millimeter* (mm.) = $\frac{1}{1{,}000}$ of a meter (m.); 1,000 mm. = 1 m.

See full chart on page 347.

If you need a refresher on multiplying and dividing by 10 and powers of 10, see Lesson 1.

Example: Convert 40 meters to kilometers.

We can see in the chart that 1 km. = 1,000 m.

$40 \cancel{m.} \times \frac{1 \text{ km.}}{1{,}000 \cancel{m.}} = 0.04 \text{ km.}$

This division by 1,000 could have been done mentally by moving the decimal place to the *left* 3 places.

Three 0s

40 ÷ 1,000 = 0.040 **Decimal place moves to the left 3 places.**

Converting Between Systems

Page 347 in the **Reference Section** lists how some measurements in the system typically used in the United States (called the **U.S. Customary System**) relate to their equivalent in the metric system.

Reminder, a conversion ratio is a fraction where both the numerator and denominator represent the same amount, only in different units.

>> Converting between systems is simply a matter of writing a conversion ratio worth 1, as we have been doing with other conversions.

Example: Convert 20 inches to centimeters.

Conversion ratio from chart on page 347: 1 in. = 2.54 cm.

$20 \cancel{in.} \times \frac{2.54 \text{ cm.}}{1 \cancel{in.}} = 50.8 \text{ cm.}$

>> You can round the conversion ratio to the nearest unit to get approximate conversions on the go.

Example: Approximately how many gallons is 40 liters?

3.79 l. = 1 gal.

Round the 3.79 liters that equal a gallon to 4.

We are converting from a smaller unit (liters) to a larger unit (gallons), so we need to divide.

Dividing 40 by 4 gets 10, so 40 liters is about 10 gallons.

KEEPING PERSPECTIVE

God gave people the creativity to do things differently. We do not all have to use the same measurement systems. The creativity amongst people reflects God's creativity, as we are all in His image.

O Lord, how manifold are your works! In wisdom have you made them all; the earth is full of your creatures (Psalm 104:24).

Unit Conversion: Currency

Not every country uses dollars and cents. If you travel out of the United States or support missionaries who travel, understanding how to convert between currencies (i.e., money systems) is an important skill.

- Converting between currencies works much the same way as converting between units of measure.
- The main difference is that with currency conversion, **the relative worth of one currency compared to another (i.e., the exchange rate) is constantly changing**, meaning the conversion ratio will too.
  - An online search can easily determine the current exchange rate. Make sure to get a parent's permission before searching online.
  - If you are in a different country, when you go to exchange dollars for that currency, you will find out the current exchange rate.

>>> Once you know the exchange rate, you can set up a conversion ratio and multiply.

Example: Convert 50 rupees to dollars if 1 Indian rupee = \$0.012.

Conversion ratios: $\frac{1 \text{ rupee}}{\$0.012}$ or $\frac{\$0.012}{1 \text{ rupee}}$

We want to use the ratio that will cause the rupees to cancel out.

$50 \cancel{\text{rupees}} \times \frac{\$0.012}{1 \cancel{\text{rupee}}} = \$0.60$

Example: Convert \$50 to rupees.

This time, we will use the other conversion ratio so the dollars cancel: $\$50 \times \frac{1 \text{ rupee}}{\$0.012} \approx 4{,}166.67$ rupees

Converting between currencies is really the same as converting between other units of measure.

While there are various tools that can convert between currencies for you, it is still helpful to know how to do it on your own, especially if you ever travel abroad.

Converting on the Go

Now, the method we just looked at is not that easy to do in your head. How would you figure out on the go how much a shirt that costs 4,167 rupees is costing you in dollars?

>>> To convert a different currency to dollars mentally, think of the equation $\textit{currency} \times \frac{\$}{\textit{currency}}$ and complete the multiplication mentally, rounding and using other mental math strategies as needed.

You might find a different mental technique works better for you, especially on some problems. That is fine.

Example: Approximate 4,167 rupees in dollars if 1 rupee = $0.012.

Round $0.012 to $0.01.

$4{,}167 \text{ } \cancel{\text{rupees}} \times \frac{\$0.01}{1 \text{ } \cancel{\text{rupee}}} = \$41.67$

(We could perform this multiplication mentally by remembering that multiplying by 0.01 is the same thing as dividing by 100 (0.01 means $\frac{1}{100}$), which can be done by moving the decimal point to the left 2 places.)

How Accurate Is Your Approximation? The accuracy of your mental approximations is determined by how much you had to round and approximate.

Example: Approximate 2,589 Australian dollars (AUD) in U.S. dollars if 1 AUD = $1.51.

Round $1.51 to $1.50 and 2,589 to 2,600.

$2{,}600 \text{ } \cancel{\text{AUD}} \times \frac{\$1.50}{1 \text{ } \cancel{\text{AUD}}}$

Find whole portion: 2,600 × $1 = $2,600

Find decimal portion: 2,600 × $0.50; 0.50 is half of $1, so if we divide $2,600 (what we would get if we multiply by $1) by 2, we will find what half of it is.

$2,600 ÷ 2 = $1,300

Add whole and decimal portion together: $2,600 + $1,300 = $3,900

Example: Approximate 345 of Currency A if 1 Currency A = $0.20.

Round 345 to 300.

$300 \text{ } \cancel{\text{A}} \times \frac{\$0.20}{1 \text{ } \cancel{\text{A}}}$

Think about how $0.20 = $0.10 + $0.10. We can multiply by 0.10 by moving decimal to the left one place and then add that to itself to find 0.20.

300 × $0.10 = $30; $30 + $30 = $60

Example: Approximate 490 of Currency B if 5 Currency B = $1.

Round 490 to 500.

$500 \text{ } \cancel{\text{B}} \times \frac{\$1}{5 \text{ } \cancel{\text{B}}} = \$100$

Notice in the last example, where we were given what equals $1 instead of how many dollars equal 1 of the other currency, we ended up dividing by 5 instead of multiplying. Always mentally think through the relationship $\textit{currency} \times \frac{\$}{\textit{currency}}$ to figure out how to solve.

KEEPING PERSPECTIVE

We are simply continuing to apply the same basic skills over and over again in different ways.

- Since creation operates consistently, multiplying by fractions worth 1 consistently does not change the value.
- We can use that truth about how God holds creation together to convert between units of measure or currencies by setting up a conversion ratio.

Unit Conversion: Temperature & Other Conversions

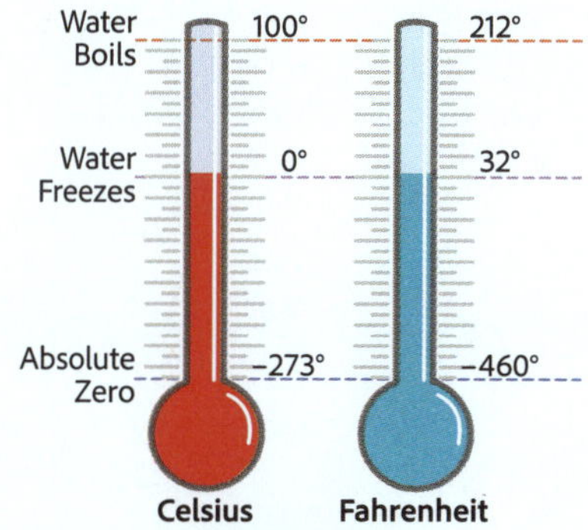

If you travel abroad, not only will you need to convert money and measurements, but you will also need to convert temperature.

- In the United States, we use a unit called Fahrenheit (°F) for most everyday temperature readings.
- In contrast, many other countries use Celsius (°C) for their everyday temperature readings.

Because both Fahrenheit and Celsius scales start 0° at different places, you can not just use a simple conversion ratio to convert between them; instead, use one of these formulas.

Formulas for Converting Between Celsius and Fahrenheit

$C = \frac{5}{9}(F - 32°)$ $F = \frac{9}{5}C + 32°$ (C = °C and F = °F)

ALGEBRA Connection

These two formulas are really the same formula, just rearranged using algebra.

Example: Convert 52° C to °F.

Plug in °C into formula for finding F: $F = \frac{9}{5}(52°) + 32°$

Perform multiplication and then addition:

$F = 93.6° + 32° = 125.6°$

We are multiplying by a fraction in these formulas. $\frac{9}{5}(52°)$ means $\frac{9}{5} \times 52°$. Enter it into a calculator as $9 \times 52 \div 5$.

Example: Convert 125.6° F to °C.

Plug in °F into formula for finding C: $C = \frac{5}{9}(125.6° - 32°)$

Solve inside the parentheses: $C = \frac{5}{9}(93.6°)$

Multiply 93.6 by 5, and then divide by 9: $C = 52°$

We are following the order of operations, solving inside the parentheses first.

Metric in the Kitchen

In the metric system, **dry goods are measured by mass (in grams) rather than by cups**. Cooks have scales in their kitchen to measure the ingredients because a gram is measured by weight, not size.

Because different ingredients weigh different amounts, the conversion ratio from grams to cups varies by ingredient. Once you know that relationship, then you can easily convert.

There are online calculators that will automatically convert various ingredients for you, as well as sites that list the relationship for various ingredients. Make sure to get a parent's permission before searching online.

Become familiar with the decimal equivalent of common fractions listed on page 346.

Example: How many cups of flour should you put into a recipe calling for 312.5 grams if 1 cup equals 125 grams? Give your answer in common kitchen measurements, rounding as needed.

$$312.5 \cancel{\text{g.}} \times \frac{1 \text{ c.}}{125 \cancel{\text{g.}}} = 2.5 \text{ c.} = 2\frac{1}{2} \text{ c.}$$

We rewrote our answer from decimal to fractions since measuring cups are labeled as fractions. This is another example of how memorizing the decimal equivalent of common fractions, which we mentioned on Day 138, proves useful.

Sometimes, you will not get an even conversion. In that case, you will have to fill up your measuring cup a little more or less to get it as close as you can.

Example: Approximately how many cups of flour should you put into a recipe calling for 400 grams if 1 cup equals 125 grams? Give your answer in common kitchen measurements, rounding as needed.

$$400 \cancel{\text{g.}} \times \frac{1 \text{ c.}}{125 \cancel{\text{g.}}} = 3.2 \text{ c.} \approx 2\frac{1}{4} \text{ c.}$$

We ended up with 0.2, which is a little less than the 0.25 that equals $\frac{1}{4}$ c. We rounded to the nearest measurer. If cooking this recipe, we should try not to fill that $\frac{1}{4}$ c. all the way to adjust for the rounding, especially if we had rounded more substantially.

>>> Once again, round to the nearest common kitchen measurer when needed.

KEEPING PERSPECTIVE

Different temperature scales are testimony to the creativity and ingenuity God gave mankind. He gave us the ability to develop different ways to record the temperature. He designed us to rule over His creation, to "subdue" this incredible world He put under our stewardship (Genesis 1:28).

It is an amazing thing to be made in God's image. Animals can not design temperature scales, nor can they know their Creator personally. Through Jesus Christ, we can. Now that is super exciting news.

Health & Nutrition

You might be surprised how often math can come in handy when understanding health and nutrition numbers. Following are a few examples.

Nutritional Facts

Nutritional labels can help you understand what is in the food you are eating. A little math can help you explore them further.

- **Want to see what the recommended daily value is for any of these values?** For example, what is the recommended daily value for sodium, given that 150 mg. is 6% of that value?

  % *of* base *is* percentage

  $0.06 \times base = 150 \text{ mg.}$

  Dividing both sides by 0.06: $base = \frac{150 \text{ mg.}}{0.06} = 2{,}500 \text{ mg.}$

- **What if you eat a double serving?** The answer is a simple *multiplication*. If one serving has 6%, then 2 servings would have 2 × 6%, or 12%.
- **Counting calories?** You can *add* up the number of calories from everything you eat to calculate the total.

Nutrition Facts

8 servings per container

Serving size **1 cup (68g)**

Amount per serving

Calories **370**

	% Daily Value*
Total Fat 5g	**7%**
Saturated Fat 1g	**5%**
Trans Fat 0g	
Cholesterol 0mg	**0%**
Sodium 150mg	**6%**
Total Carbohydrate 48g	**15%**
Dietary Fiber 5g	**14%**
Total Sugars 13g	
Includes 10g Added Sugars	**20%**
Protein 12g	
Vitamin A 10mcg	20%
Vitamin C 1mg	100%
Vitamin D 1mcg	50%
Vitamin E 2mcg	100%
Riboflavin 5mcg	75%
Folic Acid 200mcg	60%
Thiamin 2mcg	35%
Vitamin B12 5mcg	100%
Zinc 7mg	50%
Biotin 300mcg	100%
Calcium 50mcg	25%
Phosphorus 90mcg	90%
Magnesium 400mcg	100%
Chromium 75mcg	80%
Potassium 5g	100%

* The % Daily Value (DV) tells you how much a nutrient in a serving of food contributes to a daily diet. 2,000 calories a day is used for general nutrition advice.

Understanding Bloodwork

In bloodwork, there is typically a range of values that are considered normal. However, what is normal for you might be different than for someone else. You may find it helpful to keep a spreadsheet of your own levels so you can see how your bloodwork changes over time. You could even insert a graph to help explore that visually.

For example, here is a spreadsheet and graph of someone's hypothetical sodium readings from their yearly bloodwork.

	Year 1	Year 2	Year 3	Year 4	Year 5	Year 6	Year 7	Year 8	Year 9
Sodium	135	140	142	139	142	141	140	138	150

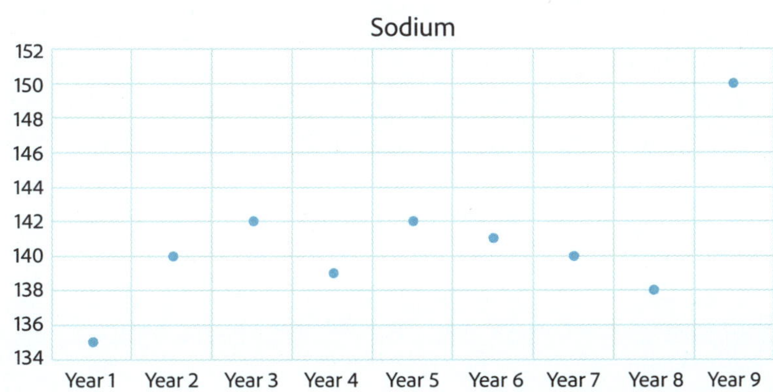

You may be able to view your bloodwork, along with graphs that show the readings over time, online. Make sure to get a parent's permission first.

Notice how a chart makes it very obvious that the last reading was abnormally high.

Doubling Up

Just as you might sometimes wish to make a different amount of a food recipe, so you might sometimes wish to make a different amount of a nutritional recipe. You can double or halve a nutritional recipe the same way you would a regular recipe.

Example: You have a recipe for an essential oil rub that calls for 2 ml. of an oil, and you want to halve the recipe. How much should you put in?

$\frac{1}{2} \times 2 \text{ ml.} = \frac{2}{2} \text{ ml.} = 1 \text{ ml.}$

Sometimes, though, you might want to modify a recipe so that it uses a certain amount of one ingredient. In that case, you will have to first figure out by how much you want to increase or decrease the recipe, similar to what we did with pan sizes.

» *amount to multiply each ingredient by* = $\frac{\textit{amount you want to use of one ingredient}}{\textit{amount recipe calls for of same ingredient}}$

» Once you know the amount to multiply each ingredient by, multiply all the other ingredients in the original recipe by that amount.

Example: You have found the following recipe for an essential oil rub. If you want to make a batch that will use 10 ml. of the coconut oil instead, how many drops of each oil should you put in?

3 drops Oil A **5 drops Oil B** 5 ml. coconut oil

amount to multiply each ingredient by = $\frac{10 \text{ ml.}}{5 \text{ ml.}} = 2$

Oil A: 2 × **3 drops** = 6 drops

Oil B: 2 × **5 drops** = 10 drops

We would need 6 drops of Oil A and 10 drops of Oil B for 10 ml.

KEEPING PERSPECTIVE

You might be surprised at how often a math concept will come in handy in life. Our bodies remind us of our need for God's redemption — sin has made a mess of this world, and our bodies feel the effects of sin each day. But there is hope. Jesus suffered and died so that death might be defeated and that eternal life would become the ultimate destination for all who believe in Him.

Jesus said to her, "I am the resurrection and the life. Whoever believes in me, though he die, yet shall he live" (John 11:25).

Review & Case Study: Mission Work & the Dollar's Buying Power

I hope you have enjoyed seeing a few other ways basic math skills apply. As you continue through life, you will encounter many other situations where math applies. When you face a real-life math problem, stop and think about the basic operations.

- Are you needing to *combine* quantities to calculate a *total*? Use addition.
- Do you need to *take away* a quantity from another quantity or *compare* quantities to calculate the *difference* between them? Use subtraction.
- Are you trying to *add repeatedly* or calculate a portion *of* another value? Use multiplication.
- Do you need to *separate* a quantity or amount into piles or calculate a price *per* some unit of time or measure? Use division.

You might need to look up a formula in order to figure out how to make a calculation. **Just know that you have the basic skills you need to help you through most daily life math applications.**

In this lesson, we also looked a lot at different units of measure. The existence of different measuring systems reminds us of two important truths:

1. God gave man creativity, which is a good thing.
2. Much of math is like a language system — and that language can vary.

But no matter what system of measure we use, quantities still operate consistently because behind the language of mathematics, there is a consistent universe held together by a faithful, never-changing God.

"For I the Lord do not change" (Malachi 3:6).

Those investing in businesses abroad also need to keep exchange rate fluctuations in mind.

Case Study: Mission Work & the Dollar's Power

Since we covered currency rates this lesson, we will talk about the effect that changes in exchange rates have on missionaries. When I went to Thailand in the early 2000s, the equivalent of a mere $8 was enough to hire a personal cab for the afternoon, while just a few dollars yielded a meal at an up-scale restaurant. I loaded my suitcase with goodies yet hardly spent much at all.

Imagine if you were a missionary to a country where a few dollars a day was enough to live off of — and then the value of the dollar went down. Suddenly, what used to cost you a few dollars costs you much more. You would need more support in a hurry.

Understanding exchange rates can help you comprehend the challenges missionaries may face so you can better support them.

Savings: Understanding Saving Options

income = giving + ***saving*** *+ spending*

Hopefully by now you have seen the importance of saving money. In this lesson, we are going to touch on some different ways to invest what you save so that your savings can grow and multiply. People spend lifetimes studying investing; we are only going to scratch the surface here, hopefully giving you some general knowledge that can help you either study further or better evaluate what a financial advisor might suggest.

Key Investment Concepts

Following are some basic investment concepts we will revisit a lot.

Reward – When you are investing, you are looking to gain money — that is, for a **return on investment** (**ROI**). You need to know what you earned to calculate an ROI. Calculate earnings by looking at how the investment has changed value. On some investments we will look at later, you will also add in anything you have received for owning the investment along the way.

- *earnings = ending value – $ invested*
- $ROI = \frac{earnings}{\$\ invested}$ List ROI as a percent.

The ROI formula is the same as the percent formula rearranged to solve for the percent $\left(\% = \frac{percentage}{base}\right)$ and with descriptive words for investments. ROI finds the % (the *ROI*) of the base (the *$ invested*) the percentage (the *earnings*) is.

Example: What ROI did you make if you spent $500 on stocks and later sold them for $600?

Calculate earnings: $600 – **$500** = **$100**

Calculate ROI: $\frac{\$100}{\$500} = 0.20 = 20\%$

- A negative ROI indicates a loss.

Example: What ROI did you make if you spent $500 on stocks and later sold them for $400?

Calculate earnings: $400 – **$500** = **–$100**

Calculate ROI: $\frac{(-\$100)}{\$500} = -0.20 = -20\%$

You lost, not gained, 20% this time.

Risk – Think of **risk** as attempting to measure how much your anticipated ROI varies. Look at the graphs. Notice that in the risky one, the value (which would in turn affect your ROI) changes drastically. If you invest at point A and then sell at point B, you would lose money (i.e., have a negative ROI). If you invest at point B and sell at point C, you would gain money (i.e., have a positive ROI). Typically, the more risk, the more potential for reward, as well as for losing money.

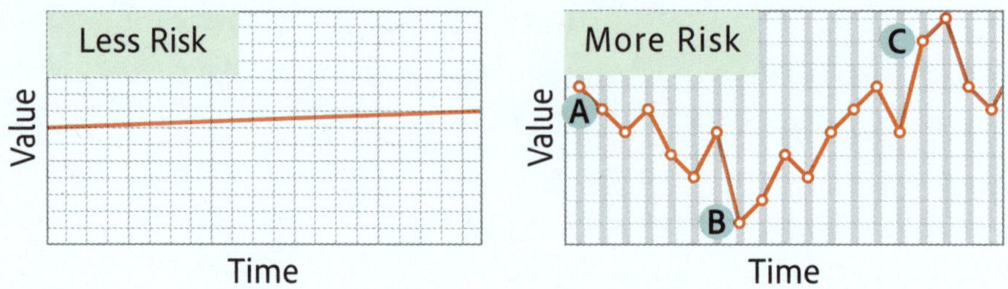

Liquidity – **Liquidity** refers to the ease of taking money out of an investment. Some investments make you promise to keep your money in them a certain amount of time. On risky investments, even if you do not have to keep your money in long term, you might need to, as you would not want to take it out at a low point, since you would lose money if you did. Typically, the harder it is to take money out of an investment (i.e., the *less* liquid it is), the more the reward.

Eternal purpose – You are supporting what you invest your money in. Without funding, companies can not stay in business. As you invest, think about the eternal impact your money is having too.

Faith-based investment options exist. Some of these use **negative screens**, eliminating companies that support certain activities, such as abortion. Others use **positive screens**, looking for companies that benefit people or have Christian values or management.

KEEPING PERSPECTIVE

When considering different investments, **stop and ask yourself, "What is the purpose for this money?"** That question can help you navigate through the many options.

- Think through the everyday purpose — when do you need the money and how risky can you be with it?
- Think through the eternal purpose — does the investment make sense in light of eternity?

Ask God to renew your mind when it comes to investing so you will be able to "discern what is the will of God" (Romans 12:2).

Supply & Demand

Imagine yourself stranded in a desert with several other people. You find one person selling water bottles, but it is only one person and only a few bottles — not nearly enough for everyone stranded. How much would you be willing to pay to make sure you got one of those bottles? More than you would be willing to pay normally at the grocery store, right?

- **Supply** indicates how much of a resource is available. In the desert scenario, there is a low supply of water bottles. There are not enough for everyone.
- **Demand** refers to how many people want a resource. In the desert scenario, there is huge demand for water bottles. Everyone needs the water.

>>> Low supply and/or high demand drives prices up.

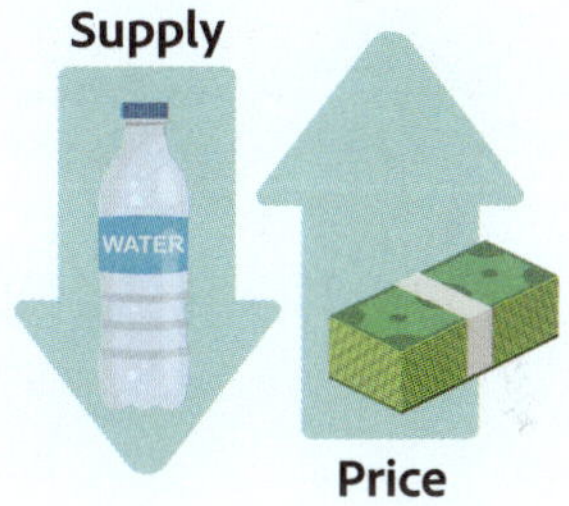

- In the desert scenario, you would end up paying a lot for a water bottle.
- If you buy your home when lots of people want to buy and few want to sell (high demand and low supply), you will end up paying extra.

>>> High supply and/or low demand drives prices down.

- If you were somewhere where no one wanted a water bottle and there were plenty, you would end up paying very little for water.
- If you go to sell your home during a time when few people want to buy and lots want to sell (low demand and high supply), you might have to sell for less than you paid.

The same holds true for any investment. The price depends on the supply and demand.

- There is the **supply and demand of individual investments**. A stock price rises when lots of people want it and lowers when only a few do.
- There is also a **more global supply and demand.** How many people want a specific type of investment? How many people are willing to lend money versus those who want to borrow it?

When thinking of investing, many people think of a **stock market**, which is where **stocks**, or small ownerships in a company, can be bought or sold, which is called trading stocks. The graph shows how the Dow Jones Industrial Average, which can be an indicator of how the stock market as a whole is doing, went up and down during 10 sample years.

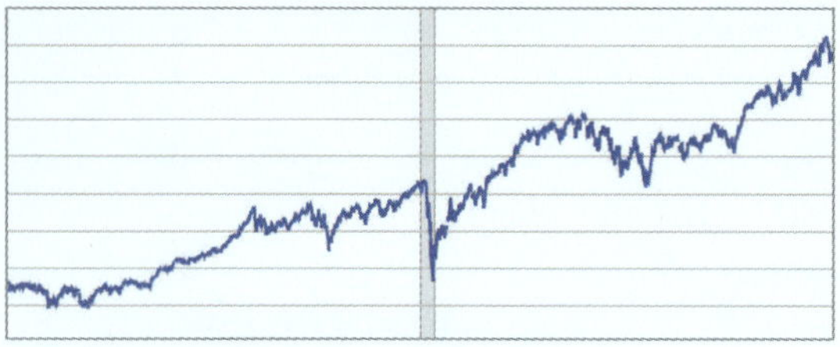

Dow Jones Industrial Average Over 10 Sample Years

There are times when stocks have been in higher demand, driving prices up ... and times when more people have chosen not to invest, causing prices to fall (low demand).

For another example, the interest banks offer depends on the following:

- how many people want loans (the demand)
- the money that is available to borrow (the supply)

The supply and demand (and thus the interest rates) are affected by various factors, including the following:

- **inflation** (think the changing value of the dollar)
- how much the Federal Reserve charges for their loans to banks.

Think of banks like gas stations. As the price of gasoline goes up and down, gas stations adjust their prices. But some sell for less than others, so it still pays to shop around. In a similar way, while the global supply and demand of money sets the general price of interest and certain types of investments like CDs and bonds, every bank still sets its own rate, so it pays to shop around.

Applying Supply & Demand

Understanding supply and demand is key to investing.

- **If you can buy when the prices are lower and sell when the prices are higher, you will make money.** If you do the reverse, you will lose money.
- **What investment is best will vary as the global supply and demand varies.** As we look at different investments, keep in mind that the supply and demand for them fluctuates, and thus which one is the best for you may fluctuate too.

KEEPING PERSPECTIVE

Figuring out when the prices are low and on their way up is not easy, and no one, not even those who do it for a living, does it perfectly. Complex computer algorithms try, but they also can not predict the future. Only God knows what will happen. In the end, we can seek wise counsel and do reasonable research to make what appear to be wise choices, but we can not expect to figure out what will happen. In light of our limited abilities, we will look at ways to invest in different categories so that hopefully if one loses value, another will increase to make up for it.

Many are the plans in the mind of a man, but it is the purpose of the Lord that will stand (Proverbs 19:21).

Variable Investments with Positive ROI (Savings Accounts)

As their name implies, **variable investments** vary in what they earn.

- **Some variable investments only vary in how much they earn.** For example, savings accounts vary in the interest they pay, but you are always earning something.
- **Other variable investments vary in whether they earn *or* lose money.** Stocks and funds are investments in companies. If the company makes money, so do you. If they lose money, so do you. There is much more risk with these investments but also a potential for greater reward.

We will start by looking at a type of variable investment that, while their returns vary, always yields something: savings accounts.

Savings Accounts

Savings accounts come with different names and details — there are **savings accounts**, **high-yield savings accounts**, and **money markets**. But the basic idea is the same. The name is not as important as seeing what the account actually offers.

We already looked at how interest in savings accounts can add up and at how you can calculate your ending balance at any point using the Compound Interest Formula from Lesson 3. Only it is even easier with savings accounts, as they advertise their interest rate as an **Annual Percentage Yield (APY)** — a rate that already takes any compounding done during the year into account. Unlike yearly rates that needed divided by the number of times per year the compounding occurred, **plug the APY directly into the formula**. Since the APY is the rate with the compounding for a year, **use the number of years as *t*.**

These accounts work well for money you need soon, as you can often take money out at any time, knowing you will have gained, not lost, money.

Compound Interest Formula $P = P_i(1 + r)^t$

$$\textit{ending balance} = \textit{beginning balance}(1 + \textit{growth rate})^{\textit{number of times compounded}}$$

Example: Calculate the ending balance of $1,000 invested at 6.5% APY after 10 years.

Plugging in values: $P = \$1{,}000\ (1 + 0.065)^{10} \approx \$1{,}877.14$

Notice we used the APY as r and the years as t.

»» As we saw in Lesson 3, here is how to calculate the total interest (which would also be your earnings):

$$\textit{total interest} = \textit{ending balance} - \textit{beginning balance}$$

Example: What was the interest earned (or earnings) in the previous example?

$\$1{,}877.14 - \$1{,}000 = \$877.14$

You can then use that to calculate an ROI.

Example: Calculate the ROI for the previous example ($1,000 invested at 6.5% APY after 10 years).

$$ROI = \frac{\text{earnings}}{\$\text{ invested}} = \frac{\$877.14}{\$1{,}000} \approx 0.8771 = 87.71\%$$

Start saving early.

When you are young, it is easy to put off the idea of saving for the future. Yet we saw in the preceding examples that 10 years of savings netted an ROI of 87.71%. That means that after 10 years, an APY of 6.5% would let you earn 87.71% of what you invested.

The earlier you start saving, the longer you can keep your savings earning interest before you need it. That allows a little bit to grow into a lot more.

Account Considerations

- **Make sure your account is FDIC (banks) or NCUA (credit unions) insured**; so long as the bank or credit union is insured by one of these government agencies, the government will cover your money if the bank or credit union goes under.
- **Check if there are any fees.**
- **Look at what the minimum balance requirements are** — that is, how much money has to stay in the account.
- **Consider the bank or credit union's convenience.** Some have local branches; others are online only.
- **Look at the rate offered.** Rates are competitive.

KEEPING PERSPECTIVE

Then you ought to have invested my money with the bankers, and at my coming I should have received what was my own with interest (Matthew 25:27).

When addressing the servant who hid his master's money, the master used a bank (or a similar concept of the day) to point out that the servant could have at least earned interest. Bank accounts are not where you are going to multiply your savings greatly; rather, they serve as a safe holding place with growth for the money you need to access.

Variable Investments with Positive *or* Negative ROI (Stocks)

While savings accounts vary in interest, they only vary in what you make and save. Other investments, such as stocks, funds, and assets, have the potential to either make *or* lose money.

- The risk of these investments is greater, but so is the potential for reward.
- Generally, variable investments that tend to gain or lose are best for **money you do not need for awhile**. That way, you have flexibility to wait to take money out at a high point rather than a low point. But be aware that a stock may not necessarily go back up — companies go out of business too. So variable investments are also **best for money you can afford to lose**.
- It is best to not put all your money in variable investments.

Example of the Value of a Hypothetical Variable Investment Changing Over Time (notice ups and downs)

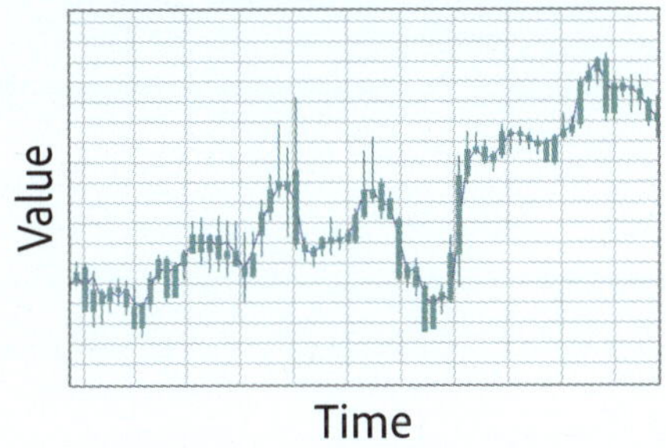

Stocks

Stocks are a tiny percent of ownership in a company and are a common variable investment. When you invest in stocks, you can buy just one or multiple **shares** (ownership) in the company; the stock is priced by the share.

>>> To calculate the total $ invested in a stock, multiply the price per share by the quantity of shares purchased.

Example: What is the total for 5 of a stock that is $40.56?

5 × $40.56 = $202.80

>>> Divide the total $ invested by the price per share to calculate the quantity of shares purchased.

Example: How many shares did you buy if you bought $202.80 worth of a share that cost $40.56 each?

$\frac{\$202.80}{\$40.56} = 5$

>>> *Gain or loss of a stock = selling price – $ invested*

If you want to make money with stocks, plan to invest a lot of time researching companies or, as you get more money, hire a financial advisor to do so for you. Even then, know there is a huge risk. Seemingly stable stocks can suddenly dip, resulting in significant loss.

Example: If you bought a share for $15 and sold it for $20, what did you gain or lose?

$20 – $15 = $5

You gained $5.

Example: If you bought a share for $15 and sold it for $10, what did you gain or lose?

$10 – $15 = –$5

You lost $5.

Day Trading: Stocks change prices throughout the day. Day traders will seek to capitalize on this, but it is more risky and time consuming.

⋙ Your total gain or loss (i.e., your earnings) would be your gain or loss per share times the number of shares you bought.

Example: If you bought 10 shares that gained $40 each, what were your earnings?

10 × $40 = $400

Some stocks also offer **dividends**, which is a share of the company's profits with those who own its dividend-paying stock.

Having some dividend stocks may help provide some earnings when stocks go down, although dividends are not guaranteed.

⋙ Add the dividends received to the ending value when calculating your earnings, as it is something you earned that is not included in the stock's ending value.

The idea behind finding your earnings is to subtract your investment from everything you have earned. That includes how much the investment itself has gained (its *ending value – $ invested*), as well as any other earnings (like dividends) not included in that.

Example: If you buy a dividend-paying stock for $100 and later it is worth $50, but you got $10 in dividends, what was your earnings?

($50 + $10) – $100 = –$40

Without the dividend, you would have lost $50; with it, you only lost $40.

Stock prices rise and fall — sometimes drastically. Your ROI with them thus varies greatly. You might more than double your investment or lose all of it within a few days.

Researching Stocks

Ideally, you want to buy a stock for a low price and sell it for a higher one. How do you know what a stock will do? You do not. That is why stocks are risky. There are, however, numbers you can look at to make better-informed choices. Earnings per share (EPS) is one of many numbers you could examine.

Earnings Per Share (EPS) is basically the business' income divided by the number of shares of its stock in existence $\left(\frac{income}{\#\ of\ shares}\right)$. You typically will not need to calculate the value but rather compare a company's EPS with that of other similar companies. **Higher is better.**

KEEPING PERSPECTIVE

When you think about investing, remember that *no one* — not the best financial analyst, yourself, or a computer algorithm — but God knows the future. Remember the Bible's caution that "Wealth gained hastily will dwindle" (Proverbs 13:11), and be careful.

Fixed Income Investments (Certificate of Deposits/CDs)

A **fixed income** investment stays steady, as you are promised a *fixed* reward (i.e., a fixed income) in return for the use of your money, sometimes for a *fixed* amount of time.

You buy a fixed income investment that earns 5.3%.

- If interest rates dip to 4%, you will still earn 5.3%.
- If interest rates rise to 8%, you will still earn 5.3%.

Because of their fixed rate, fixed income investments have little risk. Because of the fixed time commitment, they tend to earn more than a savings account, although less than riskier investments.

CDs

One common fixed income investment is a CD. A **CD** is a certificate of deposit. In a CD, the following holds true:

- You are investing your money for a fixed period of time (often months to years — called the **term**). The finish date is called its **maturation date**. Your interest rate is only fixed until the maturation date.
- You may have a minimum investment amount. CDs typically require at least a $500 to $1,000 investment.

Tip: If you think the interest rates are about to dive, fixed investments might make good purchases, as they will keep you earning a higher rate of interest.

At the maturation date of a CD, you get your deposit ***plus*** the interest you earned at the rate you agreed to. Ever since Lesson 3, we have used the Compound Interest Formula.

Compound Interest Formula $P = P_i(1 + r)^t$

$$\text{ending balance} = \text{beginning balance}(1 + \text{growth rate})^{\text{number of times compounded}}$$

Since CDs earn compound interest, we can use this formula with them too. Only as we work with investments, we will rename *ending balance* as *future value* and *beginning balance* as *present value*.

>>> **Future value** (ending amount at a future date) and **present value** (the amount you currently have) are common investing terms.

$$FV = PV(1 + r)^t$$

$$\text{Future Value} = \text{Present Value}(1 + \text{growth rate})^{\text{number of times compounded}}$$

>>> A CD at 2.3% APY means your deposit will earn 2.3% each year.

- While compounding frequency can vary with CDs, the APY used to advertise their rate adjusts for that. You can compare APYs like you would for savings accounts. A CD earning 3% APY will earn more than one earning 2.5% APY, regardless of compound frequency.
- Calculate the future value, then the earnings and ROI from that information.
- For CDs with terms less than a year, the *t* will be a portion *of* 365, and you would technically need to use an annual rate rather than an APY, as the APY is calculated for yearly increments. We will not be asking you to find future values for CDs with terms less than a year in this course. If you need to in real life, there are an abundance of future value calculators you could use.

Example: You buy a \$1,000 CD with a 3-year term at 2.3% APY. How much do you make at the end of 3 years (i.e., what is the *future value* after 3 years)? What were your earnings? What is your ROI?

Remember, APYs are already adjusted for compounding, so you can use them as the *r* and the years as the *t*.

Calculate the future value:

$$FV = PV\,(1 + r)^t = \$1{,}000\,(1 + 0.023)^3 \approx \$1{,}070.60$$

Earnings after 3 months: **\$1,070.60** – **\$1,000** = **\$70.60**

$$ROI = \frac{\$70.60}{\$1{,}000} = 0.0706 = 7.06\%$$

It does not matter if you use *P* or *FV* to represent the ending amount or P_i or *PV* to represent the initial amount. Symbols can vary. In the course, though, **we will start using future value (*FV*) and present value (*PV*) so you will be familiar with the terminology**.

KEEPING PERSPECTIVE

CDs can be helpful, steady investments to have as part of your investment plan. Because no one but God knows the future, having some fixed income investments can be prudent.

Fixed Income Investments (Bonds Plus More on CDs)

We will now explore a different type of fixed income investment — bonds — and look at how they differ from CDs.

>>> A bond is a loan.

Comparing CDs and Bonds

CD	Bond
Typically requires $500 to $1,000.	Typically requires $500 to $1,000.
A type of bank or credit union account that can be FDIC or NCUA insured.	A loan to the government or a business (so if the business goes under, you may not get your money back).
Money earns compound interest.	Money earns simple interest.
You do not receive interest until the end.	You receive interest in payouts every 6 months.
At maturation date, you get original investment and interest.	At maturation date, you get original investment back.

Considerations:

- **Liquidity** – What happens if you need your money back before the maturation date? CDs purchased through banks or credit unions are not sellable; you can also buy them through brokerage accounts, and those can be sold. Bonds can be sold.

- **Risk** – The main risk is if whoever is offering the CD or bond goes out of business. To mitigate, look for FDIC-insured banks or NCUA-insured credit unions you think will stay in business (for CDs), and businesses or government entities you think will do well (for bonds). Bonds are rated, and the ratings can help you pick safer bonds.

One key thing to note about bonds is that **they earn simple, not compound, interest**. That means that you do not earn earnings on your earnings — only on your initial investment. To earn more of a compound effect on your bond interest, reinvest the 6-month interest payments (called **payouts**) in other investments each time so it earns interest.

>>> The **coupon rate** is the simple interest percent the bond pays each year. A bond with a 6% coupon rate earns 6% simple interest each year.

Tip: While hard to find and sometimes riskier, there are bonds out there that are loans to churches or organizations seeking to make a difference in the world.

The example on the next page illustrates how bonds work; you can consult it when figuring out how to find various values for a bond.

Example: The calculations are shown for a $1,000 bond; 3-year term; coupon rate of 5.3%.

For yearly interest, multiply coupon rate by total bond value.	$0.053 \times \$1{,}000 = \$5.30$
For 6-month payout, divide yearly interest by 2.	$\frac{\$5.30}{2} = \$2.65$
For earnings over life of the bond (i.e., total earnings), multiply yearly interest by years of bond.	$\$5.30 \times 3 = \$15.90$
For what you receive back at maturation, just look at the original value.	$1,000
For ROI, use formula.	$\frac{\$15.90}{\$1{,}000} = 0.0159 = 1.59\%$

U.S. Treasury Bonds

U.S. Treasury bonds **work differently than other bonds**.

There are also different types of treasury bonds, so their details can vary.

- Unlike other bonds, you can purchase in small dollar amounts.
- Some are fixed income investments while others are variable investments in which the interest varies.
- While some pay out every 6 months, others act more like CDs, earning compound interest that is not paid out until the bond is **redeemed**, that is, turned in for the payment.
- Unlike with other bonds, you have to redeem the bond; the government will not automatically send you your money.
- Unlike other bonds, U.S. Treasury bonds can be redeemed before their maturation date, although if you do so too soon, you may have a penalty.

You do not earn any interest on U.S. Treasury bonds after their maturation date, so **do not forget to cash them in.**

As an illustration of how these bonds can vary, if you are given a $25 U.S. Treasury I bond (a specific series of treasury bonds) when you are born, it will not mature for 20 years. All along the way, though, it earns compound interest at a rate that changes every 6 months. Any time after 5 years, you could redeem it without a penalty for $25 plus whatever compound interest it has earned.

KEEPING PERSPECTIVE

The Lord will open to you his good treasury.... And you shall lend to many nations, but you shall not borrow (Deuteronomy 28:12).

A bond is a loan where you lend money in exchange for interest. Having money to lend is a gift from God. He told the Israelites it was part of His blessing to them if they followed His way. While nothing in the Bible promises Christians financial blessing (the Apostles often found themselves in prison), when you have it, know that whatever you have is entrusted to you by God.

Diversification

Diversification means that you do *not* just pick one option; instead, you pick several. By doing so, you limit what you will lose if an investment does not do well. To explore the concept, we are going to calculate actual dollars gained or lost from investments and look at some hypothetical examples.

>>> Multiply dollars invested by percent gained or lost to calculate dollars gained or lost. Use negative numbers for losses.

$ gained or lost = ***percent gained or lost*** × ***dollars invested***

Example: Grayson invests all $10,000 of his savings in Company A, of which he has heard great things. Company A's stock prices drop by 50%. How much money will he have lost if he sells his stocks now?

−0.50 × **$10,000** = −$5,000

This example is not diversified.

Example: Instead of investing all $10,000 in Company A, Grayson invests in the companies shown, who then have the gain or loss shown. How much money will he now have lost or gained if he sells his stocks?

This example shows diversification — investing in a variety of options.

	$ Invested	% Gain or Loss	$ Gained or Lost
Company A	**$2,000**	**50%** *loss*	**−0.50** × **$2,000** = −$1,000
Company B	**$2,000**	**10%** *gain*	**0.10** × **$2,000** = $200
Company C	**$2,000**	**5%** *gain*	**0.05** × **$2,000** = $100
Company D	**$2,000**	**5%** *loss*	**−0.05** × **$2,000** = −$100
Company E	**$2,000**	**8%** *gain*	**0.08** × **$2,000** = $160

Total earnings = −$1,000 + $200 + $100 + −$100 + $160 = −$640

The examples illustrate the importance of diversification. In the diversified example, even though one investment lost 50%, Grayson only lost $640 overall. When he did not diversify, he lost $5,000. **Diversification helps decrease risk.**

Even with diversification, there are times when the whole market crashes and *all* investments that are not fixed income investments crash too. **Be careful about investing everything in variable investments.**

Funds Help Diversify

Funds provide an easy way to diversify your investments.

Mutual Funds:	ETFs (Exchange-Traded Funds):
Money is pooled together and professionally managed for a fee.	Money in the fund is invested in other investments; the ETF is traded like a stock itself; you buy shares in the ETFs just as you do with stocks.
The money is invested in investments (such as stocks or bonds) that match a specific category (such as small, mid, or large companies) or the fund's objective (such as long-term growth).	ETFs traditionally buy 1 share of every investment (such as stocks or bonds) in a category (or buy selected category investments they think will result in similar rewards). If the category does well, you should too ... and vice versa.
Tend to have higher management fees than ETFs.	

Example of the Value of a Hypothetical Fund Changing Over Time (notice more moderate ups and downs than the stock chart on page 283)

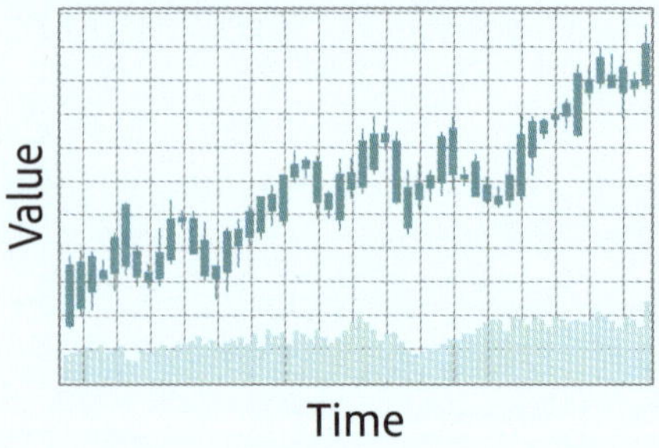

Ways to diversify:

- **Owning funds over individual stocks helps you diversify**, as the funds invest in different stocks within a category.
- **Owning several different categories of funds** diversifies further, as you are then investing in different *categories* of stocks.
- **Adding in other more stable investments like CDs and bonds (or funds that invest in these)** further diversifies and lends protection against losing too much.

More about funds:

- Funds vary in their risk based on the type and category of investments they buy.
- Funds have management fees; these fees vary by the fund, so they are something to look at when picking a fund.
- The name of the fund typically indicates what they are investing in.
- Before you pick a fund, look at all the different types and think through what best fits with the purpose of the savings you are investing.
- The **percent equity** of the fund means the percent invested in stocks (as opposed to fixed income investments). The higher the value, the more is invested in stocks.
- Biblically based funds exist that seek to invest in companies doing good or not involved in objectionable practices.

KEEPING PERSPECTIVE

Give a portion to seven, or even to eight, for you know not what disaster may happen on earth (Ecclesiastes 11:2).

We do not know what catastrophe lies around the corner. Only God knows that. By spreading money into different investments both within each type of investment you choose and by incorporating different types of investments, you can help shield against catastrophic loss.

More Variable Investments with Positive *or* Negative ROI (Assets & Direct Investments)

Assets are another way to invest money. An **asset** is anything of value, such as the following:

- real estate
- gold
- cybercurrency
- art, coins, baseball cards, gemstones, and antiques, among many other possibilities.

Assets are variable investments. You could make or lose money on them. Some have more risk of loss than others. Gold, for example, has historically held its value (although it has lost some at times), cybercurrency does not have much history, and housing markets go up and down. In general, do your research and get counsel before investing in an asset.

Asset consideration:

Keep in mind that some assets may require work. Fixing up a run-down house to sell at a higher value is labor intensive.

There is a risk of theft or damage to many assets. Think through how to keep your assets as safe as you can.

Assets tend to be harder to liquidate (i.e., turn into cash), with the exception of precious metals like gold.

Real Estate Investments

Now we will explore real estate a little further, as houses are common investments. Some of the same principles would apply to other assets.

- Due to realtor fees and closing costs, you typically have to expect a substantial increase in home value for buying and selling homes to yield a high ROI.
- If the housing market crashes, your equity in the home can *decrease* instead, meaning you will lose money.
- Renting can make money on the home or lose it if it stays vacant too long or is damaged.
- Real-estate investing can give you an opportunity to steward resources and create opportunities for families.

Every investment or job is an opportunity to serve the Lord.

For rentals, simple subtraction can help you figure out what you made.

- Subtract your yearly costs from the rent you receive:
  rental earnings = ***rent received*** – ***costs***
- Be consistent in the period of time you use — subtract monthly costs from monthly rent, or yearly costs from yearly rent. This may require additional steps.

Mortgage: $1,000 per mo.

Maintenance: $200 per mo.

Property tax: $1,200 **per yr.** (Notice this is per year and the others are per month.)

Rent received: $2,000 per mo.

Example: Calculate the yearly amount you make on your rental home given the cost and rent shown.

We will first calculate the property tax per month so as to use the same period of time for all: $\frac{\$1,200}{12} = \$100$

Now that we have the rent and all the costs in the same period of time, we can calculate the earnings.

rental earnings = **$2,000** – (**$1,000** + **$200** + **$100**) = **$700**

Yearly earnings: 12 × **$700** = $8,400

Direct Investments

While publicly traded companies sell stocks, not all businesses are funded that way. There are other ways to invest in businesses, such as through **private equity investments** (investments in a business not on the stock market) and **venture capital** (riskier investment in a startup). As with stocks, these make you a part owner in the company.

- Investing directly in a business can give the investor a significant say in how the business is run, giving those with business knowledge a chance to influence a company in a godly way.
- Has a huge potential for reward if the company does well, but also for loss if the company does not succeed.
- There are Christian companies that specialize in this type of investing.
- While there are some limited opportunities at low dollar amounts (such as through crowd funding), a lot of direct investing requires meeting certain monetary or financial training requirements.

Another way to invest directly in a business is by **investing in your own business**. Owning a business can be a wonderful way to shine God's light in the world — but it is also a huge commitment, and one not to be embarked on lightly.

KEEPING PERSPECTIVE

As you consider investments of any kind, remember that wealth can disappear in a moment. While wise investing can lead to steady growth of wealth, many people have lost their entire life savings with investments gone amiss. *Always* make sure your eternal treasures are safe in heaven.

Do not lay up for yourselves treasures on earth, where moth and rust destroy and where thieves break in and steal, but lay up for yourselves treasures in heaven, where neither moth nor rust destroys and where thieves do not break in and steal. For where your treasure is, there your heart will be also (Matthew 6:19–21).

Review & Investing in Eternity

As you think about investing, always remember to think through the purpose for the money you have saved.

- **Everyday purpose** – When do you need the money and how risky can you be with it?
- **Eternal purpose** – Does the investment support why you are here on the earth?

Do not forget to also look at the big picture: the stock market as a whole, interest rates, and other factors. **Look for the best options considering the current investing climate, and remember to diversify.**

Investing in Eternity

More and more, Christians are questioning how to use not only their giving but *all* their money for God's Kingdom. When you make any kind of investment in a stock or fund, you are really investing in a business. As we saw in the last section, if you have more to invest, there are ways to carefully invest in privately owned companies that are not on the stock market (private equity). Or if you open up your own company, you are investing in that.

No matter how the investment is made, **the businesses you invest in can either help others or hurt others in terms of what they offer and how they operate**.

While you might never have enough money, talent, or desire to start your own business or invest directly in a company, know that small investments you make matter in a bigger picture too. **Think through not only what you can *earn* through an investment but also *what your investment is accomplishing*.**

Page intentionally blank

18

Savings: Mechanics & Retirement

*income = giving + **saving** + spending*

Now that you know some of the options available for investing money you have saved, we will explore some of the mechanics of actually making and managing those investments. We will also look at planning for your old age, as when you are young is the best time to start.

Investing Mechanics

So how do you actually go about making an investment?

Some investments, such as savings accounts, CDs, and sometimes gold, can be made at a bank or credit union. Others, like real estate, are obtained through other means such as a realtor. But many investment options trade on a stock market. People commonly go through either an online **brokerage account** or a financial advisor.

Great care needs taken when creating a brokerage account or hiring a financial advisor. Get a parent's permission before signing up for anything. Brokerage accounts are often tied to your bank accounts, so it is important they be safe and legitimate. Financial advisors, likewise, have access to your money and need to be trustworthy. Even then, you should continue to do some research on your own and pray through their recommendations.

Researching and understanding the investments you make is critical, no matter how you invest.

Good financial advice can save you money in the end; bad advice can cost you dearly.

Ratings/Assessments – Various advisors and firms evaluate stocks and mutual funds, giving their assessment of the investment on various websites.

Company Numbers – We briefly mentioned in the last lesson that earnings per share are helpful to look at. There are many other aspects of a company's financial information that can help you decide if a company looks like a good investment.

Graphs – There are many kinds of graphs to help you see how an investment is doing. For example, in the graph shown, if the fund's category is the green and the blue is the fund, the fund is doing better than most funds in that same category.

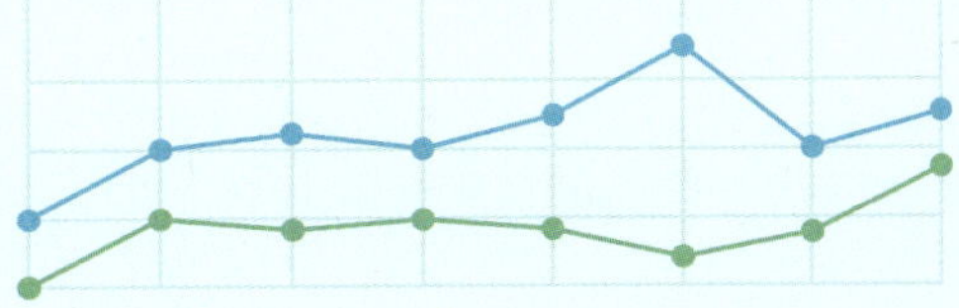

Hiring an Advisor

What you buy or sell is driven by what you *think* will happen. While no one *knows* what will happen, researching a business, their management, their history, the market, and other factors can all help you make better-informed guesses. Good financial advisors do the bulk of the research and then either advise you on what to do or do it for you.

If you hire someone to help you manage your money, here are a few things to check out:

- **Their beliefs** – You are hiring an advisor for their advice, so make sure he or she is giving you godly advice. After all, you want godly wisdom, not a worldly perspective. There are lists of Christian financial advisors. But look carefully at an advisor's beliefs and their philosophy. Do they care about what their investments support in the same way you do? Ask lots of questions before you choose them.
- **Fee structure** – How are they getting paid? Are they taking a percent of your earnings? Are they paid a set fee? Does their pay structure incentivize them to use certain funds or stocks? Some financial advisors will have agreements with funds to get some payment from them when they invest in those funds. In that case, be aware that they are financially incentivized to invest in those funds, even if that might not be the best investment at the time.

Most advisors will want to assess your entire financial situation to make sure you have a solid plan set up. That is because if you have not thought of something, it can lead to different investments.

KEEPING PERSPECTIVE

Where there is no guidance, a people falls, but in an abundance of counselors there is safety (Proverbs 11:14).

- The information and ratings you find serve as counselors, providing data that can help you make wise investments.
- If you have a financial advisor, they are another counselor. They may take the counsel of others too.

Annualized ROIs & Rule of 72

In this section, we are going to take a look at a couple of formulas that can help you evaluate options. Know that there are many others — and many use math skills you know.

Annualized ROI

In the last lesson, we looked at calculating the return on an investment, or ROI. It can be helpful to calculate an **annualized** (i.e., adjusted to be *annual*, or *yearly*) **ROI** — that is, a rate that tells you what % the investment is averaging per year. The formula, which can be expressed different ways, is very simple:

>>> $\text{annualized ROI} = (1 + ROI)^{\left(\frac{1}{\text{years}}\right)} - 1$

Example: Calculate the annualized ROI of an ROI of 60% after 6 years.

$\text{annualized ROI} = (1 + 0.60)^{\left(\frac{1}{6}\right)} - 1 \approx 0.0815 = 8.15\%$

On a calculator, $\frac{1}{6}$ gets input as (1 ÷ 6).

An annualized ROI lets you compare how much an investment makes with other investments that advertise a yearly, or annual, rate. Keep in mind, though, that you also want to compare other aspects, such as the risk and liquidity of different investments. Stocks that earned 8% one year might not earn 8% the next.

Example: Which earns more: an investment with an annualized ROI of 8.15% or one with an APY of 9%?

9%, as 9% is more than 8.15%

More on Present and Future Value

In the last lesson, we looked briefly at rewriting the Compound Interest Formula using the terms *future value* and *present value.*

$$\text{Future Value} = \text{Present Value}\,(1 + \text{growth rate})^{\text{number of times compounded}}$$

$$FV = PV\,(1 + r)^t$$

>>> You can use an annualized ROI as the growth rate in the Future Value Formula.

Example: Approximate the future value of a \$2,000 investment if it has an annualized ROI of 6% after 10 years.

$FV = \$2{,}000(1 + 0.06)^{10} \approx \$3{,}581.70$

Keep in mind that this finds an *approximate* value, since the annualized ROI is really an average, not the exact amount being earned each year. **Use yearly compounding,** as you can not assume the growth will happen steadily throughout the year.

Rule of 72

>>> It is easy to double an amount: just multiply it by 2. Double $1,000 is $2,000, as 2 × $1,000 = $2,000.

The Rule of 72 estimates the years needed to **double** an investment's value (i.e., have it equal 2 times its original value) in situations where earnings are invested back into the investment, such as with compound interest or when you keep reinvesting earnings on an investment.

>>> **Rule of 72** $\text{years to double} = \frac{72}{\text{annual growth rate}}$

- The *annual growth rate* could be an APY, an annualized ROI, or some other yearly rate. **Input it into the formula without the % sign (e.g., input 5% as 5).**
- This rule finds an **estimate, not an exact, number of years**. After all, your growth rate might vary, and even if exact, this formula approximates the answer.

Notice that we did not list 6% or 0.06 in the denominator; instead, we used 6 to stand for 6%.

Example: About how long will it take a $10,000 investment earning 6% APY to double?

$\text{years to double} = \frac{72}{6} = 12$

ALGEBRA Connection

We arrived at the *Rule of 72 Rearranged* by applying basic algebra to the *Rule of 72* equation, as shown using *y* for the *years to double* and *r* for the *annual growth rate*.

Rule of 72: $y = \frac{72}{r}$

Multiplied both sides by *r*:
$y \times r = 72$

Divided both sides by *y*:
$r = \frac{72}{y}$

Sometimes, you might want to figure out what annual, or yearly, growth rate is needed to double an investment. You can do so with the *same* formula, just rearranged.

>>> **Rule of 72 Rearranged** $\text{annual growth rate} = \frac{72}{\text{years to double}}$

In this formula, the answer you get is already in % form. Do not multiply it by 100 when you add a % sign. For example, an answer of 5.2 means the % annual growth is 5.2%.

Example: Calculate the *annual growth rate* needed to double an investment in 10 years.

$\text{annual growth rate} = \frac{72}{10} = 7.2$, which in this formula means 7.2%

KEEPING PERSPECTIVE

The first written record of the Rule of 72 was by an Italian friar (similar to a monk) named Luca Pacioli back in the late 1400s. This same friar was known as the "Father of Accounting" for his contributions to the field.

Always remember that the formulas you encounter in books did not just show up; people used their God-given abilities to find shortcuts for solving real-life problems and describing God's creation.

Mechanics of Checking Accounts

While checking accounts often do not earn interest, they typically make it easy to get money out (i.e., **withdraw** it) and in (i.e., **deposit** it). Thus, **they are helpful places to store temporary savings you plan to use to cover bills.**

Withdrawals & Deposits

- **Cash** – Some banks or credit unions have or let you use certain ATMS and/or offices to withdraw money (and, in the case of an office, to deposit it).
- **Checks** – Checks can be ordered from the bank, credit union, or an external supplier (which is often less expensive, although be careful to find one you trust, as you have to give your bank account information).
- **Electronic Withdrawals Through Your Bank or Credit Union** – It is common today for people to pay bills through an online account by entering in the information for the person/company you wish to pay. The bank or credit union then either mails a check or deposits that money into that person's/company's account.
- **Electronic Withdrawals or Deposits Through Other Sites or Services** – Many websites have a way to pay them for their services. Electronic withdrawals can also be set up to enable other sites or businesses such as your employer to automatically deposit into your account using the account's **routing** and **bank numbers**. These numbers can be found on a check or often in an online account.
- **Transfer** – A **transfer** is a withdrawal and deposit combined — where you withdraw from one of your accounts and deposit to another. Electronic transfers from accounts at the same bank are often instantaneous, while those to accounts at other banks may take a few days. A **wire transfer** is instantaneous. However, there is often a fee for a wire transfer on both the outgoing and/or incoming bank or credit union's side. They are typically used when moving large amounts of money, such as for a home down payment, from a savings account to a checking account.

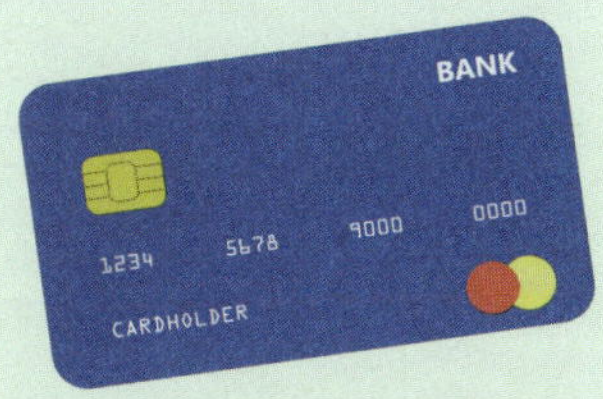

Debit cards are another way to get money out of a bank. They work similar to credit cards, except the money comes automatically out of your bank account each time. If you have one, be careful not to lose it.

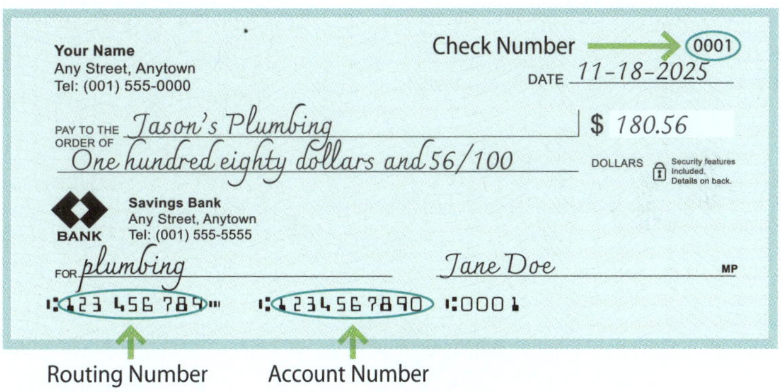

Be sure to get a parent's permission if you want to pay bills online or use online banking. It is important to discern which sites are reputable since you will be giving out your bank information.

When depositing a check from someone else, make sure you sign the back.

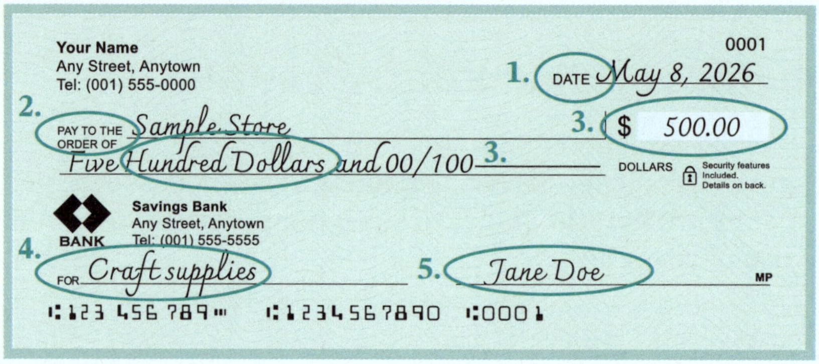

Parts of a Check

- **(1) Date** – Put the date you are writing the check here. Be careful to get the year right — the check may not be cashable if it is too old or written with a future date.
- **(2) Pay to the Order Of** – Here is where you put the person or business to which you are writing the check.
- **(3) $ Box and $ Line** – Write the amount you are paying, written with both numerals and words. When writing with words, be sure to include the cents. If it is an even, or whole, amount, use the word "even," as in "five dollars even." Or you can write "and 00/100" or "and No/100." You can even use a line afterward. That prevents anyone from adding to the check.
- **(4) For (or Memo)** – The *For* or *Memo* box is optional, but it gives you a spot to mark what the check was for. It can be helpful to notate the purpose, which can later help you and the recipient remember its purpose. You can also use it to designate funds to a specific project or area of a charity.
- **(5) Signature Line** – The last line is for your signature.

KEEPING PERSPECTIVE

Discussions of withdrawals and deposits remind me of an analogy for salvation I once heard and like to use with people. When we trust in Jesus, it is like He not only pardons an incredible debt but also deposits all His righteousness, wisdom, and other spiritual riches to our account. Picture a spiritual bank account going from negative 1 trillion to positive 1 trillion, and you have a tiny glimpse into what Jesus has done for those who believe Him.

And because of him you are in Christ Jesus, who became to us wisdom from God, righteousness and sanctification and redemption (1 Corinthians 1:30).

In whom are hidden all the treasures of wisdom and knowledge (Colossians 2:3).

Checking Your Checking Account

It is important to keep track of the money going into and out of your account. Otherwise, you could end up initiating a withdrawal or writing a check for more money than you have. This can lead to both fees and embarrassment.

There are different ways to keep track of your account to avoid **overdrafting**, which is pulling more money out than you have.

Checkbook Register

- **Checkbook Register** – Checks come with checkbook registers in which you can write down withdrawals or **payments** (money out) and **deposits** (money in). Years ago, when most transactions were done by paper checks, checkbook registers proved to be a handy way of keeping track of your account's balance. The checkbook register could then be compared every month to the checking account's monthly statement to make sure nothing had been forgotten and that there had been no errors (by you or the bank or credit unit). The balance column was kept current by subtracting the payment or adding the deposit.

Check Number	Date	Memo	Payment Amount	Deposit Amount	$ Balance
	3/1	Opening Balance			2,500
172	✓ 3/2	Cell Phone Company	45		2,455
173	3/9	My Favorite Charity	400		2,055
	✓ 3/18	Withdrawal	30		2,025
	✓ 3/18	Deposit — gift		50	2,075

✓ entries the statement also shows

Checkbook registers can still prove an effective way to track your account balances, only **be sure to also record in the register any electronic deposit or withdrawal you make.**

Check Number	Date	Memo	Payment Amount	Deposit Amount	$ Balance
	3/18	Paycheck Auto Deposit		1,000	3,075
	3/19	Online Purchase	89 50		2,985 50

- **Online Account** – With the advent of online banking, there are now other ways to keep track of checking account balances. If you have an online account, you can, with a parent's permission, simply log in to see the balance at any time. But beware: That balance does not include checks you have written that have not been cashed or electronic withdrawals or deposits that are scheduled for the future on other sites. **You still need to find a way that works for you to keep track of those.**

Tracking Transactions

Your bank or credit union does not always know about every withdrawal you have made. We say the transaction **clears** when the money actually moves out of (or into) your account. You can track transactions that have not cleared yet in a spreadsheet (with a parent's permission) or on paper. Once a transaction is showing up in the bank or credit union, you can either delete it from your list or put an X or a ✔ in a "Cleared Bank" column so you know to ignore it. Basic math can then help you adjust for whatever has not cleared the bank.

	A	B	C	D
1	**Cleared Bank**	**Date**	**Amount**	**Payee**
2		1/20/24	$ 121.50	Check 58 — Water Company
3	X	1/29/24	$ 300.00	Church (electronic withdrawal)
4		1/30/24	$ 50.00	Check 59 — Gift

Spreadsheets will also let users filter columns (as shown); here is the same sheet filtered for rows that have not cleared the bank.

With a parent's permission, you can learn more about filters in your spreadsheet's help tool.

	A	B	C	D
1	**Cleared Bank**	**Date**	**Amount**	**Payee**
2		1/20/24	$ 121.50	Check 58 — Water Company
4		1/30/24	$ 50.00	Check 59 — Gift

Example: Your bank shows a balance of $3,001.83, but you see from a list that number does not include $552.82 of scheduled withdrawals. After these transactions clear, how much money will you have in the account?

Subtracting withdrawals on the list that have not cleared:
$3,001.83 – $552.82 = $2,449.01

»» Writing a check you do not have enough money in the account to cover is financially costly. Since we live in a fallen world where our memories do not always work well, you would do well to do the following:

1. Have a little extra money in the account as a sort of bill buffer.
2. Have a system for tracking withdrawals.

KEEPING PERSPECTIVE

Whatever your system, you should look over checking account statements or online transaction listings (with a parent's permission) periodically to make sure there are not any transactions that should not be there. While it is rare, should you experience some kind of error or theft, you want to catch it sooner rather than later.

IRAs & 401ks

Now we will look at two types of accounts that help with retirement savings: IRAs and 401ks. View an IRA or 401k as a container within which you can put money into savings, CDs, bonds, stocks, and other investments. For example, if you had $10,000 in your 401k, you might have $5,000 of that money invested in one type of investment and $5,000 invested in another.

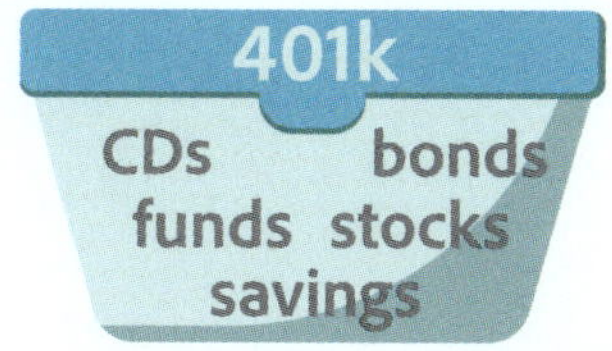

>>> Making investments within an IRA or 401k container adds certain rules and benefits.

Rules

- 401ks are only available through certain employers.
- Money invested in 401ks or IRAs can not be withdrawn without a steep penalty, except for certain situations, until you reach a government-set age. These "containers" are designed to contain money whose purpose is for retirement.
- You are limited in the amount you can put in these accounts each year, and required to start withdrawing money at a certain age (72 in a sample year).

Self-Employed? There are accounts that, like 401ks, can be used to save money tax-free.

Benefits

Many 401ks and IRAs (such as **traditional** ones) have some nice perks.

Employer Matching

As we saw on Day 106, some employers will match what you contribute up to a percent of your income.

Savings on Taxes on Money Deposited

401ks and some IRAs allow you to not have to pay taxes on what you deposit, called your **contributions**.

>>> Use your marginal tax bracket to calculate the taxes you would have paid on your deposit.

This assumes all the contribution would be at your marginal tax rate; if it is enough to bump into the higher tax bracket, you would need to calculate some at a higher rate.

Example: Savings on $5,000 with a marginal tax rate of 10%.

0.10 × $5,000 = $500

Example: Savings on $5,000 with a marginal tax rate of 22%.

0.22 × $5,000 = $1,100

>>> **This benefit only applies to certain IRAs.** Some IRAs, such as Roth IRAs, do *not* allow you to deposit tax free. Make sure you understand the IRA you are getting.

Saving Taxes on Earnings

In bank accounts and regular investing accounts, you have to include whatever you earn on your investments as income on your taxes. But you do not have to do this for earnings in 401ks and IRAs, which results in additional savings, which you can again calculate by multiplying by your marginal tax bracket (assuming the earnings would not have been enough to bump you into the next bracket).

> ***Example:*** Calculate the savings on earnings of $5,000 if earned in a 401k instead of in a taxable account if the Stinders are in the 32% tax bracket.
>
> $0.32 \times \$5{,}000 = \$1{,}600$

You will have to eventually pay taxes on the money saved tax free whenever you withdraw it from the 401k or IRA. The thinking, though, is that you will owe taxes when you are not earning as much money and are thus in a lower tax bracket.

> ***Example:*** Years later, the Stinders withdraw $5,000 from their IRA for the year. They are in the 12% tax bracket now. How much will they have to pay on it in tax?
>
> $0.12 \times \$5{,}000 = \$600$

Notice that the Stinders paid $1,000 less in taxes in the second example than in the first example. By saving when you are earning more money, you can lower your taxes when you are in a higher tax bracket and put away more money for the future. These savings can add up. Look at how $1,000 saved on taxes that is invested at 6% APY grows in 30 years.

$$FV = \$1{,}000(1 + 0.06)^{30} \approx \$5{,}743.49$$

Every $1,000 you save when you are young compounds into much, much more over time.

KEEPING PERSPECTIVE

Diligently working and setting aside money now for the future is one way you can plan ahead. Remember to plan ahead for your eternal future too. The Bible says our lives here are but "a mist that appears for a little time and then vanishes" (James 4:14). Eternity, on the other hand, is eternal. Let that reality affect how you live in *every* area, including your financial decisions.

The plans of the diligent lead surely to abundance, but everyone who is hasty comes only to poverty (Proverbs 21:5).

Saving Enough Plus Inflation Basics

One of the sad realities of how sin has messed up the world is that our bodies are now in the process of decaying and dying. As you get older, there will likely come a point where you can not do everything you can do now. For some, that means they are not able to earn income like they used to. At the same time, expenses may rise, as medical needs often increase as you age.

We are going to shift our focus to how to use investments to help save for the more distant future. You might think **retirement** (i.e., when you stop holding a job) is a long way off, but the best time to start preparing for it is when you are young.

Inflation

Have your parents or an older adult ever mentioned how much less something cost in their childhood? We call the rise in price of goods and services over time **inflation**.

The thing about inflation is that it is constantly happening. As of the writing of this course, the Federal Reserve's target has been to have 2% inflation in a year, but inflation varies from that, with some years having much more inflation and others less. Assuming the inflation rate is positive, prices rise each year.

>> Calculate the cost with inflation by multiplying by 100% *plus* the percent of the inflation.

Remember to move the decimal to the left 2 places to rewrite the percent as a decimal, just as you have been (e.g., 104% = 1.04%).

Example: If inflation rises by 4%, how much would a gallon of milk that used to cost $2 cost?

Total with inflation: 100% + 4% = 104%

Calculating 104%: 104% × $2 = 1.04 × $2 = $2.08

>> If you want to estimate the value of something over multiple years given average inflation instead, use the Future Value Formula with the average annual inflation rate as the growth rate, compounded yearly.

Example: About what would a $900 monthly rent be after 40 years at 2% inflation per year?

$FV = PV\,(1 + r)^t$

$FV = \$900(1 + 0.02)^{40} = \$1{,}987.24$

Because of inflation, everything will likely cost quite a bit more money when you retire than it does now. If your annualized ROI matches the rate of inflation, then your wealth is not really growing;

you are just preserving that money's current buying ability in an inflated world. For your wealth to truly grow, **you want your investments to yield an ROI that is greater than the rate of inflation at any given time**.

We will refer to the rate that is adjusted for inflation as the **real rate of return**. The rate can be approximated by this formula:

real rate of return = growth rate – inflation rate

Example: Approximate the real rate of return of $10,000 if it has an annualized ROI of 25% but inflation has been 5%.

real rate of return ≈ 25% – 5 % = 20%

Saving Enough

How much do you need to save for retirement? The idea is to save enough to live at your chosen lifestyle, accounting for inflation and any changes in those costs you expect in old age. Insurances and social security benefits can also help you reach that amount. **While savings needed may be quite significant, keep in mind that this is the *future value* of the savings, not the actual amount saved.** Saving $1,000 a year, at the end of 30 years at 6% APY, yields about $80,000. You do not have to save what you will need but rather what will grow into what you need.

Paying off a home is another way to save for retirement. If your mortgage is paid off and you own a home before you retire, you will not have a housing expense (except maintenance and taxes). Should you need to move, you would have the money from the home to help.

To calculate the future value when you are making ongoing deposits (like $1,000 a year), you have to use a different future value formula than the one we covered, which is why we did not show how we got $80,000. There are future value formulas (and future value calculators that do the calculations for you) for different situations.

You are probably years away from having to figure out how much to actually save, but remember the following:

- Start reasonably saving as soon as possible.
- Try to get an ROI that is higher than the inflation rate.

KEEPING PERSPECTIVE

While you might retire one day, you do not ever stop the work God has put you here on earth to do. Once you start working, it may be tempting to live for retirement; instead, live for your eternal future *beyond* retirement, making the most of each day God gives you here on Earth.

For we are his workmanship, created in Christ Jesus for good works, which God prepared beforehand, that we should walk in them (Ephesians 2:10).

Life & Long-Term Care Insurance

While hopefully death and long-term care needs are a while away for you, they are important to think about as you plan to save for the future, especially once you have a family dependent upon you.

Life Insurance

Life insurance offers whoever you pick a lump sum of money if you were to die. As with other insurance, **the more coverage you get, the more you pay in premiums**, which means you have to think through how much to really get. Many insurance calculators may estimate high amounts, but the estimates may not be affordable.

When deciding how much to buy, think about paying off any debts/obligations (including burial) and helping anyone dependent on you until they can get on their own feet. Keep in mind that a family's needs will change after someone dies. The family might not need two cars anymore, the house might be able to be paid off with the life insurance lump sum (thereby eliminating that ongoing expense), a single dad might now need to hire help for the children, children might get some Social Security money — the list of changes could continue.

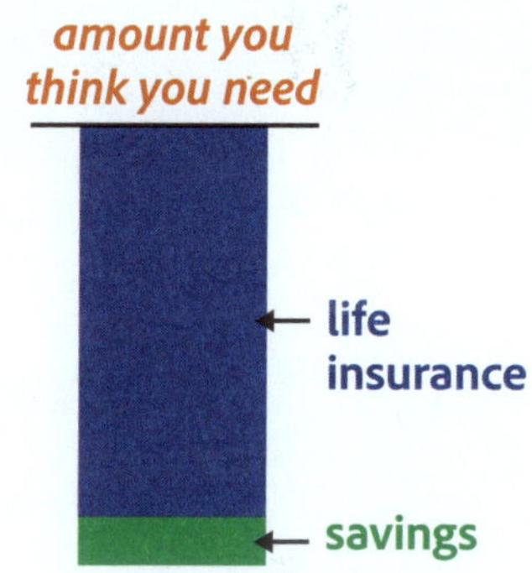

>>> Once you have decided how much is needed if you were to die, subtract what you have saved from that to help you decide how much life insurance to buy.

life insurance = *amount you think you need* – *savings*

Example: Mr. Johnson wants to have enough for his wife, if he were to die, to pay off his $200,000 home loan and $15,000 burial expenses and have $80,000 a year for 2 years to live on as she adjusts and grieves. If he has $60,000 saved, how much life insurance would he need?

life insurance = **$200,000 + $15,000 + (2 × $80,000)** – **$60,000** = $315,000

The amount of life insurance needed will likely change throughout your life. Life insurance is more important while you are working on building savings, especially if you are supporting a family. As you build up more cash in savings and pay off your home — or if you do not have people relying on your income — you probably do not need as much life insurance. This is good, as **life insurance goes up in price as you age**.

There are different types of life insurance, including these:

Term	Whole
Only lasts while you are subscribing to it.	Continues your *whole* life; more of an investment

>>> Some employers also offer **accidental death and dismemberment insurance**. This is not the same as life insurance, in that it only applies if you are seriously injured or die in a qualifying accident. It would not, for example, cover someone who dies of cancer.

Long-Term Care Insurance

Caregiving costs are extensive. **Long-term care insurance** is an insurance designed to help cover long-term care.

>>> The benefit of long-term care insurance is the payout received minus all of the premiums paid for the insurance; if a negative value, you paid more than you received.

$$\text{benefit} = \text{payout} - \text{premiums}$$

Premium is the periodic fee for an insurance. We covered this term in Lesson 7.

Payout refers to what the insurance pays on your behalf.

We multiplied \$300 by 12 to find the yearly premiums, and by 20 since paid for 20 years.

Example: How much did the Kylers *save* from long-term care insurance if they had \$300 a month premium for 20 years but the insurance company paid out \$300,000 for them?

$\text{benefit} = \$300{,}000 - (20 \times 12 \times \$300) = \$300{,}000 - \$72{,}000 = \$228{,}000$

Example: How much would the Kylers have *spent* that they did not need to if they had this insurance and never needed long-term care?

$\text{benefit} = \$0 - (20 \times 12 \times \$300) = \$0 - \$72{,}000 = -\$72{,}000$

A negative benefit is the same as what they spent that they did not need to. They spent \$72,000 extra.

- Long-term care insurance saves a significant amount of money if you use it, but it costs a significant amount if you do not.
- Long-term care insurance premiums are higher the later in life you buy it. The earlier you buy, though, the longer you have to pay premiums.
- Some companies offer long-term care in combination with life insurance, where if you do not need the care, there is some benefit to whoever you leave behind upon your death.

KEEPING PERSPECTIVE

We all like to think that we will stay healthy forever. But the Bible warns us that this world is "groaning" with the effects of the curse (Romans 8:22). As Christians, we are not exempt from this.

Whether or not you buy insurances, your faith should ultimately be in God's provision. While we can do what we can to prepare for life in this fallen world, He is ultimately our "Shepherd" who promises to care for both us and our loved ones.

Other Long-Term Saving Tools & Financial Preparations

After you stop earning a paycheck, how do you get money?

Well, hopefully you have saved up money in a 401k or IRA from which you can begin to withdraw money. In this section, we are going to take a look at two other ways to get extra income and at some other important financial preparations.

Social Security

As we saw back on Day 31, you and your employer pay Social Security tax on what you make. In exchange, when you reach a certain age (which changes as the program changes) or have a disability or circumstance that lets you withdraw early, you should be able to start getting an amount back from Social Security — a paycheck, so to speak.

- The amount of your Social Security check is based on how much you made and for how many years. If you have a lower income during your working years, you will likely have a lower income in retirement years from Social Security than someone who made more while working.
- The Social Security administration has an Online Benefits Calculator or Quick Calculator that people use to get an idea of what they can expect based on their income for each year. The calculator offers a current estimate — the government had to adjust some aspects of the program in 1983 when it neared bankruptcy, and it may need to do so again, depending on the ratio of people paying into the program versus pulling out of it.

Get a parent's permission before using online calculators.

Annuities

Annuities are sort of like privately funded Social Security available through insurance companies. They vary greatly. In some, you pay a fixed amount, and then the annuity will pay you a fixed payout in your older age. Annuities also provide a way to not have to pay taxes on earnings until later, when you will likely be at a lower tax bracket.

Be careful to get advice before getting an annuity. There are a lot of different types of annuities, and their helpfulness greatly varies based on both the annuity and the purchaser's situation.

Charitable Annuities
Some charities allow you to set up annuities with them, where your money will go toward supporting the organization while you get a fixed payout each month or year.

Other Financial Preparations

While few want to think about their illness or death, there are some financial documents that are wise to have in order before any unexpected tragedy, especially if you have people depending on you for provision. Here are some common ones.

- A **power of attorney** assigns permission to someone to make decisions for you should you become unable. There are different types of power of attorneys — some cover medical and some cover financial decisions.
- A **will** lets everyone know what should be done with all the resources God has entrusted to your care. Having a will in place makes life much easier for those you leave behind and makes sure that your resources go where you meant them to. A will is especially important if you have children one day, as you will want to specify who will care for them and think through financially providing for them.
- Rather than immediately giving money to others or charities, some wills set up a **trust**, where money is managed and distributed slowly or after a time. Trusts can be a way to care for children until they are old enough to manage money themselves; they can also be used to continue charitable giving.

Math Helps Evaluate

Basic math can help throughout the process of planning for retirement. For example, basic addition and subtraction can help you figure out what year you are planning to retire. **Rather than learning rules for each situation, just think through what you are trying to find and how to find it.**

Example: You are 15 and you plan to retire at 60. In what year will you retire if it is 2024?

To calculate the years left, compare the ages with *subtraction*, finding the *difference*: 60 − 15 = **45**.

Add that number of years to the current year: 2024 + **45** = 2069.

KEEPING PERSPECTIVE

There is a lot to think through when it comes to planning for your retirement and death, especially once you have people dependent on you that you want to make sure are taken care of. Again, there are tools to help, along with godly financial planners that specialize in helping you work through it all.

While it is important to think through the financial side, do not forget to focus on the most important preparations. Do you know your eternal destination? You can, through Christ Jesus, God's Son.

> *Whoever has the Son has life; whoever does not have the Son of God does not have life* (1 John 5:12).

If heaven is your destination by faith in Jesus, are you living in such a way to lay up eternal treasures when you get there?

Review

I hope this lesson has encouraged you to think ahead. Be purposeful about how you live in every area, including your finances. As God prospers you and you are able, make preparations for the future.

As you think about planning for the future, also think about investing in the eternal future. The Bible warns that there will be some who get to heaven but find all their earthly work burnt up, so to speak — meaningless in eternity. Others will have an eternal reward from what they allowed the Lord to do through them here on earth. There are eternal investments to be made in how we live today.

> *For no one can lay a foundation other than that which is laid, which is Jesus Christ. Now if anyone builds on the foundation with gold, silver, precious stones, wood, hay, straw—each one's work will become manifest, for the Day will disclose it, because it will be revealed by fire, and the fire will test what sort of work each one has done. If the work that anyone has built on the foundation survives, he will receive a reward. If anyone's work is burned up, he will suffer loss, though he himself will be saved, but only as through fire* (1 Corinthians 3:11–15).

You might think you will never have wealth to worry about investing money. But you can still think about how you are investing your time and talents. And you never know what God has for you down the road. Plus, even if you don't get great riches, you might have a 401k or other retirement accounts in the future to invest. Do not think that just because you have a small amount it does not make a difference — think of what Jesus did with just 5 loaves and 2 fish (Matthew 14:17–19).

We are all on a journey with the Lord. Investing well — whether of time, talent, or treasure, and whether the amounts be large or small — is part of that journey.

Page intentionally blank

Estimating Taxes, Business Math, & Review

It is time to review a lot of the concepts we have looked at by applying them to business settings, starting with estimating taxes. We will see how concepts we have looked at in this course can help you in business settings as well, touching on how work is part of God's plan and a place you can minister.

>>> On your exercises in this lesson, you are going to start reviewing for the year. Look back in the book as needed to answer questions.

Estimating Taxes

While taxes are filed once a year, they are paid throughout the year. It is important to be able to somewhat accurately estimate them.

- **Overpaying** taxes throughout the year means the government essentially borrowed money from you without interest.
- **Underpaying** taxes throughout the year could result in a penalty.

If you have an employer, you will use a W-4 form to estimate your taxes. While we will look at that next, we will first walk through a sample 1040-ES Worksheet that is used to estimate taxes for self-employed income. Understanding the worksheet will help you review taxes and better understand what the W-4 is doing behind the scenes.

Review tax terms and concepts in Lesson 4 if needed. You can also view the entire Sample Estimated Tax Worksheet on page 343 in the **Reference Section**.

Line 1: Estimate your adjusted gross income. Here is where you list all the income you plan to earn.

Sample Estimated Tax Worksheet Keep for Your Records

1	Adjusted gross income you expect in Sample Tax Year (see instructions)	1	46,000

Line 2: List either the standard deduction or your estimated itemized deductions. 2b is for qualified business income deductions, but we will not go into that. 2c is the sum of 2a and 2b, so if 2b is empty, it will be the same as 2a.

2a	Deductions . . . . • If you plan to itemize deductions, enter the estimated total of your itemized deductions. . . . . • If you do not plan to itemize deductions, enter your standard deduction.	2a	13,850
b	If you can take the qualified business income deduction, enter the estimated amount of the deduction . . .	2b	
c	Add lines 2a and 2b	2c	13,850

Line 3: Subtract the estimated deductions from your estimated adjusted gross income.

3	Subtract line 2c from line 1	3	32,150

On the form, subtract line 2c from line 1. This finds your estimated taxable income. The form is having you do the same thing as this formula we used back on Day 33:

taxable income = *adjusted gross income* – *deductions*

$46,000 – $13,850 = $32,150

Over —	But not over —				of the amount over —
$0	$11,000	---------	+	10%	$0
11,000	44,725	$1,100.00	+	12%	11,000

Lines 4–6: These lines use your taxable income from line 3 to calculate your estimated taxes using a table that is included with the form and that we worked with back in Lesson 4. For this example, the income is between $11,000 and $44,725, so we will look at that line and use it to calculate the tax.

Finding income over $11,000: **$32,150** – $11,000 = **$21,150**

Calculating tax: $1,100 + 0.12 × **$21,150** = **$3,638**

Inserting result on line 4 and line 6:

4	**Tax.** Figure your tax on the amount on line 3 by using the **Sample Tax Rate Schedules for Estimating Taxes** . **Caution:** *If you will have qualified dividends or a net capital gain, or expect to exclude or deduct foreign earned income or housing, see Worksheets 2–5 and 2–6 in Pub. 505 to figure the tax*	4	3,638
5	Alternative minimum tax from **Form 6251**	5	
6	Add lines 4 and 5. Add to this amount any other taxes you expect to include in the total on Form 1040 or 1040-SR, line 16.	6	3,638

For all cases in this course, lines 4 and 6 will be the same.

Lines 7–8: Next, subtract any nonrefundable tax credits. You might look up a list of available nonrefundable tax credits to see if you qualify for any. In the example, there were not any, so line 8 is the same as line 6.

7	Credits (see instructions). **Do not** include any income tax withholding on this line	7	
8	Subtract line 7 from line 6. If zero or less, enter -0-.	8	3,638

Lines 9–11: Add other taxes and subtract refundable credits you expect to receive. In this case, lines 9 and 10 are blank since there are not any other taxes. We have put an expected $500 refundable credit in line 11b, which gets subtracted from line 11a, with the result in line 11c, which is the estimated taxes for the year: $3,138.

9	Self-employment tax (see instructions).	9	
10	Other taxes (see instructions)	10	
11a	Add lines 8 through 10	11a	3,638
b	Earned income credit, additional child tax credit, fuel tax credit, net premium tax credit, refundable American opportunity credit, and section 1341 credit	11b	500
c	**Total 2023 estimated tax.** Subtract line 11b from line 11a. If zero or less enter -0-.	11c	3,138

Now, the 1040-ES goes on to help you figure out how much you should pay in estimated taxes each quarter, which would be important if self-employed. We are not going to cover that part of the worksheet.

KEEPING PERSPECTIVE

While taxes are complicated, estimating them is simply a matter of walking through a worksheet, performing basic addition and subtraction.

- If you have an employer, you will use a different form, but understanding the basic process we walked through in this section can help you better understand that form.
- Estimating somewhat accurately saves money.

Walking Through the W-4 Form

When you start a job, your employer will have you fill out a form called the **W-4**. This form determines how much the employer will withhold from your paycheck and send to the IRS for federal income taxes each pay period.

The form collects the following key information that affects your taxes:

- your filing status
- your credits
- your deductions
- your income (other than what the employer is paying you)

There is no spot on the form to list your estimated adjusted gross income, as the employer knows the salary they have given you already. **Make sure you list any other income on the form.**

Form **W-4** — Department of the Treasury, Internal Revenue Service

Employee's Withholding Certificate — OMB No. 1545-0074

Complete Form W-4 so that your employer can withhold the correct federal income tax from your pay.
Give Form W-4 to your employer.
Your withholding is subject to review by the IRS.

Step 1: Enter Personal Information
(a) First name and middle initial | Last name | (b) **Social security number**
Address
City or town, state, and ZIP code
Does your name match the name on your social security card? If not, to ensure you get credit for your earnings, contact SSA at 800-772-1213 or go to *www.ssa.gov*.
(c) ☐ **Single** or **Married filing separately**
☐ **Married filing jointly** or **Qualifying surviving spouse**
☐ **Head of household** (Check only if you're unmarried and pay more than half the costs of keeping up a home for yourself and a qualifying individual.)

Complete Steps 2–4 ONLY if they apply to you; otherwise, skip to Step 5. See page 2 for more information on each step, who can claim exemption from withholding, and when to use the estimator at *www.irs.gov/W4App*.

Step 2: Multiple Jobs or Spouse Works
Complete this step if you (1) hold more than one job at a time, or (2) are married filing jointly and your spouse also works. The correct amount of withholding depends on income earned from all of these jobs.
Do **only one** of the following.
(a) Use the estimator at *www.irs.gov/W4App* for most accurate withholding for this step (and Steps 3–4). If you or your spouse have self-employment income, use this option; **or**
(b) Use the Multiple Jobs Worksheet on page 3 and enter the result in Step 4(c) below; **or**
(c) If there are only two jobs total, you may check this box. Do the same on Form W-4 for the other job. This option is generally more accurate than (b) if pay at the lower paying job is more than half of the pay at the higher paying job. Otherwise, (b) is more accurate . . . . ☐

Complete Steps 3–4(b) on Form W-4 for only ONE of these jobs. Leave those steps blank for the other jobs. (Your withholding will be most accurate if you complete Steps 3–4(b) on the Form W-4 for the highest paying job.)

Step 3: Claim Dependent and Other Credits
If your total income will be $200,000 or less ($400,000 or less if married filing jointly):
Multiply the number of qualifying children under age 17 by $2,000 $ ______
Multiply the number of other dependents by $500 . . . . . $ ______
Add the amounts above for qualifying children and other dependents. You may add to this the amount of any other credits. Enter the total here . . . . **3** $

Step 4 (optional): Other Adjustments
(a) Other income (not from jobs). If you want tax withheld for other income you expect this year that won't have withholding, enter the amount of other income here. This may include interest, dividends, and retirement income . . . . **4(a)** $
(b) Deductions. If you expect to claim deductions other than the standard deduction and want to reduce your withholding, use the Deductions Worksheet on page 3 and enter the result here . . . . **4(b)** $
(c) Extra withholding. Enter any additional tax you want withheld each **pay period** . . **4(c)** $

Step 5: Sign Here
Under penalties of perjury, I declare that this certificate, to the best of my knowledge and belief, is true, correct, and complete.
Employee's signature (This form is not valid unless you sign it.) | **Date**

Employers Only
Employer's name and address | First date of employment | Employer identification number (EIN)

For Privacy Act and Paperwork Reduction Act Notice, see page 3. Cat. No. 10220Q Form **W-4**

See **Reference Section** for sample W-4.

> The W-4 form calculates the federal income tax due, not Social Security and Medicare taxes. Those are calculated and withheld separately on your paycheck based on a flat rate.

We will explore the form section by section.

Step 1: You are asked for some basic personal information, then for your filing status (highlighted).

Step 1: Enter Personal Information	(a) First name and middle initial	Last name	**(b) Social security number**
	Address		**Does your name match the name on your social security card?** If not, to ensure you get credit for your earnings, contact SSA at 800-772-1213 or go to www.ssa.gov.
	City or town, state, and ZIP code		
	(c) ☐ **Single** or **Married filing separately** ☐ **Married filing jointly** or **Qualifying surviving spouse** ☐ **Head of household** (Check only if you're unmarried and pay more than half the costs of keeping up a home for yourself and a qualifying individual.)		

Step 2: This part only applies if you have more than one job or a spouse that works. It is designed to help your employer figure out how much of your tax liability this employer versus another should be withholding.

Step 2: Multiple Jobs or Spouse Works	Complete this step if you (1) hold more than one job at a time, or (2) are married filing jointly and your spouse also works. The correct amount of withholding depends on income earned from all of these jobs. Do **only one** of the following. (a) Use the estimator at www.irs.gov/W4App for most accurate withholding for this step (and Steps 3–4). If you or your spouse have self-employment income, use this option; **or** (b) Use the Multiple Jobs Worksheet on page 3 and enter the result in Step 4(c) below; **or** (c) If there are only two jobs total, you may check this box. Do the same on Form W-4 for the other job. This option is generally more accurate than (b) if pay at the lower paying job is more than half of the pay at the higher paying job. Otherwise, (b) is more accurate . . . . ☐

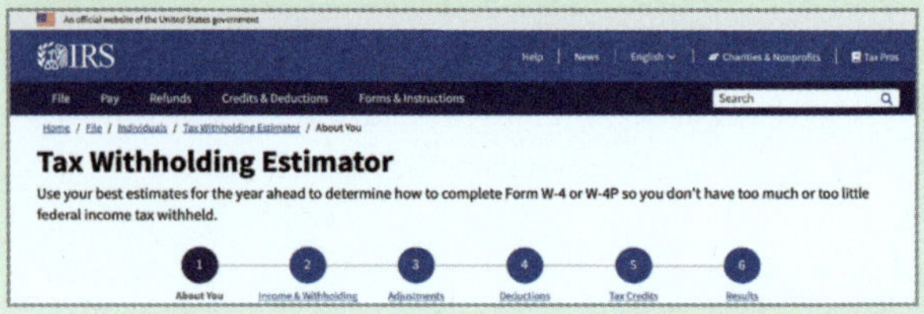

The IRS has an online Tax Withholding Estimator that walks users through questions and then shows them exactly what to put in each box of the W-4 form. This estimator can be particularly helpful for those with multiple sources of income or complicated taxes. If you wish to try it, get permission from a parent first.

Step 3: This section is where you list any credits you might qualify for. Children and dependents (people that qualify by the IRS definition as dependent on you) are a common one, so the form helps you calculate them, but you can add any others on line 3.

Step 3: Claim Dependent and Other Credits	If your total income will be $200,000 or less ($400,000 or less if married filing jointly):			
	Multiply the number of qualifying children under age 17 by $2,000 . . .	$		
	Multiply the number of other dependents by $500	$		
	Add the amounts above for qualifying children and other dependents. You may add to this the amount of any other credits. Enter the total here		3	$

The government gives guidelines as to when you get a penalty. Typically, so long as you paid 100% of what you owed the year before (110% for high income brackets), 90% of what you owe that year, or owe less than $1,000, they are not going to penalize you.

Step 4:

a. Your employer only knows about the income they are giving you. 4(a) lets you list extra income not from this job.

b. Your employer will assume you are taking the standard deduction unless you put a different amount in 4(b).

c. You can use 4(c) to specify an extra amount each pay period you want withheld if your estimate of how much you need to withhold is different than your employer calculates.

Step 4 (optional): Other Adjustments	(a) **Other income (not from jobs).** If you want tax withheld for other income you expect this year that won't have withholding, enter the amount of other income here. This may include interest, dividends, and retirement income	4(a)	$
	(b) **Deductions.** If you expect to claim deductions other than the standard deduction and want to reduce your withholding, use the Deductions Worksheet on page 3 and enter the result here	4(b)	$
	(c) **Extra withholding.** Enter any additional tax you want withheld each **pay period**	4(c)	$

KEEPING PERSPECTIVE

As you can see, the form is just collecting the information you would need to estimate your taxes.

Remember, our goal is, "render to Caesar the things that are Caesar's" (Matthew 22:21). Accurately withholding the right amount will help pay what is due and not too much extra, thereby both avoiding late fees as well as not tying up money that could be invested or used elsewhere. While it can be unfeasible to estimate precisely, getting reasonably close will serve you well.

Checking the W-4 Form

The W-4 Form does not always result in accurate tax withholding. It is easy to forget about something that affects your taxes when filling out the W-4 — especially if you have the new-job jitters. Complicated tax situations can also make it more challenging.

It can be helpful to look at your paycheck stub to see how much your employer is withholding. You should see something like this:

Taxes Withheld	Current	Year-to-Date
Federal Tax	$200	$3,000

The stub above shows that $200 was withheld for federal taxes for the current pay period, and $3,000 has been withheld so far this year.

>>> To check if your employer is withholding the right amount, follow these steps:

Your stub will also show other taxes that are being withheld, such as Social Security tax. You can ignore those when checking your W-4 form, as the form only affects federal income tax.

1. **Estimate your taxes** using the steps on Day 166.
2. **Subtract anything that has already been withheld or any estimated taxes you have paid.** The result is what you have left to pay, or, if you get a negative value, the amount you have overpaid.

Example: You estimate $4,458 taxes due, but $3,000 has already been withheld. How much is left to pay?

$4,458 – $3,000 = $1,458

Example: You estimate $4,458 taxes due, and $5,000 has already been withheld. How much have you overpaid?

$4,458 – $5,000 = –$542

The negative here means you overpaid $542; what was paid was more than what you estimated owing.

3. **Divide your answer to step 2 by the number of paychecks left to see how much should be withheld in each paycheck.** Use basic arithmetic and what you know about how often you are paid to figure out how many paychecks you have left.

Example: You are paid twice a month and there are 5 months left. If you still owe $1,458, how much should be withheld each paycheck?

Calculating paychecks left: $2 \times 5 = \mathbf{10}$

Calculating amount to withhold each paycheck:

$$\frac{\$1{,}458}{\mathbf{10}} = \$145.80$$

4. Look at how much is **currently withheld** from each paycheck and compare it to your answer to step 3.
5. If step 4 shows the need to make changes, **fill out a new W-4.** Ideally, you will change what actually changed or was inaccurate the first time you filled out the form. You may also need to think through what would decrease or increase your taxes. Things that increase them will also increase your withholding and vice versa. (See green boxes below.)
6. **After you make a change**, watch your paycheck to make sure the change to the W-4 really resulted in the correct adjustment to your withholdings.

Things that decrease taxes withheld:
decrease additional income
increase credits
increase deductions
decrease additional tax withheld in box 4(c)

Things that increase taxes withheld:
increase income
decrease credits
decrease deductions
increase additional tax withheld in box 4(c)

Additional Withholding (Box 4(c)) – Assuming the box had nothing in it previously, putting $45 in this box would result in $45 more being withheld each pay period.

(c) Extra withholding. Enter any additional tax you want withheld each **pay period** . .	**4(c)**	$ 45

Tip: If you own a side business, you can use additional withholding to withhold the extra you need to be paying in taxes for that.

KEEPING PERSPECTIVE

The W-4 can be updated at any time.

Whenever you have a major life or financial change (like getting married, having children, or getting another job), it is a good idea to revisit your W-4 and make sure the numbers there are still accurate.

You might also estimate your taxes and check your W-4 if you are uncertain if you filled it out correctly to avoid being surprised by anything come tax time.

Income Statement: Revenue, Expenses, & Profit (or Income)

Running a business shares a lot of similar concepts with running a household. For example, businesses use budgets. However, businesses often use different terminology to describe concepts. **Knowing these terms can help you understand what your employer is looking at — or help you run your own business.** While there are whole courses on business finance, we are only going to look at a few key highlights.

Back in Lesson 6, we talked about gross income (the full amount you make before paycheck deductions) and net income (the actual amount of your paycheck after all deductions). Businesses have similar concepts.

Gross profit is like personal gross income in the sense that it is not what the business actually makes in the end. It is what the business earns from selling its product or services minus what it costs to do so. We saw on Day 108 that *profit = selling price – costs*. **Gross profit** is basically the total profit on *all* a business' products or services. It is calculated by taking what was made (the **sales**) minus the costs (the **cost of goods sold**, or **COGS**).

On a product or service: *profit = selling price – costs*

For the whole business: ***gross profit*** = ***sales*** – ***COGS***

Businesses might have other income (such as interest income) and other expenses (such as operating costs and advertising) that are not tied to a specific product or service. **Net income** (or **net profit**) is what the business is actually making after all **revenue** (i.e., what was made) and expenses (including COGS, although that is calculated separately) are accounted for, similar to how personal net income is what you actually make on your paycheck.

net income = (*total of all revenue*) – *total of all expenses*

Income Statement

Net income, as well as gross profit, is shown on an aptly named business statement called the **income statement**. The one shown for a hypothetical business is color coded to help you understand it.

Income Statement for Sarah's Soaps	
Sales	$ 2,500
COGS	$ 1,000
Gross Profit	$ 1,500
Interest Income	$ 60
Expenses	
Web Expense	$ 50
Credit Card Processing	$ 40
Booth Expense	$ 300
Advertising	$ 200
Total Expenses	$ 590
Net Income	$ **970**

gross profit = $2,500 – $1,000 = $1,500

net income = ($2,500 + $60) – ($1,000 + $590) = **$970**

Income statements can get more complicated than the one shown, as a business may have different sections of revenue and expenses. But the same principle still applies: **the net income is the result of subtracting all the expenses (including COGS) from all the revenue (also called sales or income).**

When evaluating businesses (such as considering investing in one), it can be helpful to **look at their net income over several years**. By changing the income statement to include 3 years' worth of information, we can see that Sarah's business seems to be going in a good direction.

- The *gross profit* has steadily increased even while the *COGS* has stayed the same. That means the business is making more for the same costs.
- The *net income* dipped one year due to more expenses but rebounded the next year.

Things to look at on income statement:

- Is the net income going up or down?
- What has changed in sales, COGS, and expenses over the years?

Income Statement for Sarah's Soaps			
	Current Year	1 Year Ago	2 Years Ago
Sales	$ 2,500	$ 2,000	$ 1,800
COGS	$ 1,000	$ 1,000	$ 1,000
Gross Profit	$ 1,500	$ 1,000	$ 800
Interest Income	$ 60	$ 60	$ 50
Expenses			
Web Expense	$ 50	$ 0	$ 0
Credit Card Processing	$ 40	$ 0	$ 0
Booth Expense	$ 300	$ 500	$ 300
Advertising	$ 200	$ 300	$ 100
Total Expenses	$ 590	$ 800	$ 400
Net Income	$ 970	$ 260	$ 450

Use of some of the terms on an income statement varies.

KEEPING PERSPECTIVE

When running a business or working for an employer, it is easy to get caught up in the bottom line — that is, in increasing net income. Net income is listed on the *bottom line* of the income statement, so the saying makes sense. And a business' net income *is* important. If you do not make money, the business can not continue. However, **there are other considerations besides net income**. Is the way you are conducting business honoring to God? Are you building up people, both your employees and your customers?

For-profit businesses, as well as non-profit ones (businesses who do not try to make a profit), can be vehicles for serving the Lord when we remember that Jesus is to be over *every* area of our lives — that He is our true boss. The Bible's admonition to masters can be applied to employers/managers as well.

> *Masters, treat your bondservants justly and fairly, knowing that you also have a Master in heaven* (Colossians 4:1).

Business Budgeting

Businesses, like individuals, need to have a budget. Otherwise, they might not wind up with enough money to cover their expenses.

- The budget shown is the same as the income statement, except there is now *projected*, *actual*, and *difference* columns. *Projected* means what is anticipated (i.e., the budgeted amount), *actual* is what actually happened, and *difference* is the *projected* minus *actual*.
- The *actual* values listed are what get declared for the current year on the income statement.
- Businesses might also have budgets for specific projects or events.

Budget for Sarah's Soaps						
	Projected		Actual		Difference	
Sales	$	3,000	$	2,500	$	500
COGS	$	1,000	$	1,000	$	0
Gross Profit	$	2,000	$	1,500	$	500
Interest Income	$	100	$	60	$	40
	$		$		$	
Expenses	$		$		$	
Web Expense	$	50	$	50	$	0
Credit Card Processing	$	40	$	40	$	0
Booth Expense	$	250	$	300	$	(50)
Advertising	$	100	$	200	$	(100)
Total Expenses	$	440	$	590	$	(150)
	$		$		$	
Net Income	$	1,660	$	970	$	690

As we explore business budgets, we are going to apply a lot of what we have already looked at for individual budgets. Look back through the *Student* book as we go if you need a refresher on various concepts.

Good to know:

- Business budgets, like individual budgets, are often based on previous values. For example, if $2,000 was spent in a category the previous year, the business might budget $2,000 again this year — or more or less, depending on changes they are making.
- When starting a new business, there are not prior values to use, so research needs to be done and guesses need to be made. Be careful about overestimating net income.
- Business budgets are not necessarily monthly. Many times, they are yearly.

Evaluating a Business Budget

Businesses need to keep a tight watch on their budget. If actual revenues are not matching the projected revenue, **businesses, like individuals, need to make adjustments** — cutting expenses or finding other revenue streams. The same tools you looked at for evaluating personal budgets apply to business budgets.

For example, you could use graphs to explore a business budget and see pictorially where the business is spending or making money. In fact, you might find it helpful to present budget data (or other data) on the job in a graphical way. The pie chart shows the percent of total revenue coming from each of a business' products.

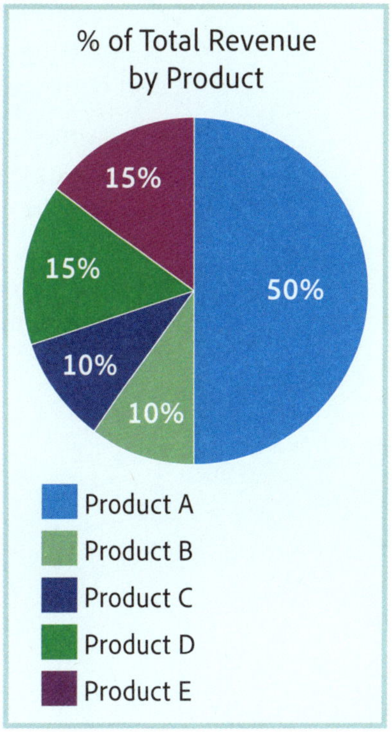

Making Changes

What would you do if you noticed that you are spending a lot more than you are making on your personal budget? What if you see that your income is a lot less than planned? You would have to find a way to either earn more money or cut spending, right?

The same is true for businesses. If a business notices that their expenses are overrunning or their revenue underrunning, they need to find a way to either make more or cut expenses.

> Remember, **overrun** means having actual amounts that are more than the projected amounts; **underrun** means having actual amounts that are less than the projected amounts.

Another option would be to take out a loan to cover their extra expenses or lack of revenue. The same principles we have explored in other lessons regarding loans apply in business too — borrowing makes you a "slave of the lender" (Proverbs 22:7) and leads to paying more in the end. While there are times loans may make sense, be careful and cautious about borrowing money, whether as an individual or a business.

> *The rich rules over the poor, and the borrower is the slave of the lender* (Proverbs 22:7).

KEEPING PERSPECTIVE

When budgeting and running a business or helping your employer run a business, always remember that there is more to consider than earthly profit. As we saw in previous case studies, businesses can be an amazing way to show God's love to those around us — to both employees and customers.

Remember too that work can be hard, but God promises to not give up on His work in us. He can strengthen us to complete whatever work He has assigned for us to do.

> *And I am sure of this, that he who began a good work in you will bring it to completion at the day of Jesus Christ* (Philippians 1:6).

Balance Sheet: Equity, Assets, & Liabilities

- What is a business worth?
- How much money or other property does the business own?
- How much does it owe people?

Think back to calculating the owner's equity in a home. You took the home value (i.e., the *asset* price) minus the principal on the loan (i.e., the *liability*). The same concepts apply in personal and business finance.

To answer these questions, look at what is called the balance sheet. The **balance sheet** shows the business' assets, liabilities, and owner's equity.

We talked about assets and liabilities on Day 115 when finding net worth. Assets are what you have, and liabilities are what you owe.

net worth = *assets* – *liabilities*

It is the same idea for a business. Only *net worth* is called *equity*.

equity = *assets* – *liabilities*

Equity is also referred to as *owner's equity* or *shareholder's equity* (if the company has stocks).

Balance Sheet for Sarah's Soaps	
Assets	
Cash	$ 700
Accounts Receivable	$ 100
Inventory	$ 500
Total Assets	$ **1,300**
Liabilities	
Accounts Payable	$ 200
Unearned Revenue	$ 50
Total Liabilities	**$ 250**
Owner's Equity	$ **1,050**

Here is a hypothetical balance sheet for Sarah's Soaps, showing the business' assets, liabilities, and equity. The equity equals the assets minus the liabilities.

equity = **$1,300** – **$250** = **$1,050**

The boxes show some examples of common business assets and liabilities.

Assets are anything of value, such as the following:

- cash
- equipment
- money customers owe the business (*accounts receivable*)
- inventory
- trade names (business or product name)

Some assets, like inventory and money customers owe, are easily converted to cash. Other assets, like a building, are not as easily turned to cash.

It may seem odd to list a business name as an asset. But think about the name of a big store everyone knows. That name offers value to the company since people recognize it, even though it is not really a physical item you could touch.

Liabilities are what the business owes. You can think of it as **a business' debts**.

- Loans and credit card balances are examples of liabilities.
- Sometimes businesses will purchase supplies on account, where they pay later (*accounts payable*).
- A business may also have products/services that have been paid for but have not delivered yet (*unearned revenue*).

Analyzing the Numbers

As with income statements, by **looking at multiple years** of balance sheets, you can see how assets, liabilities, and owner's equity are changing. **You can also look at different relationships between values to see how a business is doing.** A few are shown here. The desired ratios depend on the industry, so look up what is normal for that industry when evaluating a business.

While D/E ratio could be expressed as a percent, it is typically not. This is a key ratio to look at when evaluating stocks.

»» *Debt to Equity Ratio (D/E)* $= \frac{liabilities}{equity}$.

This ratio shows how much debt (liabilities are debt) is owed compared to equity. Looking at this lets you know proportionately how much debt the business has in a way the numbers themselves can not show. For example, a company with $10,000 of debt does not really have much debt proportionally if their equity is $1 million, but $10,000 of debt would be a lot for a company with only $4,000 of equity. A D/E ratio under 1 means there is more equity, or value, in the company than debt. Lower is better.

Example: For Sarah's Soaps, $\frac{liabilities}{equity} = \frac{\$250}{\$1{,}050} \approx 0.24$

»» *Return on Equity Ratio (ROE)* $= \frac{net\ income}{average\ equity}$.

Typically written as a percent, ROE shows you how much the business is making compared to the amount of money invested in it. Higher is better.

Note that *average* equity and assets are not shown on the financial statements we looked at. They are a calculated average.

Example: Approximating for Sarah's Soaps if the *average equity* was $1,000 and using the net income value from the income statement from Day 170.

$$ROE = \frac{net\ income}{average\ equity} = \frac{\$970}{\$1{,}000} = 0.97 = 97\%$$

»» *Return on Assets Ratio (ROA)* $= \frac{net\ income}{average\ assets}$.

Typically written as a percent, ROA shows you how much the company is making compared to assets. This can be helpful to compare with other companies. Higher is better.

Example: Approximating for Sarah's Soaps if the *average assets* were $1,066 and using the net income value from the income statement from Day 170.

$$ROA = \frac{net\ income}{average\ assets} = \frac{\$970}{\$1{,}066} \approx 0.91 = 91\%$$

KEEPING PERSPECTIVE

As we look at math for businesses, remember that work is a gift from God. We were designed to work. God works. In fact, "... it is God who works in you" (Philippians 2:13).

Self-Employment Taxes & Starting a Business

If you ever own a business or are self-employed, you will need to think through paying medicare and social security taxes.

- Business owners need to pay the employer portion for their employees, plus withhold the employee portion from their employees' paychecks.
- Self-employed individuals need to pay both the employer and employee portions themselves, called a **self-employment tax**.

Sample Tax Year Values

Medicare tax = 1.45% for both employer and employee = 1.45% + 1.45% = **2.9%** total

Social Security tax = 6.2% for both employer and employee = 6.2% + 6.2% = **12.4%** total

Self-employment tax = **2.9%** + **12.4%** = **15.3%**

Since it is a flat tax, self-employment tax is easy to calculate. Just note that in the sample tax year **it was only due on 92.35% of the self-employment net income.**

>>> Estimate self-employment taxes by taking 15.3% on 92.35% of self-employment net income — that is, multiply net income by 15.3% *and* 92.35%.

Example: You are an independent consultant and earned $1,200 for that work. How much self-employment tax will you owe given the self-employment tax rate of 15.3%?

$0.153 \times 0.9235 \times \$1{,}200 \approx \$170$

Since self-employment tax is only due on 92.35% of income rather than on 100%, it comes to slightly less than the total of the medicare and social security taxes an employer and employee normally pay. But it is still a significant expense to plan for if self employed. To help, in the sample tax year, **self-employed individuals could deduct half of what they pay in self-employment taxes on their federal income taxes.**

Example: In the scenario above, how much could you deduct for your self-employment taxes in the sample tax year?

$\frac{\$170}{2} = \$85$

Starting a Business

There are many ways to own a business. Here are four main categories of business ideas:

- **Render a service**, such as teaching an instrument, tutoring, mowing lawns, designing websites, delivering food, driving people, or babysitting.
- **Sell a product that you produce**, such as a book, vegetables from the garden, or hand-carved masterpieces.
- **Sell a product that someone else produces**, such as signing on to be a consultant for companies that sell through individuals and home-based parties instead of retail stores.
- **Rent something out for a fee**, such as a room, tool, or truck.

Whatever business idea you have, math can help you evaluate it to better understand what you are getting into before you start.

For example, you are at a friend's product party and wondering about becoming a consultant. It sounds like you could make a lot of money easily. A little math can help you better evaluate it.

Example: A company offers you 20% of what you sell, plus other perks. Ignoring the perks, how much would you have to sell each month to earn $1,000?

20% *of* sales *is* $1,000

Wrote as equation: $0.20 \times s = \$1{,}000$

Divided both sides by 0.20: $s = \frac{\$1{,}000}{0.20} = \$5{,}000$

ALGEBRA Connection

We used an *s* to stand for the total sales. Letters are just placeholders for values.

Now, selling $5,000 might take some effort. You could also use math to look into any discounts you would get on products you buy as a consultant, the minimums you might have to buy each month (which you would need to budget for), and any other aspects of the potential business.

In a similar way, math can help you evaluate any businesses you might want to start. It can help you think through anticipated revenue and expenses and calculate what your net income might be.

You may find you have to supplement your own business at first, get investors, or take out a loan in order to pay bills as your business gets off the ground.

KEEPING PERSPECTIVE

Starting your own business can be an exciting adventure and an opportunity to bless others, but it takes work and time that you could use to serve in other ways. Be sure you do the math and count the cost first — and above all, seek God's direction and wise counsel from people you can trust.

Gambling

We have been looking all lesson at various things connected to work. We will end by talking about a trap that many fall into in an effort to avoid hard work: gambling.

As you go through life, you will encounter many advertisements saying you can get rich quickly by winning the lottery, betting on the right team, or winning big at the casino. Let us look at what the Bible has to say about gambling, and at how a little math shows that gambling is a great way to lose, not gain, money.

A Biblical Perspective

While the Bible never says "thou shalt not gamble," it gives principles that can help inform how we view gambling. In contrast to gambling's get-rich-quick appeal, the Bible encourages gaining wealth little by little.

> *Wealth gained hastily will dwindle, but whoever gathers little by little will increase it* (Proverbs 13:11).

The Bible also emphasizes work as a blessing and the way to earning wealth, not something to be avoided through a get-rich-quick scheme.

> *Do you see a man skillful in his work? He will stand before kings; he will not stand before obscure men* (Proverbs 22:29).

> *A slack hand causes poverty, but the hand of the diligent makes rich* (Proverbs 10:4).

> *For even when we were with you, we would give you this command: If anyone is not willing to work, let him not eat* (2 Thessalonians 3:10).

We are also warned to beware of the love of money, which is, at its core, what pulls many people towards gambling.

> *Keep your life free from love of money, and be content with what you have* (Hebrews 13:5a).

The Bible also warns us to be controlled by the Spirit, not other things. Gambling, like wine, has a tendency to be addictive and seek to control those who dabble in it.

> *And do not get drunk with wine, for that is debauchery, but be filled with the Spirit* (Ephesians 5:18).

The Bible also teaches that all we have comes from God, and we are a steward of it. Our wealth (or lack thereof) is in His hands.

> *The earth is the LORD's and the fullness thereof, the world and those who dwell therein* (Psalm 24:1).

Looking at the Math

Every year, billions of dollars are made in the gambling industry. If the industry makes money, that means the people gambling are losing that money. Gambling in any form is not an investment — on average, it yields a negative, not positive ROI. Casinos and other gambling setups are designed to make sure the company makes money, which means they are set up so the user will, on the whole, lose money.

What about the lottery? If we were to use math (which we will not go into here) to explore the odds of winning and losing, we would find the odds of losing are incredibly high, and those of winning a sizeable amount so low as to be almost nonexistent.

Here is a stark comparison:

- Suppose someone lost just \$100 a year gambling. They would lose 30 × \$100 = \$3,000 over 30 years.
- If that \$100 was deposited in a savings account earning 5% APY instead, after 30 years, they would have about \$7,000.

Which one seems like the wise investment to you? Now, you might think you would gain instead of losing at gambling. But remember, even if you could gain at times, the industry is set up so that you lose overall. That is why we factored in losing, not gaining.

KEEPING PERSPECTIVE

Remember that getting rich does not solve life's problems. It is contentment, not riches, that brings "great gain" (1 Timothy 6:6). As you evaluate different ways you could spend your money, view work as God's gift and money as a stewardship. Let biblical principles inform your decisions.

Review & Business Can Be a Blessing

I hope this lesson has helped you see that the skills you have learned in this course apply in other areas too, such as on the job. Hopefully this lesson has also helped you review some of the concepts we covered earlier like taxes and budgeting. We will continue to review more in the next lesson in preparation for both the quarter and final tests.

Business Can Be a Blessing

As we talked about business in this lesson, I want to end with a reminder that business, as with the rest of life, is an opportunity to build God's Kingdom.

> *And whatever you do, in word or deed, do everything in the name of the Lord Jesus, giving thanks to God the Father through him* (Colossians 3:17).

The Bible does not compartmentalize between spiritual things and secular. *All* things serve God.

> *By your appointment they stand this day, for all things are your servants* (Psalm 119:91).

A business can be used to generate money to give to Kingdom work. It can also be a blessing to communities, providing jobs and needed products or services. It can be an opportunity to invest in the lives of employees. Businesses have even been used to gain access to closed countries with the gospel.

As an employee, business can provide you the opportunity to share Christ with your coworkers. It can give you a chance to set an example of hard work, working "for the Lord and not for men" (Colossians 3:23). It allows you to earn money to provide for a family and give.

Page intentionally blank

Review

As we wrap up Consumer Math, we will spend our remaining time reviewing and taking the quarter test and final exam. As we do, there is time for some more case studies and a recap of some key biblical principles I hope you will keep in mind as you move forward into adulthood and manage whatever resources God sends your way.

Quarter Review & Case Study: Partnering with God (R.G. LeTourneau)

Today, your exercise is going to help you review the quarter in preparation for the quarter test. First, though, here is a case study of a man who stewarded his time, treasure, and talent for God's glory. May it encourage you to faithfully steward whatever God has and will entrust to you.

Case Study: Partnering with God (R.G. LeTourneau)

This case study is based (with permission) primarily on R.G. LeTourneau: Mover of Men and Mountains.[45]

> *"Only God holds the future, and when you have found Him, you might not know what your future holds, but you will have the blessed security of knowing Who holds you, now and forever." — R.G. LeTourneau*[46]

Today's case study is an amazing story of a man who truly took to heart what it meant to live for God's Kingdom. R.G. LeTourneau excelled at construction equipment to move dirt. He did not start out looking like a success, though. He quit high school and started work at an iron foundry.[47] He then bounced around to various construction or mechanical-type jobs, including clearing land, becoming a car mechanic, owning a construction company, and more. The struggles he experienced and the skills he learned on these jobs and on his own later combined to give him both a knowledge of problems that needed solved and the know-how to design machines to solve those problems.

At one point early in his career, LeTourneau experienced spiritual revival and considered becoming a missionary. His pastor prayed with him and said, "God needs businessmen as well as preachers and missionaries."[48]

continued

A 1959 version of the LeTourneau Sheep's Foot Roller, used to compact dirt. The original version was built in 1929.

LeTourneau began viewing himself as a partner with God, wanting God to be glorified by his business. One time, after failing to honor the Sabbath and his commitment to tithe, he resolved to do both, even while digging his way out of debt.[49] He once chose to lose a lot of money instead of taking a brother in Christ to court, leaving the wrong of it in God's hands. He says in his autobiography, "I would not lightly refer to God as my partner. He is my Lord and Savior, and I am His servant. But by His grace He has made us members of His family, and we can refer to Him as our Father which art in Heaven. More than that, He has let us be 'workers, together with Him.' "[50]

LeTourneau was not an instant success. He persevered through many setbacks — the loss of a child, a major car accident, and times he almost went bankrupt. He trusted God in the good *and* bad times.

When LeTourneau started to experience incredible success, he did not let the wealth change him. While he had 299 patents in his lifetime[51] and his machines revolutionized the earth-moving industry, he used that success to further God's Kingdom. He and his wife decided that they did not need all the income they were making. They had a foundation set up, giving 50% of the company profits plus 50% of their own income to it each year. Later, they increased to giving 90% of both their common stock (a type of ownership in their business) and their personal income to the foundation.[52] Through the foundation, they supported Christian work and blessed numerous people. They even started a school to help educate WWII veterans that later became a Christian university.

LeTourneau was also able to use his business to further God's Kingdom in how he ran it. He opened a factory in an area where people needed work and training, bringing both in the process.[53] He started two colonies abroad in underdeveloped areas, showing the people there how they could harness the natural resources they had to make a living and bringing the gospel as he did. He went from being afraid of public speaking to buying his own airplane and traveling around the country and world talking to businessmen about God. He still worked hard too, even designing machines on the plane as he went.

According to his autobiography, LeTourneau once had a customer say, "I shovel it out and God shovels it back, but God has a bigger shovel."[54] That phrase has since often been attributed to LeTourneau, as it describes beautifully in earth-moving language how LeTourneau gave his money and time to God. One biographer said, "Although he has made and spent millions, he is remarkably detached from money as such; he is only concerned about what it can accomplish. He often says, 'The question is not how much of my money I give to God, but rather how much of God's money I keep for myself.' "[55]

"Some people think I am all mixed up — that you can not serve the Lord and business, too, but that is just the point. God needs businessmen as partners as well as preachers." — R.G. LeTourneau[56]

Two Masters

No one can serve two masters, for either he will hate the one and love the other, or he will be devoted to the one and despise the other. You cannot serve God and money (Matthew 6:24).

We have talked a lot in this course about managing money. I hope you are feeling better equipped to manage what you have now as well as what you will get as you move into adulthood and greater responsibilities.

- The skills we have looked at can help you serve God with whatever wealth God entrusts to you. They can help you budget and use your resources, big or small, wisely.
- You could also use many of the *same* tools and principles to merely accumulate wealth for yourself. Cutting expenses, avoiding debt, saving money, and budgeting help you avoid debt and reach goals, whatever they are.

Who or what you serve with the tools you have learned is up to you. Jesus warned that God and money are like two different masters you could serve.

Your "master" tells you what to do.

- **If money is your master**, you will make decisions based on getting as much of it as you can, which means it will not ever be enough. When you die, you will not get to keep any of what you have worked for.
- **If God is your master**, you will want to make decisions based on God's Kingdom, purposes, and what He leads you to do. You can live for an eternal Kingdom where treasures last forever. You know what? God can give you *contentment* with whatever life He calls you to lead. You can have joy in watching God provide for you.

Do not lay up for yourselves treasures on earth, where moth and rust destroy and where thieves break in and steal, but lay up for yourselves treasures in heaven, where neither moth nor rust destroys and where thieves do not break in and steal. For where your treasure is, there your heart will be also (Matthew 6:19–21).

Look at the birds of the air: they neither sow nor reap nor gather into barns, and yet your heavenly Father feeds them. Are you not of more value than they? ... Therefore do not be anxious, saying, "What shall we eat?" or "What shall we drink?" or "What shall we wear?" For the Gentiles seek after all these things, and your heavenly Father knows that you need them all. But seek first the kingdom of God and his righteousness, and all these things will be added to you (Matthew 6:26, 31–33).

Jesus did not offer a merge between these two masters. We are going to live for one of two kingdoms: this world's (where we need money to buy things) or the next. We are all born living for this world's kingdom. We want our way. It is as if we pretend life is about us and we are our own gods. This sin leads to eternal death and separation from God.

But God loves us so much that "he gave his only Son," Jesus, to "be sin who knew no sin" (John 3:16; 2 Corinthians 5:21). He made a way for us to enter into God's Kingdom — to be declared righteous by simple faith in what Jesus did. When we do that, we become "children of God" (John 1:12). God becomes our Father and our new Master — the One for whom we want to live.

Yet the choice of kingdoms is not a choice we only make once. Even after we receive the forgiveness and new life Jesus offers, making God our Master, we often live as if we are still a slave to this world's ways. We need the Holy Spirit's continual ministry to help us learn to live as servants of God. We also need other believers to "exhort" us (Hebrews 3:13).

We have the privilege as God's children and servants to *grow in living for the new Kingdom* we are now a part of through what Jesus did for us. And as we do, we discover the joy of working with God for His purposes in this world.

KEEPING PERSPECTIVE

The Apostles viewed themselves as servants of God. How about you?

This is how one should regard us, as servants of Christ and stewards of the mysteries of God (1 Corinthians 4:1).

James, a servant of God and of the Lord Jesus Christ ... (James 1:1).

Begin Now & Case Study: Examples from the Bible

Throughout this course, we have looked at a lot of skills. It may be tempting to think that you do not need these skills for a while, but that is not the case. **Begin wise stewardship now.**

- As you begin giving generously, you will experience the joy of giving and grow in trusting God to provide your needs. Remember that just a handful of rice can make a difference (see the case study in Lesson 1).
- As you learn to budget with small amounts, it will help you prioritize the goals God lays on your heart and avoid the love of money.
- As you make small decisions, it will build your faith muscles, preparing you for bigger decisions.

We can look back in the Bible at how men and women used whatever God gave them, whether for good or evil. What do you want the write-up on your life to be?

Know that whatever God has given you, whether much or little, it is given to you for His purposes. You are part of His master plan. If you are His child, look for the "good works" He has planned for you to do with the time, talent, and treasure He has given you.

> *For we are his workmanship, created in Christ Jesus for good works, which God prepared beforehand, that we should walk in them* (Ephesians 2:10).

Case Study: Examples from the Bible

Here are a few examples of people in Scripture and how they used the resources God gave them.

Abraham – When three visitors (two angels and the Lord Himself) came to Abraham, he hastened to prepare a costly feast. He gave generously of what he had to the Lord. While the Lord might not show up to your house, Jesus said that gifts to His people are really given to Him. Abraham, along with many New Testament verses, reminds us to be hospitable with what we have.

Then the King will say to those on his right, "Come, you who are blessed by my Father, inherit the kingdom prepared for you from the foundation of the world. For I was hungry and you gave me food, I was thirsty and you gave me drink, I was a stranger and you welcomed me, I was naked and you clothed me, I was sick and you visited me, I was in prison and you came to me." Then the righteous will answer him, saying, "Lord, when did we see you hungry and feed you, or thirsty and give you drink? And when did we see you a stranger and welcome you, or naked and clothe you? And when did we see you sick or in prison and visit you?" And the King will answer them, "Truly, I say to you, as you did it to one of the least of these my brothers, you did it to me" (Matthew 25:34–40).

Israelites – When the Israelites left Egypt, God moved the Egyptians to give them gold, silver, and clothing (Exodus 12:35). Sadly, they used some of that gold to build a golden calf. Thankfully, God did not give up on them, and they later used some of the gold and other supplies to build the tabernacle. The Israelites remind us that God's gifts can be used wrongly, such as for idolatry or worship.

Solomon – God gave King Solomon great wealth (1 Kings 10:23). Solomon used his wealth and the peace of his reign to build God's temple. Yet his blessings ensnared him, as he began pursuing pleasure and all this world has to offer. Solomon's conclusion was that wealth — and everything here on earth apart from living content in the Creator — is meaningless. His life and words remind us not to live for money or what money can buy.

> *Vanity of vanities, says the Preacher, vanity of vanities! All is vanity* (Ecclesiastes 1:2).

> *He who loves money will not be satisfied with money, nor he who loves wealth with his income; this also is vanity* (Ecclesiastes 5:10).

> *There is nothing better for a person than that he should eat and drink and find enjoyment in his toil. This also, I saw, is from the hand of God, for apart from him who can eat or who can have enjoyment?* (Ecclesiastes 2:24–25).

The Boy with 5 Loaves and 2 Fish – While we do not know his name, we know that a boy once gave up the food he had and watched Jesus multiply it into enough to feed thousands. His life reminds us that God can take what little we have and do amazing things with it.

Lydia – We are told Lydia was "a dealer in purple cloth" (Acts 16:14; NIV). Purple dye was pricey, so this meant she was a woman God had given wealth to. What did she do with it? After accepting Jesus, she invited Paul to stay with her (Acts 16:15). We see Paul going to her house to encourage her and the believers in her house (Acts 16:40). She is an example of someone using her resources and her home for the Lord and a reminder to us to do the same.

Reference Section

Page intentionally blank

Taxes and Tax Forms

Disclaimer: This information is from the sample tax year. The brackets and deductions get re-released each year.

You can find current information on brackets, deductions, and other tax topics, by searching on the IRS website (with a parent's permission); finding current tax forms at a library, IRS Taxpayer Assistance Center, or mail (call 1-800-TAX-FORM); or by utilizing a tax program or accountant.

- **Medicare Tax Rate:** 2.9% (1.45% for both the employer and employee for all examples given; additional 0.9% for employee on incomes over a certain amount)
- **Social Security Tax Rate:** 12.4% (6.2% for both the employer and employee up to $160,200 of income)
- **Self-Employment Tax:** The sum of the employer and employee taxes for both Medicare and Social Security (1.45% + 6.2% = 15.3%) on 92.35% of the self-employment income

Sample Federal Tax Brackets

This table represents information from the sample tax year.

Tax Rate	Single	Married Filing Jointly or Qualifying Surviving Spouse	Married Filing Separately	Head of Household
10%	$0 to $11,000	$0 to $22,000	$0 to $11,000	$0 to $15,700
12%	$11,001 to $44,725	$22,001 to $89,450	$11,001 to $44,725	$15,701 to $59,850
22%	$44,726 to $95,375	$89,451 to $190,750	$44,726 to $95,375	$59,851 to $95,350
24%	$95,376 to $182,100	$190,751 to $364,200	$95,376 to $182,100	$95,351 to $182,100
32%	$182,101 to $231,250	$364,201 to $462,500	$182,101 to $231,250	$182,101 to $231,250
35%	$231,251 to $578,125	$462,501 to $693,750	$231,251 to $346,875	$231,251 to $578,100
37%	$578,126 and up	$693,751 and up	$346,876 and up	$578,101 and up

Sample Tax Rate Schedules for Estimating Taxes

Sample Tax Rate Schedules

Schedule X — Use if your **Sample Tax Year** filing status is **Single**

If line 3 is: Over—	But not over—	The tax is:			of the amount over—
$0	$11,000	---------	+	10%	$0
11,000	44,725	$1,100.00	+	12%	11,000
44,725	95,375	5,147.00	+	22%	44,725
95,375	182,100	16,290.00	+	24%	95,375
182,100	231,250	37,104.00	+	32%	182,100
231,250	578,125	52,832.00	+	35%	231,250
578,125	---------	174,238.25	+	37%	578,125

Schedule Z — Use if your **Sample Tax Year** filing status is **Head of household**

If line 3 is: Over—	But not over—	The tax is:			of the amount over—
$0	15,700	---------	+	10%	$0
15,700	59,850	$1,570.00	+	12%	15,700
59,850	93,350	6,868.00	+	22%	59,850
95,350	182,100	14,678.00	+	24%	95,350
182,100	231,250	35,498.00	+	32%	182,100
231,250	578,100	51,226.00	+	35%	231,250
578,100	---------	172,623.00	+	37%	578,100

Schedule Y-1 — Use if your **Sample Tax Year** filing status is **Married filing jointly or Qualifying surviving spouse**

If line 3 is: Over—	But not over—	The tax is:			of the amount over—
$0	$22,000	---------	+	10%	$0
22,000	89,450	$2,200.00	+	12%	22,000
89,450	190,750	10,294.00	+	22%	89,450
190,750	364,200	32,580.00	+	24%	190,750
364,200	462,500	74,208.00	+	32%	364,200
462,500	693,750	105,664.00	+	35%	462,500
693,750	---------	186,601.50	+	37%	693,750

Schedule Y-2 — Use if your **Sample Tax Year** filing status is **Married filing separately**

If line 3 is: Over—	But not over—	The tax is:			of the amount over—
$0	$11,000	---------	+	10%	$0
11,000	44,725	$1,100.00	+	12%	11,000
44,725	95,375	5,147.00	+	22%	44,725
95,375	182,100	16,290.00	+	24%	95,375
182,100	231,250	37,104.00	+	32%	182,100
231,250	346,875	52,832.00	+	35%	231,250
346,875	---------	95,300.75	+	37%	346,875

Sample Standard Deduction

Filing Status	Sample Standard Deduction
Single	$13,850
Married filing jointly; surviving spouse	$27,700
Married filing separately	$13,850
Head of household	$20,800

Sample 1040

Form **1040** Department of the Treasury—Internal Revenue Service
U.S. Individual Income Tax Return | SAMPLE TAX YEAR | OMB No. 1545-0074 | IRS Use Only—Do not write or staple in this space.

For the Sample Tax Year, or other tax year beginning ______________, ending ______________ | See separate instructions.

Your first name and middle initial	Last name	**Your social security number**
If joint return, spouse's first name and middle initial	Last name	**Spouse's social security number**

Home address (number and street). If you have a P.O. box, see instructions.		Apt. no.
City, town, or post office. If you have a foreign address, also complete spaces below.	State	ZIP code
Foreign country name	Foreign province/state/county	Foreign postal code

Presidential Election Campaign
Check here if you, or your spouse if filing jointly, want $3 to go to this fund. Checking a box below will not change your tax or refund. ☐ **You** ☐ **Spouse**

Filing Status
Check only one box.
☐ Single
☐ Married filing jointly (even if only one had income)
☐ Married filing separately (MFS)
☐ Head of household (HOH)
☐ Qualifying surviving spouse (QSS)
If you checked the MFS box, enter the name of your spouse. If you checked the HOH or QSS box, enter the child's name if the qualifying person is a child but not your dependent: ______________

Digital Assets At any time during Sample Tax Year, did you: (a) receive (as a reward, award, or payment for property or services); or (b) sell, exchange, or otherwise dispose of a digital asset (or a financial interest in a digital asset)? (See instructions.) ☐ **Yes** ☐ **No**

Standard Deduction **Someone can claim:** ☐ You as a dependent ☐ Your spouse as a dependent
☐ Spouse itemizes on a separate return or you were a dual-status alien

Age/Blindness **You:** ☐ Were born before January 2, 1959 ☐ Are blind **Spouse:** ☐ Was born before January 2, 1959 ☐ Is blind

Dependents (see instructions):
If more than four dependents, see instructions and check here . . ☐

(1) First name Last name	(2) Social security number	(3) Relationship to you	(4) Check the box if qualifies for (see instructions): Child tax credit	Credit for other dependents
			☐	☐
			☐	☐
			☐	☐
			☐	☐

Income

Attach Form(s) W-2 here. Also attach Forms W-2G and 1099-R if tax was withheld.
If you did not get a Form W-2, see instructions.

1a	Total amount from Form(s) W-2, box 1 (see instructions)	1a	
b	Household employee wages not reported on Form(s) W-2	1b	
c	Tip income not reported on line 1a (see instructions)	1c	
d	Medicaid waiver payments not reported on Form(s) W-2 (see instructions)	1d	
e	Taxable dependent care benefits from Form 2441, line 26	1e	
f	Employer-provided adoption benefits from Form 8839, line 29	1f	
g	Wages from Form 8919, line 6	1g	
h	Other earned income (see instructions)	1h	
i	Nontaxable combat pay election (see instructions) 1i		
z	Add lines 1a through 1h	1z	
2a	Tax-exempt interest . . . 2a ____ b Taxable interest	2b	
3a	Qualified dividends . . . 3a ____ b Ordinary dividends	3b	
4a	IRA distributions . . . 4a ____ b Taxable amount	4b	
5a	Pensions and annuities . . 5a ____ b Taxable amount	5b	
6a	Social security benefits . . 6a ____ b Taxable amount	6b	
c	If you elect to use the lump-sum election method, check here (see instructions) ☐		
7	Capital gain or (loss). Attach Schedule D if required. If not required, check here ☐	7	
8	Additional income from Schedule 1, line 10	8	
9	Add lines 1z, 2b, 3b, 4b, 5b, 6b, 7, and 8. This is your **total income**	9	
10	Adjustments to income from Schedule 1, line 26	10	
11	Subtract line 10 from line 9. This is your **adjusted gross income**	11	
12	**Standard deduction or itemized deductions** (from Schedule A)	12	
13	Qualified business income deduction from Form 8995 or Form 8995-A	13	
14	Add lines 12 and 13	14	
15	Subtract line 14 from line 11. If zero or less, enter -0-. This is your **taxable income**	15	

Attach Sch. B if required.

Standard Deduction for—
- Single or Married filing separately, $13,850
- Married filing jointly or Qualifying surviving spouse, $27,700
- Head of household, $20,800
- If you checked any box under *Standard Deduction,* see instructions.

For Disclosure, Privacy Act, and Paperwork Reduction Act Notice, see separate instructions. Cat. No. 11320B Form **1040**

Tax and Credits

16	**Tax** (see instructions). Check if any from Form(s): **1** ☐ 8814 **2** ☐ 4972 **3** ☐			16	
17	Amount from Schedule 2, line 3			17	
18	Add lines 16 and 17			18	
19	Child tax credit or credit for other dependents from Schedule 8812			19	
20	Amount from Schedule 3, line 8			20	
21	Add lines 19 and 20			21	
22	Subtract line 21 from line 18. If zero or less, enter -0-			22	
23	Other taxes, including self-employment tax, from Schedule 2, line 21			23	
24	Add lines 22 and 23. This is your **total tax**			24	

Payments

If you have a qualifying child, attach Sch. EIC.

25	Federal income tax withheld from:				
a	Form(s) W-2	25a			
b	Form(s) 1099	25b			
c	Other forms (see instructions)	25c			
d	Add lines 25a through 25c			25d	
26	Sample tax year estimated tax payments and amount applied from previous return			26	
27	Earned income credit (EIC)	27			
28	Additional child tax credit from Schedule 8812	28			
29	American opportunity credit from Form 8863, line 8	29			
30	Reserved for future use	30			
31	Amount from Schedule 3, line 15	31			
32	Add lines 27, 28, 29, and 31. These are your **total other payments and refundable credits**			32	
33	Add lines 25d, 26, and 32. These are your **total payments**			33	

Refund

Direct deposit? See instructions.

34	If line 33 is more than line 24, subtract line 24 from line 33. This is the amount you **overpaid**			34	
35a	Amount of line 34 you want **refunded to you**. If Form 8888 is attached, check here ☐			35a	
b	Routing number	**c** Type: ☐ Checking ☐ Savings			
d	Account number				
36	Amount of line 34 you want **applied to your Sample Year estimated tax**	36			

Amount You Owe

37	Subtract line 33 from line 24. This is the **amount you owe**. For details on how to pay, go to *www.irs.gov/Payments* or see instructions			37	
38	Estimated tax penalty (see instructions)	38			

Third Party Designee

Do you want to allow another person to discuss this return with the IRS? See instructions ☐ **Yes.** Complete below. ☐ **No**

Designee's name		Phone no.		Personal identification number (PIN)	

Sign Here

Under penalties of perjury, I declare that I have examined this return and accompanying schedules and statements, and to the best of my knowledge and belief, they are true, correct, and complete. Declaration of preparer (other than taxpayer) is based on all information of which preparer has any knowledge.

Joint return? See instructions. Keep a copy for your records.

Your signature	Date	Your occupation	If the IRS sent you an Identity Protection PIN, enter it here (see inst.)
Spouse's signature. If a joint return, **both** must sign.	Date	Spouse's occupation	If the IRS sent your spouse an Identity Protection PIN, enter it here (see inst.)
Phone no.	Email address		

Paid Preparer Use Only

Preparer's name	Preparer's signature	Date	PTIN	Check if: ☐ Self-employed
Firm's name			Phone no.	
Firm's address			Firm's EIN	

Go to *www.irs.gov/Form1040* for instructions and the latest information. Form **1040**

Sample Estimated Tax Worksheet

Sample Estimated Tax Worksheet

Keep for Your Records

1	Adjusted gross income you expect in Sample Tax Year (see instructions)		1
2a	Deductions • If you plan to itemize deductions, enter the estimated total of your itemized deductions. • If you do not plan to itemize deductions, enter your standard deduction.		2a
b	If you can take the qualified business income deduction, enter the estimated amount of the deduction		2b
c	Add lines 2a and 2b		2c
3	Subtract line 2c from line 1		3
4	**Tax.** Figure your tax on the amount on line 3 by using the **Sample Tax Rate Schedules for Estimating Taxes**. **Caution:** *If you will have qualified dividends or a net capital gain, or expect to exclude or deduct foreign earned income or housing, see Worksheets 2-5 and 2-6 in Pub. 505 to figure the tax*		4
5	Alternative minimum tax from **Form 6251**.		5
6	Add lines 4 and 5. Add to this amount any other taxes you expect to include in the total on Form 1040 or 1040-SR, line 16.		6
7	Credits (see instructions). **Do not** include any income tax withholding on this line.		7
8	Subtract line 7 from line 6. If zero or less, enter -0-.		8
9	Self-employment tax (see instructions).		9
10	Other taxes (see instructions)		10
11a	Add lines 8 through 10		11a
b	Earned income credit, additional child tax credit, fuel tax credit, net premium tax credit, refundable American opportunity credit, and section 1341 credit		11b
c	**Total sample tax year estimated tax.** Subtract line 11b from line 11a. If zero or less enter -0-		11c
12a	Multiply line 11c by 90% (66 2/3% for farmers and fishermen)	12a	
b	Required annual payment based on prior year's tax (see instructions)	12b	
c	**Required annual payment to avoid a penalty.** Enter the **smaller** of line 12a or 12b. **Caution:** *Generally, if you do not prepay (through income tax withholding and estimated tax payments) at least the amount on line 12c, you may owe a penalty for not paying enough estimated tax. To avoid a penalty, make sure your estimate on line 11c is as accurate as possible. Even if you pay the required annual payment, you may still owe tax when you file your return. If you prefer, you can pay the amount shown on line 11c. For details, see chapter 2 of Pub. 505.*		12c
13	Income tax withheld and estimated to be withheld during sample tax year (including income tax withholding on pensions, annuities, certain deferred income, etc.).		13
14a	Subtract line 13 from line 12c Is the result zero or less? ☐ **Yes.** Stop here. You are not required to make estimated tax payments. ☐ **No.** Go to line 14b	14a	
b	Subtract line 13 from line 11c Is the result less than $1,000? ☐ **Yes.** Stop here. You are not required to make estimated tax payments. ☐ **No.** Go to line 15 to figure your required payment.	14b	
15	If the first payment you are required to make is due April 18, enter 1/4 of line 14a (minus any prior year overpayment that you are applying to this installment) here, and on your estimated tax payment5 voucher(s) if you are paying by check or money order		15

Form **W-4**

Department of the Treasury
Internal Revenue Service

Employee's Withholding Certificate

Complete Form W-4 so that your employer can withhold the correct federal income tax from your pay.
Give Form W-4 to your employer.
Your withholding is subject to review by the IRS.

OMB No. 1545-0074

SAMPLE TAX YEAR

Step 1: Enter Personal Information

(a) First name and middle initial	Last name	**(b) Social security number**
Address		**Does your name match the name on your social security card?** If not, to ensure you get credit for your earnings, contact SSA at 800-772-1213 or go to *www.ssa.gov*.
City or town, state, and ZIP code		

(c)
- ☐ **Single** or **Married filing separately**
- ☐ **Married filing jointly** or **Qualifying surviving spouse**
- ☐ **Head of household** (Check only if you're unmarried and pay more than half the costs of keeping up a home for yourself and a qualifying individual.)

Complete Steps 2–4 ONLY if they apply to you; otherwise, skip to Step 5. See page 2 for more information on each step, who can claim exemption from withholding, and when to use the estimator at *www.irs.gov/W4App*.

Step 2: Multiple Jobs or Spouse Works

Complete this step if you (1) hold more than one job at a time, or (2) are married filing jointly and your spouse also works. The correct amount of withholding depends on income earned from all of these jobs.

Do **only one** of the following.

(a) Use the estimator at *www.irs.gov/W4App* for most accurate withholding for this step (and Steps 3–4). If you or your spouse have self-employment income, use this option; **or**

(b) Use the Multiple Jobs Worksheet on page 3 and enter the result in Step 4(c) below; **or**

(c) If there are only two jobs total, you may check this box. Do the same on Form W-4 for the other job. This option is generally more accurate than (b) if pay at the lower paying job is more than half of the pay at the higher paying job. Otherwise, (b) is more accurate . . . . ☐

Complete Steps 3–4(b) on Form W-4 for only ONE of these jobs. Leave those steps blank for the other jobs. (Your withholding will be most accurate if you complete Steps 3–4(b) on the Form W-4 for the highest paying job.)

Step 3: Claim Dependent and Other Credits

If your total income will be $200,000 or less ($400,000 or less if married filing jointly):

Multiply the number of qualifying children under age 17 by $2,000 $ ______

Multiply the number of other dependents by $500 . . . . . $ ______

Add the amounts above for qualifying children and other dependents. You may add to this the amount of any other credits. Enter the total here . . . . **3** $

Step 4 (optional): Other Adjustments

(a) Other income (not from jobs). If you want tax withheld for other income you expect this year that won't have withholding, enter the amount of other income here. This may include interest, dividends, and retirement income . . . . **4(a)** $

(b) Deductions. If you expect to claim deductions other than the standard deduction and want to reduce your withholding, use the Deductions Worksheet on page 3 and enter the result here . . . . **4(b)** $

(c) Extra withholding. Enter any additional tax you want withheld each **pay period** . . **4(c)** $

Step 5: Sign Here

Under penalties of perjury, I declare that this certificate, to the best of my knowledge and belief, is true, correct, and complete.

Employee's signature (This form is not valid unless you sign it.) **Date**

Employers Only

Employer's name and address	First date of employment	Employer identification number (EIN)

For Privacy Act and Paperwork Reduction Act Notice, see page 3. Cat. No. 10220Q Form **W-4**

Units of Time & Measure

Time Abbreviations & Relationships

8 hours (hrs.) = 1 typical workday

5 days (ds.) = 1 typical workweek

12 months (mos.) = 4 quarters (qtrs.) = 52 weeks (wks.) = 1 year (yr.) ≈ 365 days (ds.)

1 quarter (qtr.) = 3 months (mos.)

4.33 weeks (wks.) ≈ 1 month (mo.)

(found by taking $52 \div 12 \approx 4.33$, as there are 52 wks. in a yr.)

Shape Formulas

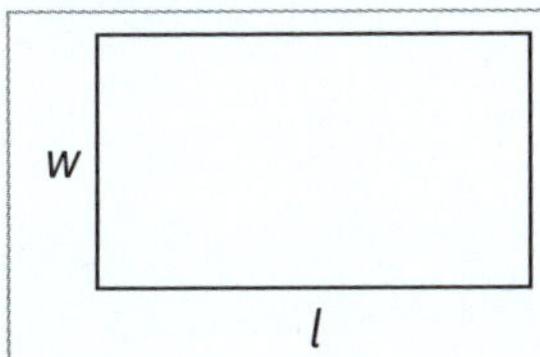

Rectangle

Area = length × width

$A = l \times w$

Perimeter = sum of all sides

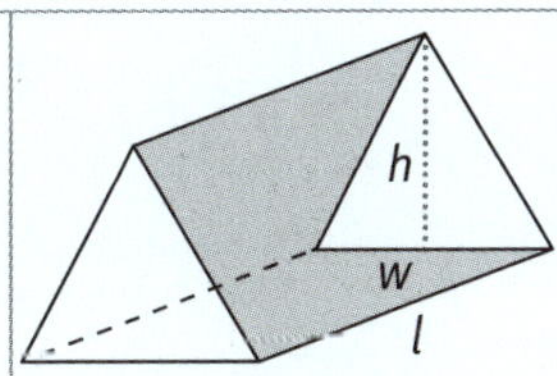

Rectangular Prism

V = volume = length × width × height

$V = l \times w \times h$

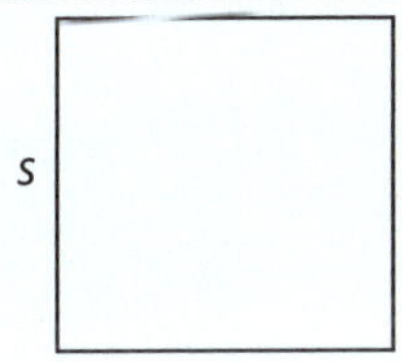

Square

Area = side^2

$A = s^2$

Perimeter = sum of all sides

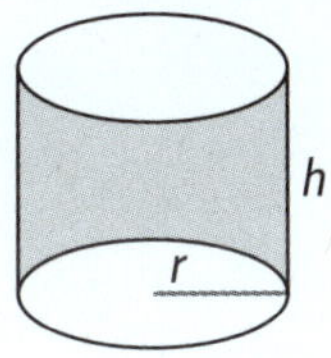

Cylinder

V = volume = 3.14 × *radius*^{2} × *height*

$V = 3.14 \times r^2 \times h$

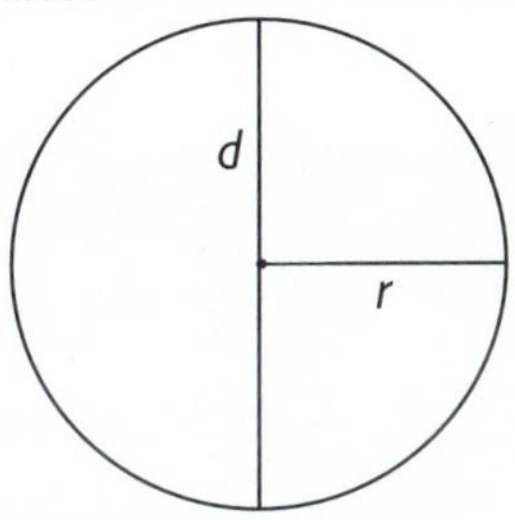

Circle

Area = $3.14 \times r^2$

Circumference = 2 × 3.14 × *radius*

$C = 2 \times 3.14 \times r$

d = diameter = $2 \times r$

r = radius = $\frac{d}{2}$

U.S. Customary Measurements

Distance Measurements

12 inches (in.) = 1 foot (ft.)

36 in. = 3 ft. = 1 yard (yd.)

5,280 ft. = 1,760 yd. = 1 mile (mi.)

Common Kitchen Measurers

1 c.	$\frac{3}{4}$ c.	$\frac{2}{3}$ c.	$\frac{1}{2}$ c.	$\frac{1}{3}$ c.	$\frac{1}{4}$ c.
1 TBSP.	$\frac{1}{2}$ TBSP.				
1 tsp.	$\frac{1}{2}$ tsp.	$\frac{1}{4}$ tsp.	$\frac{1}{8}$ tsp.		

Decimal Equivalent of Common Fractions

$\frac{1}{8} = 0.125$ $\frac{1}{4} = 0.25$ $\frac{1}{2} = 0.5$ $\frac{3}{4} = 0.75$

$\frac{1}{5} = 0.20$ $\frac{1}{3} \approx 0.33$ $\frac{2}{3} \approx 0.66$

Household Measurements

1 gallon (gal.) = 4 quarts (qt.) = 8 pints (pt.) = 16 cups (c.) = 128 fluid ounces (fl. oz.)

3 teaspoons (tsp.) = 1 tablespoon (TBSP.)

4 tablespoons (TBSP.) = $\frac{1}{4}$ cup (c.)

16 ounces (oz.) = 1 pound (lb.)

Metric Measurements

Each unit is 10x previous row

Each unit is previous row divided by 10 (i.e., $\frac{1}{10}$ of previous row)

Metric Chart

	Meaning	**Distance**	**Liquid Capacity**	**Mass**
kilo	1,000	1 *kilo*meter (km.) = 1,000 meters (m.)	1 *kilo*liter (kl.) = 1,000 liters (l.)	1 *kilo*gram (kg.) = 1,000 grams (g.)
hecto	100	1 *hecto*meter (hm.) = 100 meters (m.)	1 *hecto*liter (hl.) = 100 liters (l.)	1 *hecto*gram (hg.) = 100 grams (g.)
deka	10	1 *deka*meter (dam.) = 10 meters (m.)	1 *deka*liter (dal.) = 10 liters (l.)	1 *deka*gram (dag.) = 10 grams (g.)
	1	**meter (m.)**	**liter (l.)**	**gram (g.)**
deci	$\frac{1}{10}$	1 *deci*meter (dm.) = $\frac{1}{10}$ of a meter (m.); 10 dc. = 1 m.	1 *deci*liter (dl.) = $\frac{1}{10}$ of a liter (l.); 10 dl. = 1 l.	1 *deci*gram (dg.) = $\frac{1}{10}$ of a gram (g.); 10 g. = 1 dg.
centi	$\frac{1}{100}$	1 *centi*meter (cm.) = $\frac{1}{100}$ of a meter (m.); 100 cm. = 1 m.	1 *centi*liter (cl.) = $\frac{1}{100}$ of a liter (l.); 100 cl. = 1 l.	1 *centi*gram (cg.) = $\frac{1}{100}$ of a gram (g.); 100 cg. = 1 g.
milli	$\frac{1}{1,000}$	1 *milli*meter (mm.) = $\frac{1}{1,000}$ of a meter (m.); 1,000 mm. = 1 m.	1 *milli*liter (ml.) = $\frac{1}{1,000}$ of a liter (l.); 1,000 ml. = 1 l.	1 *milli*gram (mg.) = $\frac{1}{1,000}$ of a gram (g.); 1,000 g. = 1 mg.

Orange terms are the most common and important to know.

Selected Relationships Between U.S. Customary and Metric Systems

Units of Distance

1 inch (in.) = 2.54 centimeters (cm.)

1 foot (ft.) = 30.48 centimeters (cm.)

1 yard (yd.) ≈ 0.91 meters (m.)

1 mile (mi.) ≈ 1.61 kilometers (km.)

Units of Liquid Capacity

1 teaspoon (tsp.) ≈ 5 milliliters (ml.)

1 tablespoon (TBSP.) ≈ 15 milliliters (ml.)

1 gallon (gal.) ≈ 3.79 liters (l.)

Mass

1 ounce (oz.) ≈ 28.35 grams (g.)

1 pound (lb.) ≈ 453.59 grams (g.)

1 U.S. ton (called a short ton) ≈ 0.91 metric tons

Page intentionally blank

Index

Photo Credits

All images from shutterstock.com unless indicated.
All tax forms from IRS.gov
fred.stlouisfed.org: page 228, page 230 (both), page 280

Page intentionally blank

Endnotes

Please note that we in no way endorse all endnotes, but are rather acknowledging where a specific piece of information was obtained. Note that websites especially may have changed since examined. Please check with a parent before exploring any of these (and, parents, please use your own discernment).

For this course, we have relied on the information and tax forms which can be found on the www.irs.gov website. It is a good source for additional forms and other information. Some forms in the course have been modified or only partially shown for relevance per the topic. Generic tax years were also used.

1. *Google's English Dictionary*, Oxford Languages, s.v. "consumer," accessed August 22, 2023.
2. Howard Dayton (1973), as referenced in John Cortines and Gregory Baumer, *God and Money*, p. 13.
3. "Handful of Rice, 7 Minute Version," https://generousgiving.org/handful-of-rice-7-minute-version / ~5:40
4. "Handful of Rice," https://generousgiving.org/handful-of-rice/ at beginning.
5. "Handful of Rice, 7 Minute Version," https://generousgiving.org/handful-of-rice-7-minute-version / ~6:24.
6. Ibid., ~5:59.
7. Ibid., ~2:43.
8. Ibid., ~2:48.
9. *Merriam-Webster.com Dictionary,* s.v. "utility," accessed 2023, https://www.merriam-webster.com/dictionary/utility.
10. Charles Spurgeon, "The Bed and Its Covering," New Park Street Pulpit Volume 5, January 9, 1859, https://www.spurgeon.org/resource-library/sermons/the-bed-and-its-covering/#flipbook/.
11. Michelle Black and Robin Saks Frankel, "What Is the Average Credit Card Interest Rate This Week? November 13, 2023," (Forbes), https://www.forbes.com/advisor/credit-cards/average-credit-card-interest-rate/.
12. Elvina M. Hall and John Grape. *Sabbath Carols: A New Collection of Music and Hymns* (1868), p. 93.
13. "USDA Food Plans: Monthly Cost of Food Reports," Center for Nutrition Policy and Promotion (CNPP), USDA, https://www.fns.usda.gov/cnpp/usda-food-plans-cost-food-monthly-reports, October 2023 report.
14. "Budget share for total food increased 13 percent in 2022," Chart Detail, USDA, https://www.ers.usda.gov/data-products/chart-gallery/gallery/chart-detail/?chartId=76967.
15. Economic Research Service, USDA, "Home-Grilled Cheeseburger Costs Grew 1.8 Percent from 2023 to 2024" (June 25, 2024 update).
16. David Watsky, "Wondering if a Meal Delivery Service is Cheaper Than Takeout? I Did the Math," CNET, updated June 14, 2024, https://www.cnet.com/health/nutrition/i-priced-8-prepared-meal-delivery-services-against-the-average-cost-of-takeout-and-wow/.
17. Tobie Stanger, "15 Ways to Save on Thanksgiving Food Shopping," Consumer Reports, updated November 16, 2022, https://www.consumerreports.org/health/food-shopping/how-to-save-time-and-money-at-the-grocery-store-a2751262166/.
18. Greg shares the "Generosity Covenant" they settled on in his book by John Cortines and Gregory Baumer, *God and Money: How We Discovered True Riches at Harvard Business School* (Carol Stream, IL: Tyndale House Ministries, 2016), pp. 203–206.
19. Generous Giving, "Greg and Alison Baumer Part 1," video, 04:59, https://generousgiving.org/greg-alison-baumer/.
20. Generous Giving, "Greg and Alison Baumer Part 2," video, 07:20, https://generousgiving.org/greg-and-alison-baumer-five-years-later/.
21. Generous Giving, "Greg and Alison Baumer Part 1," 11:10.
22. Generous Giving, "Greg and Alison Baumer Part 2," 06:00.
23. "Health Expenditures 1960-2022," National Health Spending Explorer, Peterson-KFF Health System Tracker, accessed 10/21/2024, https://www.healthsystemtracker.org/chart-collection/u-s-spending-healthcare-changed-time/#Total%20national%20health%20expenditures,%20US%20$%20Billions,%201970-2022
24. "Publication 969 (2023), Health Savings Accounts and Other Tax-Favored Health Plans," Publications, IRS, s.v. "Qualifying for an HSA Contribution," https://www.irs.gov/publications/p969#en_US_2022_publink1000204121.
25. "HMO, POS, PPO, EPO and HDHP with HSA: What's the difference?" Aetna, https://www.aetna.com/health-guide/hmo-pos-ppo-hdhp-whats-the-difference.html.
26. Generous Giving, "Alan and Katherine Barnhart," video, 15:00, https://generousgiving.org/alan-and-katherine-barnhart/.

27. Monthly Labor Review, A Consumption Measure for Automobiles, (January 2024), Chart 1. Annual Depreciation Rates by Automobile Age (1996-2022), https://www.bls.gov/opub/mlr/2024/article/a-consumption-measure-for-automobiles.htm
28. Generous Giving, Tom and Bree Hsieh, video, 05:15, https://generousgiving.org/tom-and-bree-hsieh-living-below-the-median-income/.
29. Ibid., 07:05.
30. Burning Glass Institute and Strada Institute for the Future of Work, Talent Disrupted: Underemployment, College Graduates, and the Way Forward, 2024, p. 10. https://stradaeducation.org/wp-content/uploads/2024/02/Talent-Disrupted.pdf
31. Federal Reserve, Consumer Credit Outstanding (Levels), 2021-2024, https://www.federalreserve.gov/releases/g19/hist/cc_hist_memo_levels.html
32. "What kinds of scholarships are available?" under "Find and apply for as many scholarships as you can," Finding and Applying for Scholarships, Federal Student Aid, accessed July 15, 2024, https://studentaid.gov/understand-aid/types/scholarships#what-kinds-of-scholarships-are-available.
33. "How Financial Aid Is Calculated," Complete the Financial Aid Process, Federal Student Aid, accessed July 15, 2024, https://studentaid.gov/complete-aid-process/how-calculated.
34. "Federal Pell Grants," Grants, Federal Student Aid, accessed July 15, 2024, https://studentaid.gov/understand-aid/types/grants/pell.
35. "A Federal Supplemental Educational Opportunity Grant (FSEOG)," FSEOG (Grants), Federal Student Aid, accessed July 15, 2024, https://studentaid.gov/understand-aid/types/grants/fseog.
36. "Direct Subsidized and Direct Unsubsidized Loans," Subsidized and Unsubsidized Loans, Federal Student Aid, accessed July 15, 2024, https://studentaid.gov/understand-aid/types/loans/subsidized-unsubsidized.
37. "(GEN-24-01) 2024-2025 Federal Pell Grant Maximum and Minimum Award Amounts (Updated April 5, 2024)," Dear Colleague Letters, Federal Student Aid, updated April 5, 2024, https://fsapartners.ed.gov/knowledge-center/library/dear-colleague-letters/2024-01-31/2024-2025-federal-pell-grant-maximum-and-minimum-award-amounts-updated-april-5-2024.
38. "Explore Approved Occupations for Registered Apprenticeship," Apprenticeship Occupations, Apprenticeship USA, accessed July 15, 2024, https://www.apprenticeship.gov/apprenticeship-occupations.
39. U.S. Census Bureau and U.S. Department of Housing and Urban Development, Median Sales Price of Houses Sold for the United States [MSPUS], retrieved from FRED, Federal Reserve Bank of St. Louis; https://fred.stlouisfed.org/series/MSPUS, December 30, 2024. Accessed 12/30/2024.
40. Federal Housing Finance Agency, "FHFA Announces Conforming Loan Limit Values for 2025" https://www.fhfa.gov/news/news-release/fhfa-announces-conforming-loan-limit-values-for-2025?utm_source=chatgpt.com
41. "How should I use lender credits and points (also called discount points)?" Consumer Financial Protection Bureau, reviewed October 19, 2023, https://www.consumerfinance.gov/ask-cfpb/what-are-discount-points-and-lender-credits-and-how-do-they-work-en-136/.
42. Ibid.
43. Freddie Mac, "30-Year Fixed Rate Mortgage Average in the United States (MORTGAGE30US)," retrieved January 6, 2024, from FRED, Federal Reserve Bank of St. Louis, https://fred.stlouisfed.org/series/MORTGAGE30US.
44. Board of Governors of the Federal Reserve System (US), Delinquency Rate on Single-Family Residential Mortgages, Booked in Domestic Offices, All Commercial Banks [DRSFRMACBS], retrieved from FRED, Federal Reserve Bank of St. Louis; https://fred.stlouisfed.org/series/DRSFRMACBS, January 23, 2025.
45. Much of this information is from *R.G. LeTourneau: Mover of Men and Mountains* (Moody Press, Chicago: 1967).
46. Ibid., p. 267.
47. Ibid., Chapter 3.
48. Ibid., p. 109.
49. Ibid., Chapter 17.
50. Ibid., p. 110.
51. "The Machines," Museum & Archives, LeTourneau University, accessed 12/30/2024, https://www.letu.edu/library/rg-museum.html#ContentBlock-1-5.
52. LeTourneau, p. 204-205.
53. Ibid., p. 221.
54. Ibid, p. 110.
55. Ibid., p. 280, in epilogue to 1967 edition by Nels E. Stjernstrom.
56. Ibid, p. 1.

HIGH SCHOOL MATH

ELEMENTARY ALGEBRA

This high school algebra curriculum provides a full year of math in a clearly written format with guidance for teachers as well as for students who are self-directed. The full-color student text is divided into 17 sections, covering functions and graphs, integers, rational numbers, exponents, polynomials, factoring, fractions, and more.

GEOMETRY

With nearly one million students having learned geometric principles through this text, *Geometry* has been an authoritative standard for years. With the use of innovative discussions, cartoons, anecdotes, and vivid exercises, students will not only learn but will also find their interest growing with each lesson.

PRINCIPLES OF ALGEBRA 2

Algebra 2 students will get a glimpse of God's handiwork as they apply their studies. Most algebra books come across as a whole bunch of meaningless, often hard-to-understand problems. But algebra doesn't have to be that way at all. This book is full of different "tools" in algebra, seeing how they all really help us explore God's creation and complete real-life tasks.

MASTERBOOKS ACADEMY.COM
ONLINE CLASSES
A FAITH-FUELED EDUCATION
WHEREVER YOU ARE
Go online with Master Books video supplements!
Applied Engineering
Intro to Biblical Greek
MASTERBOOKS CURRICULUM
Where Faith Grows!